STUDENT'S SOLUTIONS MANUAL

MILTON LOYER
Penn State University

ELEMENTARY STATISTICS

ELEVENTH EDITION

Mario F. Triola
Dutchess Community College

Addison-Wesley
is an imprint of

PEARSON

Reproduced by Pearson Addison-Wesley from electronic files supplied by the author.

ISBN-13: 978-0-321-57062-8
ISBN-10: 0-321-57062-6

4 5 6 BBR 10

Addison-Wesley
is an imprint of

www.pearsonhighered.com

INTRODUCTION

by Milton Loyer

This *Student's Solutions Manual* contains detailed solutions to the odd-numbered exercises for each section in the text *Elementary Statistics*, Eleventh Edition, by Mario Triola, and the solutions to all of the end-of-chapter exercises: the Statistical Literacy and Critical Thinking exercises, the Chapter Quick Quiz, the Review Exercises, and the Cumulative Review Exercises. To aid in the comprehension of the calculations, worked problems typically include intermediate steps of algebraic or computer/calculator notation. When appropriate, additional hints and comments are included and prefaced by NOTE.

Many statistical problems are best solved using particular formats. Because recognizing and following these patterns promote understanding, this manual identifies and employs such formats whenever practicable.

For best results, read the text carefully before attempting the exercises, and attempt the exercises before consulting the solutions. This manual has been prepared to provide a check and extra insights for exercises that have already been completed, and to provide guidance for solving exercises that have already been attempted but have not been successfully completed.

I would like to thank Mario Triola for writing an excellent elementary statistics textbook and for inviting me to prepare this solutions manual.

TABLE OF CONTENTS

Chapter 1

Introduction to Statistics

1-2 Statistical Thinking

1. A voluntary response sample, also called a self-selected selected, is one in which the subjects themselves decide whether or not to participate in the study.

3. Statistical significance occurs when the sample data indicate that a particular conclusion is mathematically justified. Practical significance occurs when the sample data lead to a conclusion that is meaningful and useful. A conclusion may be mathematically justified (i.e., statistically significant) but of no meaningful or useful value (i.e., of no practical significance). For example, the data may lead to the conclusion that a gasoline additive improves mileage – but if the estimated improvement is only 0.1 mpg, common sense dictates that using the additive is not worth the time or the money. That would be an example of a case where there is statistical significance but not practical significance.

5. Determining whether or not the weight loss is statistically significant properly requires applying the techniques presented in future chapters. But even if one can conclude that the program is effective (i.e., that the 3.0 lb weight loss is statistically significant), such a weight loss does not have practical significance. Most people would not be willing to subject themselves for one full year to a regimen that produced such a small weight loss.

7. (b) Possible, but very unlikely. Possible because there are no physical constraints or rules of the game that would prohibit such scores.

9. (c) Possible and likely. Likely in the sense of "not unusual enough to raise suspicion," not likely in the sense of "more probable to happen than not to happen."

11. (b) Possible, but very unlikely. This would be either an amazing coincidence or the result of specific planning.

13. (a) Impossible. The highest possible total on a normal pair of dice is 6+6 = 12.

15. There is nothing in the presentation of the data to suggest that the x and y values are matched. If the x and y values are not matched as presented, it does not make sense to calculate the differences between the x and y values – moving the x values around, for example, would yield a different set of differences.

17. The data can address the issue of whether or not the two types of cigarette (menthol and king-size) contain the same amounts of nicotine.

19. The presentation of the data suggests that the weight and mpg values are matched, each pair of values referring to a particular car. Even though the x (lbs) and y (mpg) values are matched, it does not make sense to calculate their differences. In general, quantities must have the same units in order to be added or subtracted – and then the resulting sum or difference will also have those same units. No meaningful units or interpretation can be assigned to "pounds minus mpg."

21. If car manufacturers supplied the information, there is reason to suspect that the data might be biased. Since the industry profits from increased sales, and since consumers tend to look for cars with high mpg ratings, the individual manufacturers might tend to supply information only for their better models (i.e., the one with the highest mpg values).

23. The weight loss program has statistical significance (i.e., one can conclude that the program produces a weight loss) because it is very unlikely (only 3 chances in 1000) that such a weight loss could have occurred by chance alone. The program does not have practical significance because the amount of lost weigh (3.3 lbs) is so small.

25. Determining whether or not the difference between the survey results (85%) and the statement of the industry representative (50%) is statistically significant properly requires applying the techniques presented in future chapters. But common sense suggests that this is statistically significant evidence against the representative's claim because (1) the observed result of 85% is so much greater than the stated claim of the 50% and (2) the sample of 1038 randomly selected adults appears to be properly chosen and large enough to provide reliable data.

27. a. Yes. In every case the highway amount is substantially greater than the city amount. If there were no difference, we would expect the highway and city amounts to be approximately equal – and when they were not, we would expect the highway amount to be larger about half the time and the city amount to be larger about half the time.
    b. Yes. The differences appear to be substantial – generally ranging from 6 to 10 mpg higher for the highway amounts.
    c. One practical implication of the difference is the increased cost of city driving compared to highway driving. Given a choice between two otherwise equal routes, financial considerations would suggest choosing the highway route over the city one.

1-3 Types of Data

1. A parameter is a numerical value that describes a population, while a statistic is a numerical value that describes a sample.

3. Discrete data are limited to certain finite or countable number of values, while continuous data can assume any value within a specified range.

5. Statistic, since it was determined from a sample of households.

7. Parameter, since it was determined from the population of all 2223 passengers..

9. Parameter, since it was determined from the population of all 50 states.

11. Statistic, since it was determined from a sample of 40 days.

13. Discrete, since the value must be one of the integers 0,1,2,…3249,3250.

15. Continuous, since the amount of nicotine could be any value on a continuum – even though it is reported to the nearest 0.1 mg.

17. Discrete, since the number of baby girls must be one of the integers 0,1,2,…,725,726.

19. Continuous, since the weight could be any value on a continuum – even though it is reported to the nearest 0.1 kg.

21. Ratio, since differences are meaningful and there is a meaningful zero.

23. Ordinal, since the categories have a natural ordering but the differences between the categories are not necessarily uniform.

25. Nominal, since the data give category names only and there is no natural ordering.

27. Interval, since differences between years are meaningful but there is no meaningful zero.

IMPORTANT NOTE for exercises 29-32: The population and sample are determined by the intent of the researcher, which must be clearly defined at the outset of any project. Unfortunately these exercises state only what the researcher did, and do not specifically identify the intent of the researcher. Consequently there may be differences of interpretation in some of the exercises, but some general principles apply. (1) The sample is a subset of the population of interest and must have the same units as the population. If the population of interest is all households, for example, then the sample must be a selection of households and not a selection of adults – as households with more adults would have a higher chance of being included in the study, thus creating a bias. (2) The problem of nonresponse must be addressed. If 500 persons are randomly selected and asked a personal question, for example, but only 400 choose to answer the question, what is the sample? Depending on the situation, the sample could be either the 500 people randomly selected (and the 100 "no response" answers be reported as part of the sample data) or the just 400 people who actually gave data in the form of a specific answer.

29. a. The sample is the readers who completed and returned the survey.
    b. The population is all people who read *USA Today*.
    c. No, since the sample is self-selected it is not likely to be representative of the population. Only those with special interest in health matters are likely to return the survey.

31. a. The sample is the people who responded to the request.
    b. The intended population is likely all persons over 18. The actual population is actually all persons over 18 who have opportunity to receive the request. If the request went out over the Internet, for example, the population would be all Internet users over 18. If the request went out over several radio and/or TV stations, the population would be all persons over 18 who tune in to those stations. And so on.
    c. No, since the sample is self-selected it is not likely to be representative of the population. Only those with special interest in abortion issues are likely to spend the time and money to respond to the survey.

33. Temperature ratios are not meaningful because a temperature of 0° does not represent the absence of temperature in the same sense that $0 represents the absence of money. The zero temperature in the exercise (whether Fahrenheit or Centigrade) was determined by a criterion other than "the absence of temperature."

35. This is an example of ordinal data. It is not interval data because differences are not meaningful – i.e., the difference between the ratings +4 and +5 does not necessarily represent the same differential in the quality of food as the difference between 0 and +1.

1-4 Critical Thinking

1. A voluntary response sample is one in which the subjects themselves choose whether or not to participate. This is generally unsuitable for statistical purposes because it is persons with strong feelings and/or a personal interest in the topic that tend to respond – and the opinions of such people are not necessarily representative of the entire population.

3. a. No. An association between two variables does not imply that there is a cause and effect. Even if there is a cause and effect between two variables, the statistics alone cannot identify which is the cause and which is the effect – saying that an increase in the number of registered weapons caused an increase in the murder rate is no more justified than saying it was an increase in the murder rate that caused the increase in the number of registered weapons.
   b. No. Since there is not necessarily a cause and effect relationship, a decrease in the number of registered weapons will not necessarily result in a reduced murder rate.

5. College graduates tend to earn more money than non-graduates, and people who have more money are able to purchase better health care. Having more money (whether or not it resulted from having a college degree) and not studying more is the primary contributing factor toward longer life.

7. If the population of Orange County includes significantly more minority drivers than white drivers, one would expect more speeding tickets to be issued to minorities than to whites – even if the percentage of white drivers who violated the speed limit was greater than the percentage of minority drivers who did so. It is also possible that police tend to target minority drivers – so that the numbers of tickets issued to the various racial/ethnic groups does not correspond to the actual amount of speed limit violations occurring. The fact that more speeding tickets are given to minorities does not warrant the conclusion that minority persons are more likely to speed.

9. Self-reporting is not always a reliable method of gathering information. People tend to answer questions in a manner that will reflect well on them; they tend to give the answer they wish were true, or that they wish others to believe about them, rather than the true answer. The 82% rate is more likely to be accurate because it was based on objective observation rather than self-reporting.

11. The fact that the study was financed by groups with vested interests in the outcome might have influenced the conducting and/or reporting to concentrate on aspects of the study favorable to chocolate. The fact that a product contains one ingredient associated with positive health benefits may not outweigh other negative aspects of the product that were not reported.

13. There are at least two reasons why the results cannot be used to reach any conclusion about how the general population feels about keeping the United Nations in the United States. First, only viewers of the ABC "Nightline" program, which are not necessarily representative of the general population, were aware of the poll. Secondly, the 186,000 persons who responded (and there is always the possibility that some people called more than once in order to give their opinion more weight) are a voluntary response sample – which means that the responses will include an over representation of those with strong feelings and/or a personal interest in the survey.

15. People who were killed in motorcycle crashes, when a helmet may have saved their lives, were not present to testify.

17. No, this is not likely to be a good estimate of the average per capita income for all individuals in the United States. States with small populations but high wages (due to the cost of living) like Alaska and Hawaii, for example, contribute 2/50 = 4% of the average for the 50 states – but since they do not contain 4% of the US wage earners, they inflate the estimate to make it higher than the true average of all US wage earners. In general, the average of averages does not give the overall average.

NOTE: Imagine a hypothetical country of two states: one where all 4 persons earned a total of $400,000 (for a per capita income of $400,000/4 = $100,000), and one where all 96 persons earned a total of $4,800,000 (for a per capita income of $4,800,000/96 = $50,000). The true per capita income for the country is (400,000 + 4,800,000)/100 = 5,200,000/100 = $52,000, but the average of the state averages is (100,000 + 50,000)/2 = 150,000/2 = $75,000.

19. Nothing. Because the information comes from a voluntary response sample, the opinions expressed are more likely to be representative of those with strong feelings on the topic than of the general population. Even if the information had not come from a voluntary response sample, the conclusion would apply only to women who read *Good Housekeeping* magazine and not to "all women."

21. a. 5/8 = 0.625 = (0.625)(100%) = 62.5%
    b. 23.4% = 23.4/100 = 0.234
    c. (37%)(500) = (37/100)(500) = 185
    d. 0.127 = (0.127)(100%) = 12.7%

23. a. (49%)(734) = (49/100)(734) = 360
    b. 323/734 = 0.440 = (0.440)(100%) = 44.0%

25. a. (5%)(38410) = (5/100)(38410) = 1920.5
    While either 1920 or 1921 seem like reasonable answers, it is not possible to give an exact answer; any number between 1729 and 2112 rounds to 5%.
    b. 18053/38410 = 0.470 = (0.470)(100%) = 47.9%
    c. No. Because the information comes from a voluntary response sample, the opinions expressed are more likely to be representative of those with strong feelings on the topic than of the general population. Even if the information had not come from a voluntary response sample, the conclusion would apply only to those with AOL Internet access and not to the "general population."

27. If something falls 100%, there is none remaining. For foreign investment to fall 500% it would have to decline by an amount equal to 5 times as much as it started with – which is not possible.

29. If the researcher started with 20 mice in each of the six groups, and none of the mice dropped out of the experiment for any reason, then the proportions of success could only be fractions like 0/20, 1/20, 2/20, etc. – and the success rates in percents could only be multiples of 5 like 0%, 5%, 10%, etc. A success rate of 53%, for example, would not be possible.

NOTE: This appears to be a case of miscommunication rather than falsifying data. If the 120 mice had not been divided evenly among the 6 groups, the following fractions would account for all the 19+19+19+24+21+18 = 120 mice and give the 6 reported success rates:
10/19 = 53%, 11/19 = 58%, 12/19 = 63%, 11/24 = 46%, 10/21 = 48%, 12/18 = 67%.

1-5 Collecting Sample Data

1. In a random sample, every individual member has an equal chance of being selected; while in a simple random sample of size n, every possible sample of size n has an equal chance of being selected.

3. No. A random sample is likely to be representative of the population from which it is selected. The stated sample is likely to be representative of the student's friends, but not necessarily of the general population.

5. Observational study, since the researcher (Emily) merely measured whether the therapist could identify the chosen hand and did nothing to modify the therapist.

7. Experiment, since the effect of an applied treatment (in this case a zero dose of the appropriate medicine) was measured.
NOTE: There is room for disagreement – some might argue that this is an observational study, since the intent was to learn about the effects of the disease on untreated men and specific characteristics were measured on unmodified subjects. But since the patients were given a placebo and/or were part of a larger formal study in which other persons received the proper medicine, then the given statement describes an experiment. Even if there were no such larger formal study, the spirit of the statement is that the subjects were selected and monitored for the purposes of seeing what particular effect their "treatment" had and/or comparing this effect to known results when the proper treatment was given.

9. Convenience, since the sample is those who happened to be in the student's family.

11. Cluster, since all the voters at randomly selected polling stations were surveyed.

13. Stratified, since the population of wines was subdivided into 5 different subgroups (wineries), and then samples were drawn from each subgroup.

15. Systematic, since every 100th spark plug is tested.

17. Random, since every person who filed a return had an equal chance of being selected..

19. Cluster, since all the members of six randomly selected health plans were interviewed.

21. Yes, it is a random sample because each pill has an equal chance of being selected.
Yes, it is a simple random sample because each sample of size 30 has an equal chance of being selected.

23. Yes, it is a random sample because each voter has an equal chance of being selected – and that probability is the number of precincts selected divided by the total number of precincts.
No it is not a simple random sample because some samples of a given size are not possible – e.g., a sample with at least voter from every precinct is not possible.

25. No, this is not a random sample of all New Yorkers because persons who did not visit the location had not chance of being selected.
No it is not a simple random sample because it is not a random sample.

27. Retrospective. The researchers identified groups with a common characteristic (those with and without present respiratory problems) and gathered data by going <u>backward </u>in time (to see how they were involved in the events of 9/11).

29. Cross-sectional. The data examined refer to a single point in time.

31. Blinding occurs when either the subject or the evaluator doesn't know if the subject received a valid treatment or a placebo. Double-blinding occurs when neither the subject nor the evaluator knows whether a treatment or a treatment was received. Blinding is important in experiments in which there is subjectivity involved in measuring the response. In this experiment, knowledge of whether a treatment or a placebo was received might influence the subject's or evaluator's assessment of whether the "treatment" was effective in dealing with the cold.

33. To see which of two statistics textbooks results in better student understanding, have Professor A use one text in all his classes and Professor B use the other text in all her classes. At the end of the semester give all the students a common exam. Confounding will occur because it will not be possible to tell if any difference between the two groups is due to the textbook or to the professor.

Statistical Literacy and Critical Thinking

1. No. Since the responders constituted a voluntary response sample, they likely represented only those with strong feelings for or against one candidate or the other and were not necessarily representative of the entire population.

2. a. Quantitative, because the data consist of numbers representing measurements
   b. Continuous, since length can be any value on a continuum – even though the values are recorded in whole minutes.
   c. Observational study, since the data were collected without involving any treatments or modifications.
   d. Nominal, since the data give names only and do not measure or rank the films.
   e. Ratio, since differences are meaningful and there is a meaningful zero.

3. The subjects must be selected so that all samples of the same size have the same chance of being selected.

4. No, this is the average travel time to work for all individuals in the United States. States with larger populations (which would likely have th greater travel times) should be given more weight.
   NOTE: Imagine a hypothetical country of two states: one where all 4 persons had an average travel time of 10 minutes, and one where all 96 persons had an average travel time of 30 minutes. The true average travel time for the country is $[(4)(5) + (96)(30)]/100 = [2900]/100 = 29$ minutes, but the average of the state averages is $[5 + 30]/2 = [35]/2 = 17.5$ minutes.

Chapter Quick Quiz

1. True. That is a possible complete set from which one might wish to take a sample.

2. Continuous. They may take on any value within a continuum.

3. False. Every combination of n names is not possible – e.g., no combination of names containing two adjacent names is possible.

4. False. A numerical value describing a characteristic of a sample is a statistic, not a parameter.

5. Experiment. The application of a treatment makes the study an experiment.

6. False. There is no natural ordering of the colors.

7. Population. For a sample, the numerical value that describes some characteristic is a statistic.

8. Categorical. The data consist of names or labels, and they are not numbers representing counts or measurements. Since there is a natural ordering to the ratings, they are categorical data at the ordinal level of measurement.

9. Nominal. The book categories are names only, and no natural ordering scheme applies.

10. No. Since the people were randomly selected, the responders constitute a random sample of the population. The typical voluntary response sample results from those choosing to respond to surveys that are sent out (whether by print media, the Internet, on the phone, etc.) indiscriminately to broad audiences. If some of the 500 people in this particular survey refused to reply, it would still be a random sample – but the response "no reply" should be used to bring the total number of responses up to 500 in order to account for all the subjects in the survey.

Review Exercises

1. a. Since the survey was based on a voluntary response sample, its results might not be representative of the population. It could be that people who respond to such surveys tend to have characteristics (e.g., being conscientious) that would affect how they squeeze their toothpaste tubes.
   b. As stated to apply to the population of all Americans, the 72% is a parameter. In truth, the figure is based on a sample and is properly a statistic. While it may be used to estimate the value in the population, it is accurate only for the sample. It is a statistic that is used to estimate the parameter, and the article should have stated that "it is estimated that 72% of Americans squeeze their toothpaste tube from the top."
   c. Observational study, since there was no attempt to modify the behavior being observed.

2. No, results based only those who agree to respond may not be representative of the entire population. Those who refuse to respond may be fundamentally different from their more cooperative peers and may possess opinions that would become under-represented in the sample. One possible strategy for dealing with those who refuse to respond would be to call back (perhaps using a different pollster with an especially agreeable manner) and hope to find them in a more cooperative mood.

3. a. Ratio, since differences are meaningful and there is a meaningful zero.
   b. Nominal, since gender is only a category with no natural ordering.
   c. Interval, since differences between temperatures are meaningful but there is no meaningful zero.
   d. Ordinal, since the categories have a natural ordering but the differences between the categories are not necessarily uniform.

4. a. Nominal, since color is only a category with no natural ordering.
   b. Ratio, since differences are meaningful and there is a meaningful zero.
   c. Ordinal, since the categories have a natural ordering but the differences between the categories are not uniform.
   d. Interval, since differences between temperatures are meaningful but there is no meaningful zero.

5. a. Discrete. While it may be possible to own ½ shares, the number of shares cannot be any value on a continuum – e.g., a person cannot own π shares.

  b. Ratio. Differences are meaningful and there is a meaningful zero.

  c. Stratified. The population has been divided into subgroups, and subjects are selected from each subgroup.

  d. Statistic. While it may be used to estimate the true value in the population (i.e., the parameter), the average in the sample is a statistic.

  e. That would be a voluntary response sample and would likely be representative only of those stockholders with strong opinions and not of the general population of stockholders.

6. NOTE: The second part of the problem involves deciding "whether the sampling scheme is likely to result in a sample that is representative of the population." This is subjective, since the term "representative" has not been well-defined. Mathematically, any random sample is representative in the sense that there is no bias and a series of such samples can be expected to average out to the true population values – even though any one particular sample may not be representative. And so we expect a random sample of size n=50 to be representative. But how about a random sample of size n=2? It has all the mathematical properties of a random sample of size n=50, but common sense suggests that any one sample of size n=2 is not necessarily likely to be representative of the population.

  a. Systematic, since the selections were made at regular intervals. Yes, there is no reason why every 500th stockholder should have some characteristic that would introduce a bias.

  b. Convenience, since those selected were the ones who happened to attend. No, those who attend would tend to be the more interested and/or more well-off stockholders.

  c. Cluster, since the stockholders were organized into groups (by stockbroker) and all the stockholders in the selected groups were surveyed. Yes, since every stockholder has the same chance of being included in the sample (assuming each stockholder works through a single stockbroker), the sample will be a random sample – and considering the large number of stockbrokers involved (see the NOTE at the beginning of the problem), it is reasonable to expect the sample to be representative of the population.

  d. Random, since each stockholder has the same chance of being selected. Yes, since every stockholder has the same chance of being included in the sample, the scheme is likely to result in a representative sample.

  e. Stratified, since the stockholders were divided into subpopulations (zip codes) from which the actual sampling was done. No, since all the zip codes were given equal weights (5 stockholders from each) – because "significant" zip codes (with large numbers of stockholders) are counted equal with "insignificant" zip codes (with small numbers of small stockholders), stockholders from "insignificant" zip codes have a greater chance of being included and the types of stockholders that live in "significant" zip codes will be under-represented.

7. a. $12/35 = 0.343 = (0.343)(100\%) = 34.3\%$

  b. $(18\%)(4544) = (18/100)(4544) = 818$

8. a. Parameter, since it was calculated from the entire population.

  b. Discrete, since the numbers of votes must be whole numbers.

  c. $(49.72\%)(68,838,000) = (49.72/100)(68,838,000) = 34,226,000$ (rounded to agree with the accuracy used for the size of th population)

9. a. To contain 100% less fat would be to contain no fat at all – 100% of the fat has been removed. It is not physically possible to contain 125% less fat.

   b. $(58\%)(1182) = (58/100)(1182) = 686$

   c. $331/1182 = 0.280 = (0.280)(100\%) = 28.0\%$

10. The Gallup poll used randomly selected respondents, while the AOL respondents were a voluntary response sample. In addition, those with Internet access to participate in the AOL poll would include more affluent and well-educated persons (who tended to support Obama) and fewer blue-collar workers (who tended to support Clinton). The Gallup poll is more likely to reflect the true opinions of American voters.

Cumulative Review Exercises

NOTE: Throughout the text intermediate mathematical steps will be shown to aid those who may be having difficulty with the calculations. In practice, most of the work can be done continuously on calculators and the intermediate values are unnecessary. Even when the calculations cannot be done continuously, DO NOT WRITE AN INTERMEDIATE VALUE ON YOUR PAPER AND THEN RE-ENTER IT IN THE CALCULATOR. That practice can introduce round-off and copying errors. Store any intermediate values in the calculator so that you can recall them with greater accuracy and without copying errors. In general, the degree of accuracy appropriate depends upon the particular problem – and guidelines for this will be given as needed in subsequent chapters. Unless there is reason to otherwise, answers in this section are given with 3 decimal accuracy.

1. $\dfrac{1.1+1.7+1.7+1.1+...+1.1}{25} = \dfrac{31.4}{25} = 1.256$ mg

2. $\dfrac{110+96+170+125+...+119}{35} = \dfrac{4196}{35} = 119.9$ min

3. $\dfrac{85-80}{3.3} = \dfrac{5}{3.3} = 1.52$

4. $\dfrac{12.13-12.00}{0.12/\sqrt{24}} = \dfrac{0.13}{0.0245} = 5.31$

5. $\left[\dfrac{(1.96)(0.25)}{0.01}\right]^2 = [49]^2 = 2401$

6. $\dfrac{(491-513.174)^2}{513.174} = \dfrac{(-22.174)^2}{513.174} = \dfrac{491.6863}{513.174} = 0.9581$

7. $\dfrac{(98.0-98.4)^2+(98.6-98.4)^2+(98.6-98.4)^2}{3-1}$

   $= \dfrac{(-0.4)^2+(0.2)^2+(0.2)^2}{2} = \dfrac{0.16+0.04+0.04}{2} = \dfrac{0.24}{2} = 0.12$

8. $\sqrt{\dfrac{(98.0\text{-}98.4)^2+(98.6\text{-}98.4)^2+(98.6\text{-}98.4)^2}{3\text{-}1}}$

$= \sqrt{\dfrac{(-0.4)^2+(0.2)^2+(0.2)^2}{2}} = \sqrt{\dfrac{0.16+0.04+0.04}{2}} = \sqrt{\dfrac{0.24}{2}} = \sqrt{0.12} = 0.346$

NOTE FOR EXERCISES 9-12: Many calculators have a display that is limited to 8 or 10 characters, and they will not display as many digits as are shown in the following solutions. Answers with fewer significant digits are acceptable.

9. $(0.4)^{12} = 1.6777216\text{E-}05 = 0.000016777216$

10. $5^{15} = 3.0517578125\text{E}10 = 30{,}517{,}578{,}125$

11. $9^{11} = 3.1381059609\text{E}10 = 31{,}381{,}059{,}609$

12. $(0.25)^6 = 2.44140625\text{E-}04 = 0.000244140625$

Chapter 2

Summarizing and Graphing Data

2-2 Frequency Distributions

1. No. The first class frequency, for example, tells us only that there were 18 pennies with weights in the 2.40-2.49 grams class, but there is no way to tell the exact values of those 18 weights.

3. No. This is not a relative frequency distribution because the sum of the percentages is not 100%. It appears that each respondent was asked to indicate whether he downloaded the four types of material (and so the sum of the percentages could be anywhere from 0% to 400%), and not to place himself in one of the four categories (in which case the table would be a relative frequency distribution and the sum of the percentages would be 100%).

5. a. Class width: subtracting the first two lower class limits, $14-10 = 4$.
   b. Class midpoints: the first class midpoint is $(10+13)/2 = 11.5$, and the others can be obtained by adding the class width to get 11.5, 15.5, 19.5, 23.5, 27.5.
   c. Class boundaries: the boundary between the first and second class is $(13+14)/2 = 13.5$, and the others can be obtained by adding or subtracting the class width to get 9.5, 13.5, 17.5, 21.5, 25.5, 29.5.

7. a. Class width: subtracting the first two lower class limits, $1.00-0.00 = 1.00$.
   b. Class midpoints: the first class midpoint is $(0.00+0.99)/2 = 0.495$, and the others can be obtained by adding the class width to get 0.495, 1.495, 2.495, 3.495, 4.495.
   c. Class boundaries: the boundary between the first and second class is $(0.99+1.00)/2 = 0.995$, and the others can be obtained by adding or subtracting the class width to get -0.005, 0.995, 1.995, 2.995, 3.995, 4.995.

9. a. Strict interpretation: No; because there are more values at the upper end, there is not symmetry.
   b. Loose interpretation: Yes; there is a concentration of frequencies at the middle and a tapering off in both directions.

11. The requested figure is given below. Obtain each relative frequency by dividing the given frequency by 25, the total number of observations in each table. The "total" line is not necessary.

The non-filtered cigarettes have much more tar. Yes, the filters appear to be effective in reducing the amount of tar.

Relative Frequency Comparison

tar (mg)	cigarette type	
	non-filtered	filtered
2−5	0%	8%
6−9	0%	8%
10−13	4%	24%
14−17	0%	60%
18−21	60%	0%
22−25	28%	0%
26−29	8%	0%
total	100%	100%

NOTE: For cumulative tables, this manual uses upper class boundaries in the "less than" column. Consider exercise #13, for example, to understand why is done. Conceptually, weights occur on a continuum and the integer values reported are assumed to be the nearest whole number representation of the precise measure. An exact weight of 17.7, for example, would be reported as 18 and fall in the third class. The values in the second class, therefore, are better described as "less than 17.5" (using the upper class boundary) than as "less than 18" (using the lower class limit of the next class). This distinction is crucial in the construction of pictorial representations in the next section. To present a visually simpler table, however, it is common practice to follow the example in the text and use the lower class limit of the next class. Regardless of the "less than" label, the final cumulative frequency must equal the total sample size – and the sum of the cumulative frequency column has no meaning and should never be included.

13. Obtain the cumulative frequency values by adding the given frequencies.

tar (mg) in non-filtered cigarettes	cumulative frequency
less than 13.5	1
less than 17.5	1
less than 21.5	16
less than 25.5	23
less than 29.5	25

15. Obtain the relative frequencies by dividing the given frequencies by the total of 2223.

category	relative frequency
male survivors	16.2%
males who dies	62.8%
female survivors	15.5%
females who died	5.5%
	100.0%

17. The requested table is given below. The frequency distribution of the last digits shows unusually high numbers of 0's and 5's. This is typical for data that have been rounded off to "convenient" values. It appears that the heights were reported and not actually measured.

digit	frequency
0	9
1	2
2	1
3	3
4	1
5	15
6	2
7	0
8	3
9	1
	37

19. The requested table is given below.

nicotine (mg)	frequency
1.0−1.1	14
1.2−1.3	4
1.4−1.5	3
1.6−1.7	3
1.8−1.9	1
	25

21. The requested table is given below. No, the voltages do not appear to follow a normal distribution – instead of being concentrated near the middle of the distribution, the values appear to be rather evenly distributed.

voltage (volts)	frequency
123.3–123.4	10
123.5–123.6	9
123.7–123.8	10
123.9–124.0	10
124.1–124.2	1
	40

23. The requested table is given below. While over half of the screws are within 0.01 inches of the claimed value (28 of 50 fall between 0.74 and 0.76), there are over twice as many screws below that range as there are above it (15 vs. 7). It appears that there might be a slight tendency to err on the side of making the screws too small.

length (in)	frequency
0.720–0.729	5
0.730–0.739	10
0.740–0.749	11
0.750–0.759	17
0.760–0.769	7
	50

25. The requested table is given below. The ratings appear to have a distribution that is not normal. While there is a maximum score with progressively smaller frequencies on either side of the maximum, the distribution is definitely not symmetric (i.e., the maximum score is not near the middle, but at the upper end of the distribution).

FICO score	frequency
400–449	1
450–499	1
500–549	5
550–599	8
600–649	12
650–699	16
700–749	19
750–799	27
800–849	10
850–899	1
	100

27. The requested table is given below.

weight (g)	frequency
6.0000–6.0499	2
6.0500–6.0999	3
6.1000–6.1499	10
6.1500–6.1999	8
6.2000–6.2499	6
6.2500–6.2999	7
6.3000–6.3499	3
6.3500–6.3999	1
	40

29. The requested table is given below.

blood group	frequency
O	22
A	20
B	5
AB	3
	50

31. The frequency distributions including and excluding the outlier are given below. In general, an outlier can add several rows to a frequency distribution. Even though most of the added rows have frequency zero, the table tends to suggest that these are possible values – thus distorting the reader's mental image of the distribution.

0.0111 CANS (with the outlier)

weight (lbs)	frequency
200 – 219	6
220 – 239	5
240 – 259	12
260 – 279	36
280 – 299	87
300 – 319	28
320 – 339	0
340 – 359	0
360 – 379	0
380 – 399	0
400 – 419	0
420 – 439	0
440 – 459	0
460 – 479	0
480 – 499	0
500 – 519	1
	175

0.0111 CANS (without the outlier)

weight (lbs)	frequency
200 – 219	6
220 – 239	5
240 – 259	12
260 – 279	36
280 – 299	87
300 – 319	28
	174

2-3 Histograms

1. The pulse rate data have been organized into 7 classes. Examining the frequency distribution requires consideration of 14 pieces of information: the 7 class labels, and the 7 class frequencies. The histogram efficiently presents the same information in one visual image and gives all the relevant CVDOT (center, variation, distribution shape, outlier, [time is not relevant for these data]) details in an intuitive format.

3. The data set is small enough that the individual numbers can be examined; they do not require summarization in a figure. The data set is not large enough for a histogram to reveal the true nature of the distribution; the histogram will essentially be a repeat of the individual numbers.

NOTE: For exercises 5-8, the following values are used to answer the questions. It appears that the midpoints of the first 3 classes are 5,000 and 10,000 and 15,000. It appears that the heights of the 5 bars are 2, 30, 8, 15, 5.

5. a. Adding the heights of all the bars, the total number is 2+30+8+15+5 = 60.
   b. Adding the heights of the two rightmost bars, the number over 20,000 miles is 15+5 = 20.

7. a. The minimum possible miles traveled is the lower class boundary associated with the leftmost bar, 2500 miles.

b. The maximum possible number of miles traveled is the upper class boundary associated with the rightmost bar, 42,500 miles.

9. The histogram is given below. The digits 0 and 5 occur disproportionately more than the others. This is typical for data that have been rounded off to "convenient" values. It appears that the heights were reported and not actually measured.

11. The histogram is given below.

13. The histogram is given below. No, the voltages do not appear to follow a normal distribution – instead of being concentrated near the middle of the distribution, the values appear to be rather evenly distributed.

15. The histogram is given below. The true class boundaries are 0.7195, 0.7295, 0.7395, etc. The manual follows the text in presenting a histogram that communicates the information in an appropriate, though approximate, manner. While the 0.75″ label appears reasonably accurate in that all but 5 of the screws were within 0.02″ of that value, it appears that there are slightly more screws below the labeled value than above the labeled value and that the values extended farther below the labeled value than above the labeled value.

17. The histogram is given below. The true class boundaries are 399.5, 449.5, 499.5, etc. The manual follows the text in presenting a histogram that communicates the information in an appropriate, though approximate, manner. The ratings appear to have a distribution that is not normal. While there is a maximum score with progressively smaller frequencies on either side of the maximum, the distribution is definitely not symmetric (i.e., the maximum score is not near the middle, but at the upper end of the distribution).

19. The histogram is given below at the left. The true class boundaries are 5.99995, 6.04995, 6.09995, etc. The manual follows the text in presenting a histogram that communicates the information in an appropriate, though approximate, manner.

21. The back-to-back relative frequency histograms are given below. The pulse rates of the males tend to be lower than those of the females.

2-4 Statistical Graphics

1. The dotplot permits identification of each original value and is easier to construct. The dotplot gives an accurate visual impression of the proportion of the data within <u>any</u> selected range of values; while the polygon is limited to impressions concerning the specified classes (and only the heights at the class midpoints, and not the areas under the lines, give an accurate visual impression of those proportions).

3. Using relative frequencies allows direct comparison of the two polygons. When two sets of data have different sample sizes, the larger data set will naturally have higher frequencies and direct comparison of the heights of the two polygons does not give meaningful information.

5. The dotplot is given below. The Strontium-90 levels appear to have a "spread-out" normal distribution, a wide range of values clustered around 150 and occurring with less frequency at the extremes.

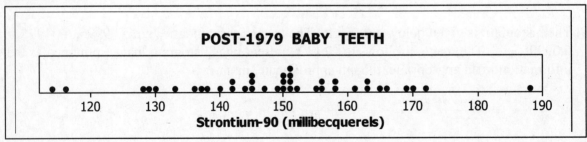

7. The frequency polygon is given below at the left.
   NOTE: The frequencies are plotted at the class midpoints, which are not integer values. The polygon must begin and end at zero at the midpoints of the adjoining classes that contain no data values.

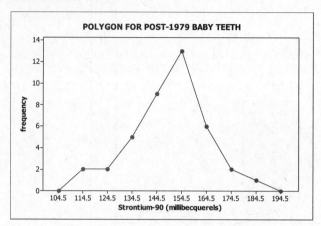

9. The stemplot is given below. The weights appear to be approximately normally distributed, except perhaps for the necessary lower truncation at zero.

   weight (pounds)
   0. | 12356677888999
   1. | 11234444444445556678
   2. | 0011111113334668888999
   3. | 9345
   4. | 36
   5. | 2

11. The ogive is given below at the left. Using the figure: move up from 4 on the horizontal scale to intersect the graph, then move left to intersect the vertical scale at 59. This indicates there were approximately 59 data values which would have been recorded as being below 4, which agrees with the actual data values.
    NOTE: Ogives always begin on the vertical axis at zero and end at n, the total number of data values. All cumulative values are plotted at the upper class boundaries.

13. The Pareto chart is given below.

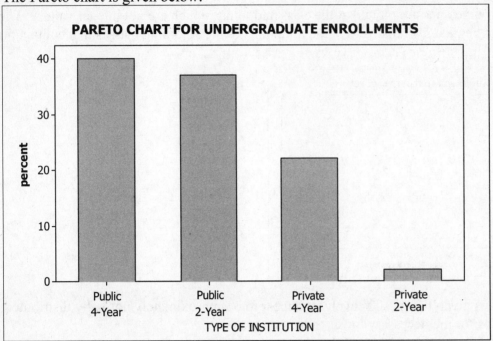

PARETO CHART FOR UNDERGRADUATE ENROLLMENTS

15. The pie chart is given below. The "slices" of the pie may appear in any order and in any position, but their relative sizes must be as shown. There were 1231 total responses, and the central angle of the pie chart for each category was determined as follows.

Interview: 452/1231 = 36.7%, and 36.7% of 360° is 132°

Resume: 297/1231 = 24.1%, and 24.1% of 360° is 87°

Reference Checks: 143/1231 = 11.6%, and 11.6% of 360° is 42°

Cover Letter: 141/1231 = 11/5%, and 11.5% of 360° is 41°

Interview Follow-Up: 113/1231 = 9.2%, and 9.2% of 360° is 33°

Screening Call: 85/1231 = 6.9%, and 6.9% of 360° is 35°

AREA OF THE MISTAKE

PIE CHART FOR JOB APPLICATION MISTAKES

17. The pie chart is given below at the left. The "slices" of the pie may appear in any order and in any position, but their relative sizes must be as shown.

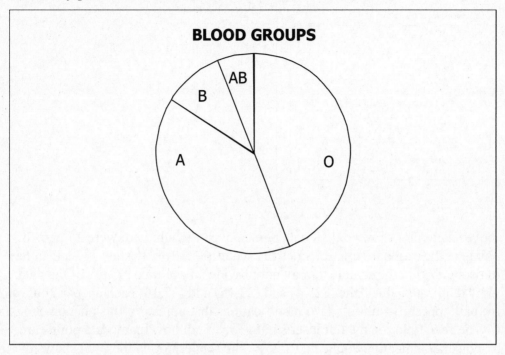

19. The Pareto chart is given below at the left.

21. The simple, unmodified scatterplot is given below. There appears to be a slight tendency for cigarettes with more tar to also have more CO.

NOTE: The above scatterplot shows only 9 data points, even though there were 25 pairs of tar/CO data points in the original sample. Since the scatterplot actually shows less than half the information contained in the sample, it may not provide an accurate picture of the data. This is caused by duplicate values: the (22,14) and (23,15) and (27,16) each appear 2 times, and the (20,16) pair appears <u>14</u> times! Two modifications that adjust for this phenomenon are shown below. The scatterplot on the left inserts numbers to tell how many data points are represented by dots that indicating duplicate values. The scatterplot on the right shows the true number of dots. The same effect can also be obtained by using dots whose size is proportional to the number of duplicate values it represents. The modified scatterplots indicate that there appears to be no relationship between the amounts of tar and CO.

23. The time series graph is given below. Note that the given years are not evenly spaced.

25. The multiple bar graph is given below. As the population increases, the <u>numbers</u> of marriages and divorces will automatically increase. To identify any change in marriage and divorce patterns, one needs to examine the <u>rates</u>. This is analogous to using percents (or relative frequencies) instead of frequencies to compare categories for two samples of different sizes. The marriage rate appears to have remained fairly constant, with a possible slight decrease in recent years. The divorce rate appears to have steadily grown, with a possible slight decrease in recent years.

27. The back-to-back stemplot is given below. The pulse rates for men appear to be lower than the pulse rates of women.

```
                 PULSE RATES
            Women|    |Men
                 |  5 | 666666
    888884444000 |  6 | 00000004444444888
 66666622222222 |  7 | 22222266
   88888000000 |  8 | 44448888
              6 |  9 | 6
              4 | 10 |
                | 11 |
              4 | 12 |
```

2-5 Critical Thinking: Bad Graphs

1. The illustration uses two-dimensional objects (dollar bills) to represent a one-dimensional variable (purchasing power). If the illustration uses a dollar bill with ½ the original length and ½ the original width to represent ½ the original purchasing power, then the illustration is misleading (because ½ the length and ½ the width translates into ¼ the area and gives the visual impression of 25% instead of 50%). But if the illustration uses a dollar bill with ½ the area (i.e, with .707 of the original length and .707 of the original width) to represent ½ the original purchasing power, then the illustration conveys the proper visual impression.

3. No. Results should be presented in a way that is fair and objective so that the reader has the reliable information necessary to reach his own conclusion.

5. No. The illustration uses two-dimensional objects to represent a one-dimensional variable (weight). The average male weight is $172/137 = 1.255$ times the average female weight. Making a two-dimensional figure 1.255 times taller and 1.255 times wider increases the area by $(1.255)^2 = 1.58$ and gives a misleading visual impression.

7. The average income for men is about 1.4 times the average income for women. Making the men's pictograph 1.4 times as wide and 1.4 times as high as the women's produces a men's image with $(1.4)^2 = 1.96$ times the area of the women's image. Since it is the area that gives the visual impression in a two-dimensional figure, the men's average income appears to be almost twice that of the women's average income. A graph that depicts the data fairly is given at the right.

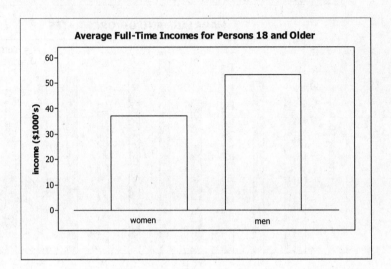

9. The graph in the text makes it appear that the braking distance for the Acura RL is more than twice that of the Volvo S80. The actual difference is about 60 feet, and the Acura RL distances is about $192/133 = 1.44$ times that of the Volvo S80. The exaggeration of differences is caused by the fact that the distance scale dies not start at zero. A graph that depicts the data fairly is given at the right.

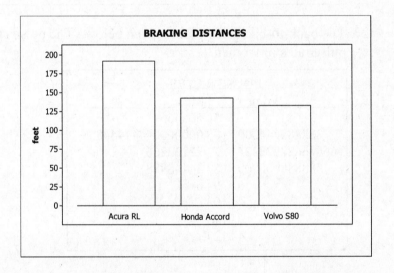

11. The given figure is misleading because the backside of the head is not visible. Categories extending to the backside of the head will not have as much area showing as comparable categories shown at the front of the head. A regular pie chart would give the relative sizes of the categories in as undistorted manner. A better graph would be a bar chart – with the vertical axis starting at 0, and the categories given in order by age. When there is a natural ordering of the categories that can be preserved with a bar chart – but it is hidden in a pie chart, which ends up placing the "first" and "last" categories side by side.

Statistical Literacy and Critical Thinking

1. When investigating the distribution of a data set, a histogram is more effective than a frequency distribution. Both figures contain the same information, but the visual impact of the histogram presents that information in a more efficient and more understandable manner.

2. When investigating changes over a period of years, a time series graph would be more effective than a histogram. A histogram would indicate the frequency with which different amounts occurred, but by ignoring the years in which those amounts occurred it would give no information about changes over time.

3. Using two-dimensional figures to compare one-dimensional variables exaggerates differences whenever the <u>areas</u> of the two dimensional figures are not proportional to the amounts being portrayed. Making the height and width proportional to the amounts being portrayed creates a

   distorted picture because it is area that makes the visual impression on the reader – and a two-fold increase in height and width produces a four-fold increase in area.

4. The highest histogram bars should be near the center, with the heights of the bars diminishing toward each end. The figure should be approximately symmetric.

Chapter Quick Quiz

1. $10 - 0 = 10$. The class width may be found by subtracting consecutive lower class limits.

2. Assuming the data represent values reported to the nearest integer, the class boundaries for the first class are -0.5 and 9.5.

3. No. All that can be said is that there are 27 data values somewhere within that class.

4. False. A normal distribution is bell-shaped, with the middle classes having higher frequencies than the classes at the extremes. The distribution for a balanced die will be flat, with each class having about the same frequency.

5. Variation.

6. 52, 52, 59. The 5 to the left of the stem represent the tens digit associated with the ones digits to the right of the stem.

7. Scatterplot. The data is two-dimensional, requiring separate axes for each variable (shoe size and height).

8. True. The vertical scale for the relative frequency histogram will be the values of the frequency histogram divided by the sample size n.

9. A histogram reveals the shape of the distribution of the data.

10. Pareto chart. When there is no natural order for the categories, placing them in the order of their frequencies shows the relative importance without losing the nature of any significant relationships between the categories.

Review Exercises

1. The frequency distribution is given at the right. The pulse rates for the males appear to be lower than those for the females.

MALE PULSE RATES	
beats per	
minute	frequency
50–59	6
60–69	17
70–79	8
80–89	8
90–99	1
	40

2. The histogram is given below. The basic shape is similar to the histogram for the females, but the male pulse rates appear to be lower.

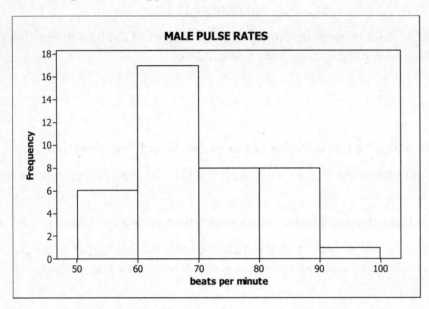

3. The dotplot is given below at the left. It shows that the male pulse rates appear to be lower than those for the females.

beats per minute
5 | 666666
6 | 00000004444444888
7 | 22222266
8 | 44448888
9 | 6

4. The stemplot is given above at the right. It shows that the male pulse rates appear to be lower than those of the females.

5. The scatterplot is given below.

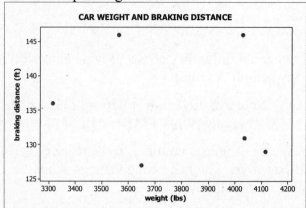

6. The time-series graph is given below.

7. The graph is misleading because the vertical axis does not start at zero, causing it to exaggerate the differences between the categories. A graph that correctly illustrates the acceleration times is given below.

8. a. 25. The difference between the first two lower class limits is $125 - 100 = 25$.
   b. 100 and 124. These are the values given in the first row of the table.
   c. 99.5 and 124.5. Values within these boundaries will round to the whole numbers given by the class limits.
   d. No. The distribution is not symmetric; the class with the largest frequency is near the right end of the distribution.

Cumulative Review Exercises

1. Yes. The sum of the relative frequencies is 100%.

2. Nominal. The yes-no-maybe responses are categories only; they do not provide numerical measures of any quantity, nor do they have any natural ordering.

3. The actual numbers of responses are as follows. Note that $884 + 433 + 416 = 1733$. .
   Yes: $(0.51)(1733) = 884$. No: $(0.25)(1733) = 433$. Maybe: $(0.24)(1733) = 416$

4. Voluntary Response Sample. A voluntary response sample is not likely to be representative of the population of all executives, but of those executives who had strong feelings about the topic and/or had enough free time to respond to such a survey.

5. a. A random sample is a sample in which every member of the population has an equal chance of being selected.
   b. A simple random sample of size n is a sample in which every possible sample of size n has an equal chance of being selected.
   c. Yes, it is a random sample because every person in the population of 300,000,000 has the same chance of being selected. No, it is not a simple random sample because all possible groups of 1000 do not have the same chance of being selected – in fact a group of 1000 composed of the oldest person in the each of the first 1000 of the 300,000 groups has no chance of being selected.

6. a. 100. The difference between the first two lower class limits is $100 - 0 = 100$.
   b. -0.5 and 99.5. Values within these boundaries will round to the whole numbers given by the class limits.
   c. $11/40 = 0.275$, or 27.5%. The total of the frequencies is 40.
   d. Ratio. Differences between the data values are meaningful and there is a meaningful zero.
   e. Quantitative. The data values are measurements of the cotinine levels.

7. The histogram is given below. Using a strict interpretation of the criteria, the cotinine levels do not appear to be normally distributed – the values appear to be concentrated in the lower portion of the distribution, with very few values in the upper portion.
   NOTE: The histogram bars extend from class boundary to class boundary. We follow the text and for clarity of presentation use the labels 0, 100, 200, etc. instead of the more cumbersome -0.5, 99.5, 199.5 etc.

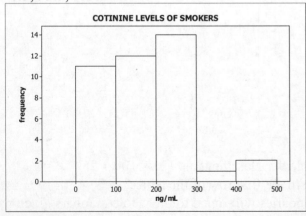

8. Statistic. A statistic is a measurement of some characteristic of a sample, while a parameter is a measurement of some characteristic of the entire population.

Chapter 3

Statistics for Describing, Exploring, and Comparing Data

3-2 Measures of Center

1. The mean, median, mode, and midrange are measures of "center" in the sense that they each attempt to determine (by various criteria – i.e., by using different approaches) what might be designated as a typical or representative value.

3. No. The price "exactly in between the highest and the lowest" would be the mean of the highest and lowest values – which is the midrange, and not the median.

NOTE: As it is common in mathematics and statistics to use symbols instead of words to represent quantities that are used often and/or that may appear in equations, this manual employs the following symbols for the various measures of center.

   mean = $\bar{x}$ mode = M
   median = $\tilde{x}$ midrange = m.r.

This manual will generally report means, medians and midranges accurate to one more decimal place than found in the original data. In addition, these two conventions will be employed.

   (1) When there is an odd number of data, the median will be one of the original values. The manual follows example #2 in this section and reports the median as given in the original data. And so the median of 1,2,3,4,5 is reported as $\tilde{x}$ = 3.
   (2) When the mean falls <u>exactly</u> between two values accurate to one more decimal place than the original data, the round-off rule in this section gives no specific direction. This manual follows the commonly accepted convention of always rounding up. And so the mean of 1,2,3,3 is reported as $\bar{x}$ = 9/4 = 2.3 [i.e., 9/4 = 2.25, rounded up to 2.3].

In addition, the median is the middle score <u>when the scores are arranged in order</u>, and the midrange is halfway between the first and last score <u>when the scores are arranged in order</u>. It is usually helpful to arrange the scores in order. This will not affect the mean, and it may also help in identifying the mode. Finally, no measure of center can have a value lower than the smallest score or higher than the largest score. Remembering this helps to protect against gross errors, which most commonly occur when calculating the mean.

5. Arranged in order, the n=10 scores are: 34 36 39 43 51 53 62 63 73 79
   a. $\bar{x} = (\Sigma x)/n = (533)/10 = 53.3$ words   c. M = (none)
   b. $\tilde{x} = (51 + 53)/2 = 52.0$ words   d. m.r. = (34 + 79)/2 = 56.5 words
   Using the mean of 53.3 words per page, a reasonable estimate for the total number of words in the dictionary is (53.3)(1459) = 77,765. Since the sample is a simple random sample, it should be representative of the population and the estimate of 77,675 is a valid estimate for the number of words in the dictionary – but not for the total number of words in the English language, since the dictionary does not claim to contain every word.
   NOTE: The exercise also asks whether the estimate for the total is "accurate." This is a relative term. A change of 1.0 in the estimate for the mean produces a change of 1459 in the estimate for the total. While a change of 1459 is only a change of 1459/77765 = 1.9%, this illustrates that accuracy in terms of absolute numbers is sometimes hard to attain when estimating totals.

7. Arranged in order, the n=5 scores are: 4277 4911 6374 7448 9051
   a. $\bar{x} = (\Sigma x)/n = (32061)/5 = \$6412.2$ c. M = (none)
   b. $\tilde{x} = \$6374$ d. m.r. $= (4277 + 9051)/2 = \$6664.0$
   No. Even though the sample values cover a fairly wide range, the measures of center do not differ very much.

9. Arranged in order, the n=10 scores are: 8.4 9.6 10 12 13 15 27 35 36 38
   a. $\bar{x} = (\Sigma x)/n = (204)/10 = \$20.4$ million   c. M = (none)
   b. $\tilde{x} = (13 + 15)/2 = \$14.0$million   d. m.r. $= (8.4 + 38)/2 = \$23.2$ million
   No, the top ten salaries provide no information (other than an upper cap) about the salaries of TV personalities in general. No, top 10 lists are generally not valuable for gaining insight into the larger population.
   NOTE: The "one more decimal place than is present in the original set of values" round-off rule given in the text sometimes requires thought before being applied. Since the data in this exercise is presented with two digit accuracy, the rule suggests reporting the appropriate measures of center with three digit accuracy.

11. Arranged in order, the n=15 scores are:
    0 73 95 165 191 192 221 235 235 244 259 262 331 376 381
    a. $\bar{x} = (\Sigma x)/n = (3260)/15 = 217.3$ hrs   c. M = 235 hrs
    b. $\tilde{x} = 235$ hrs   d. m.r. $= (0 + 381)/2 = 190.5$ hrs
    The duration time of 0 appears to be very unusual. It likely represents the Challenger disaster of January 1986, when the mission ended in an explosion shortly after takeoff.

13. Arranged in order, the n=20 scores are: 1 2 2 2 2 2 2 2 2 2 2 2 2 2 3 3 3 3 4 4
    a. $\bar{x} = (\Sigma x)/n = (47)/20 = 2.4$ mpg   c. M = 2 mpg
    b. $\tilde{x} = (2 + 2)/2 = 2.0$ mpg   d. m.r. $= (1 + 4)/2 = 2.5$ mpg
    No, there would not be much of an error. Since the amounts do not vary much, the mean appears to be a reasonable value to use for each of the older cars.

15. Arranged in order, the n=16 scores are:
    213 213 227 231 239 242 244 246 246 255 257 258 262 280 293 448
    a. $\bar{x} = (\Sigma x)/n = (4154)/16 = 259.6$ sec   c. M = 213 sec, 246 sec (bimodal)
    b. $\tilde{x} = (246 + 246)/2 = 246.0$ sec   d. m.r. $= (213 + 448)/2 = 330.5$ sec
    Yes, the time of 448 seconds appears to be very different from the others.

17. Arranged in order, the n=20 scores are:
    4 4 4 4 4 4 4.5 4.5 4.5 4.5 4.5 4.5 6 6 8 9 9 13 13 15
    a. $\bar{x} = (\Sigma x)/n = (130)/20 = 6.5$ yrs   c. M = 4 yrs, 4.5 yrs (bi-modal)
    b. $\tilde{x} = (4.5 + 4.5)/2 = 4.5$ yrs   d. m.r. $= (4 + 14)/2 = 9.5$ yrs
    Yes, it is common to earn a bachelor's degree in 4 years in that we estimate that 6/20 = 30% of the people do so – but the typical time to earn a bachelor's degree appears to be more than that.
    NOTE: The "one more decimal place than is present in the original set of values" round-off rule given in the text sometimes requires thought before being applied. In this exercise it appears that the basic measurement is being done in whole number of years, and so the appropriate measures of center should be reported with one decimal accuracy. When there are different number of decimal places in the original data, it is generally appropriate to use the least accurate data as the basis for applying the round-off rule.

19. Arranged in order, the n=12 scores are: 14 15 59 85 95 97 98 106 117 120 143 371

    a. $\bar{x} = (\Sigma x)/n = (1320)/12 = 110.0$ c. M = (none)

    b. $\tilde{x} = (97 + 98)/2 = 97.5$ d. m.r. $= (14 + 371)/2 = 192.5$

    Yes, there does appear to be a trend in the data. The numbers of bankruptcies seems to steadily increase up to a certain point and then drop off dramatically. It could be that new bankruptcy laws (that made it more difficult to declare bankruptcy) went into effect.

21. Arranged in order, the scores are as follows.

    30 days: 244 260 264 264 278 280 318   one day: 456 536 567 614 628 943 1088

    n = 7   n = 7

    $\bar{x} = (\Sigma x)/n = 1908/7 = \$272.6$ $\bar{x} = (\Sigma x)/n = 4832/7 = \$690.3$

    $\tilde{x} = \$264$ $\tilde{x} = \$614$

    The tickets purchased 30 days in advance are considerably less expensive – they appear to cost less than half of the tickets purchased one day in advance.

23. Arranged in order, the scores are as follows.

    non-filtered: 1.0 1.1 1.1 1.1 1.1 1.1 1.1 1.1 1.1 1.1 1.1 1.1 1.1
    　　　　　　 1.1 1.2 1.2 1.3 1.3 1.4 1.4 1.5 1.6 1.7 1.7 1.8

    filtered: 0.2 0.4 0.6 0.8 0.8 0.8 0.8 0.8 0.8 0.8 0.9 1.0 1.0
    　　　　 1.0 1.0 1.0 1.1 1.1 1.1 1.1 1.1 1.1 1.1 1.2 1.3

non-filtered	filtered

    n = 25   n = 25

    $\bar{x} = (\Sigma x)/n = 31.4/25 = 1.26$ mg $\bar{x} = (\Sigma x)/n = 22.9/25 = 0.92$ mg

    $\tilde{x} = 1.1$ mg $\tilde{x} = 1.0$ mg

    The filtered cigarettes seem to have less nicotine. Filters appear to be effective in reducing the amounts of nicotine.

25. The Minitab DESCRIBE results are as follows.

Variable	N	Mean	SE Mean	StDev	Minimum	Q1	Median	Q3	Maximum
BodyTemp	106	98.200	0.0605	0.623	96.500	97.800	98.400	98.600	99.600

    $\bar{x} = 98.20\ °F$ and $\tilde{x} = 98.40\ °F$. These results suggest that the mean body temperature is less than 98.6 °F.

27. The Minitab DESCRIBE results are as follows.

Variable	N	Mean	SE Mean	StDev	Minimum	Q1	Median	Q3	Maximum
home	40	123.66	0.0380	0.240	123.30	123.43	123.70	123.90	124.20
generator	40	124.66	0.0457	0.289	124.00	124.50	124.70	124.80	125.20
UPS	40	123.59	0.0487	0.308	122.90	123.53	123.70	123.80	123.90

    home: $\bar{x} = 123.66$ volts and $\tilde{x} = 123.70$ volts

    generator: $\bar{x} = 124.66$ volts and $\tilde{x} = 124.70$ volts

    UPS: $\bar{x} = 123.59$ volts and $\tilde{x} = 123.70$ volts

    The three different sources appear to have about the same voltage measurements, and within each source the mean and the median are very close to each other.

29. The x values below are the class midpoints from the given frequency table, the class limits of which indicate the original data were integers.

x	f	f·x
11.5	1	11.5
15.5	0	0
19.5	15	292.5
23.5	7	164.5
27.5	2	55.0
	25	523.5

    $\bar{x} = \Sigma(f \cdot x)/\Sigma f = 523.5/25 = 20.9$ mg

    This is close to the true value of 21.1 mg.

31. The x values below are the class midpoints from the given frequency table, the class limits of which indicate the original data were integers.

x	f	f·x
43.5	25	1087.5
47.5	14	665.0
51.5	7	360.5
55.5	3	166.5
59.5	1	59.5
	50	2339.0

$\bar{x} = \Sigma(f \cdot x)/\Sigma f = 2339.0/50 = 46.8$ mph

This is close to the true value of 46.7 mph. This is considerably higher than the posted speed limit of 30 mph – but remember that this does not estimate the mean speed of all drivers, but the mean speed of drivers who receive speeding tickets.

33. The x values below are the numerical values of the letter grades, and the corresponding weights are the numbers of credit hours.

x	w	w·x
3	3	9
2	3	6
3	4	12
4	4	16
1	1	1
	15	44

Σx $\bar{x} = \Sigma(w \cdot x)/\Sigma w = 44/15 = 2.93$

No; since 2.93 is below 3.00, the student did not make the Dean's list.

35. a. The formula $\bar{x} = (\Sigma x)/n$ implies that $n \cdot \bar{x} = \Sigma x$, the sum of the sample values.

   Since $n \cdot \bar{x} = 5 \cdot (657.054) = 3285.27$, that is the sum of the sample values.
   Since the 4 given sample values sum to 2642.76, the missing value is found by subtracting
      3285.27 – 2642.76 = 642.51 micrograms.

   b. As above, the formula $\bar{x} = (\Sigma x)/n$ implies that $n \cdot \bar{x} = \Sigma x$.

   Since n and $\bar{x}$ are given, the total Σx is determined.
   For any freely assigned n−1 values, the final value is determined to be the value
      (whether + or −) necessary to obtain the desired total.

37. a. Arranged in order, the original 100 scores are:

```
444  497  502  503  517  519  532
559  568  579  579  591  591  594  598  603  604  611  611  617  618
628  630  635  636  637  638  651  654  657  660  661  664  681  681
682  689  692  693  694  696  697  698  701  706  708  709  709  711
713  714  714  714  722  729  731  732  739  741  743  744  745  751
752  753  753  755  756  756  760  768  768  777  779  781  782  782
783  784  787  789  792  793  795  795  796  797  797  798  802  809
818  824  829  830  834  835  836  849  850
```

   $\bar{x} = (\Sigma x)/n = (70311)/100 = 703.1$

   b. Trimming the highest and lowest 10% (or ten scores), the remaining 80 scores are:

```
579  591  591  594  598  603  604  611  611  617  618  628  630  635
636  637  638  651  654  657  660  661  664  681  681  682  689  692
693  694  696  697  698  701  706  708  709  709  711  713  714  714
714  722  729  731  732  739  741  743  744  745  751  752  753  753
755  756  756  760  768  768  777  779  781  782  782  783  784  787
789  792  793  795  795  796  797  797  798  802
```

   $\bar{x} = (\Sigma x)/n = (56777)/80 = 709.7$

c. Trimming the highest and lowest 20% (or 20 scores), the remaining 60 scores are:

```
618   628   630   635   636   637   638   651   654   657   660   661   664   681
681   682   689   692   693   694   696   697   698   701   706   708   709   709
711   713   714   714   714   722   729   731   732   739   741   743   744   745
751   752   753   753   755   756   756   760   768   768   777   779   781   782
782   783   784   787
```

$\overline{x} = (\Sigma x)/n = (42824)/60 = 713.7$

The results are not dramatically different. Since the trimmed mean appears to increase slightly as more values are trimmed, the data may be negatively skewed – i.e., have smaller values that are not balanced by corresponding larger ones.

39. The geometric mean of the three values is the third (i.e., cube) root of their product:
$\sqrt[3]{(1.10)(1.05)(1.02)} = 1.056$, which corresponds to a 5.6% growth rate.

This is not the same as the mean of the growth rates:
(10% + 5% + 2%)/3 = (17%)/3 = 5.7% .

41. There are 25 scores in the table in Exercise 29. The median, x_{13} for the ordered scores, is in the 18-21 class. The relevant values are as follows:

median lower class limit = 18   n = 25 total scores

class width = 4   m = 1 total scores preceding the median class

median class frequency = 15

$\tilde{x}$ = (median class lower limit) + (class width)·[(n+1)/2 – (m+1)]/(median class frequency)

= 18 + 4·[26/2 – 2]/15

= 18 + 4·[11]/15

= 18 + 2.9

= 20.9 mg

This compares favorably with the true median value of 20.0 mg calculated directly from the original list. Since frequency tables are summaries which do not contain all the information, values obtained from the original data should always be preferred to ones obtained from frequency tables.

3-3 Measures of Variation

1. Variation is a general descriptive term that refers to the fact that all the items being measured are not identical. In statistics, variance is a specific and well-defined measure of variation that has a particular mathematical formula.

3. The incomes of the adults selected from the general population should have more variation. Statistics teachers are a more homogeneous group – they would tend to have similar salaries that would not be likely to include the extremely high or extremely low salaries possible in the general population.

NOTE: Although not given in the text, the symbol R will be used for the range throughout this manual. Remember that the range is the difference between the highest and the lowest scores, and not necessarily the difference between the last and first scores as they are listed. Since calculating the range involves only the subtraction of 2 original pieces of data, the manual follows the advice in this section that "In general, the range should not be rounded" and reports the range with the same accuracy found in the original data. In general, the standard deviation and the variance will be reported with one more decimal place than the original data.

When finding the square root of the variance to obtain the standard deviation, use all the decimal places of the variance – and not the rounded value reported as the answer. To do this, keep the value on the calculator display or place it in the memory. Do not copy down all the decimal places and then re-enter them to find the square root – as that not only is a waste of time, but also could introduce copying and/or round-off errors.

5. preliminary values: n = 10 $\Sigma x = 533$ $\Sigma x^2 = 30615$

   R = 79 – 34 = 45 words

   $s^2 = [n(\Sigma x^2) – (\Sigma x)^2]/[n(n-1)]$

   $\quad = [10(30615) – (533)^2]/[10(9)] = 22061/90 = 245.1$ words^2

   $s = \sqrt{245.1} = 15.7$ words

There seems to be considerable variation from page to page. For small samples with n = 10, there could be considerable variation in the sample mean and, therefore, considerable variation in the projected totals for the entire dictionary. It appears that there would be a question about the accuracy of an estimate based on this sample for the total number of words.

NOTE: The above format will be used in this manual for calculation of the variance and standard deviation. To reinforce some of the key symbols and concepts, the following detailed mathematical analysis is given for the data of the above problem.

x	$x-\bar{x}$	$(x-\bar{x})^2$	x^2	
34	-19.3	372.49	1156	$\bar{x} = (\Sigma x)/n = 533/10 = 53.3$ words
36	-17.3	299.29	1296	
39	-14.3	204.49	1521	by formula 3-4, $s^2 = \Sigma(x-\bar{x})^2/(n-1)$
43	-10.3	106.09	1849	$\quad = 2206.10/9$
51	-2.3	5.29	2601	$\quad = 245.1222$ words^2
53	-0.3	0.09	2809	
62	8.7	75.69	3844	by formula 3-5, $s^2 = [n(\Sigma x^2) – (\Sigma x)^2]/[n(n-1)]$
63	9.7	94.09	3969	$\quad = [10(30615) – (533)^2]/[10(9)]$
73	19.7	388.09	5329	$\quad = 22061/90$
79	25.7	660.49	6241	$\quad = 245.1222$ words^2
533	0.0	2206.10	30615	

NOTE: As shown in the preceding detailed mathematical analysis, formulas 3-4 and 3-5 give the same answer. The following three observations give hints for using these two formulas.

• When using formula 3-4, constructing a table having the first 3 columns shown above helps to organize the calculations and makes errors less likely. In addition, verify that $\Sigma(x-\bar{x}) = 0$ [except for a possible discrepancy at the last decimal due to rounding] before proceeding – if such is not the case, an error has been made and further calculation is fruitless.

• When using formula 3-5, the quantity $[n(\Sigma x^2) – (\Sigma x)^2]$ must be non-negative (i.e., greater than or equal to zero) – if such is not the case, an error has been made and further calculation is fruitless.

• In general, formula 3-5 is to be preferred over formula 3-4 because (1) it does not involve round-off error or messy calculations when the mean does not "come out even" and (2) the quantities Σx and Σx^2 can be found directly (on most calculators, with one pass through the data) without having to construct a table or make a second pass through the data after finding the mean.

7. preliminary values: n = 5 $\Sigma x = 32061$ $\Sigma x^2 = 220431831$

   R = 9052 – 4227 = $4774

   $s^2 = [n(\Sigma x^2) – (\Sigma x)^2]/[n(n-1)]$

   $\quad = [5(220431831) – (32061)^2]/[5(4)] = 74251434/20 = 3712571.7$ dollars^2

   $s = \sqrt{3712571.7} = \$1926.8$

A value is considered unusual if it differs from the mean by more than two standard deviations. Since $10,000 differs from the mean by $(10{,}000 – 6{,}412.2)/1926.8 = 1.86$ standard deviations, in this context it would not be considered an unusual value.

9. preliminary values: n = 10 $\Sigma x = 204$ $\Sigma x^2 = 5494.72$
   R = 38 − 8.4 = \$29.6 million
   $s^2 = [n(\Sigma x^2) − (\Sigma x)^2]/[n(n-1)]$
   $= [10(5494.72) − (204)^2]/[10(9)] = 13331.2/90 = 148.1$ (million dollars)^2
   $s = \sqrt{148.1} = \$12.2$ million
   No. The variation among the top ten salaries reveals essentially nothing about the variation of the salaries in general. All that can be said is that the range of salaries in the general population of TV personalities must be at least \$29.6 million.

11. preliminary values: n = 15 $\Sigma x = 3260$ $\Sigma x^2 = 865574$
    R = 381 − 0 = 381 hrs
    $s^2 = [n(\Sigma x^2) − (\Sigma x)^2]/[n(n-1)]$
    $= [15(865574) − (3260)^2]/[15(14)] = 23356010/210 = 11219.1$ hrs^2
    $s = \sqrt{11219.1} = 105.9$ hrs
    A value is considered unusual if it differs from the mean by more than two standard deviations. Since 0 hours differs from the mean by (0 − 217.3)/105.9 = -2.05 standard deviations, in this context it would be considered an unusual value.

13. preliminary values: n = 20 $\Sigma x = 47$ $\Sigma x^2 = 121$
    R = 4 − 1 = 3 mpg
    $s^2 = [n(\Sigma x^2) − (\Sigma x)^2]/[n(n-1)]$
    $= [20(121) − (47)^2]/[20(19)] = 211/380 = 0.6$ (mpg)^2
    $s = \sqrt{0.6} = 0.7$ mpg
    A value is considered unusual if it differs from the mean by more than two standard deviations. Since 4 mpg differs from the mean by (4 − 2.35)/0.75 = 2.21 standard deviations, in this context it would be considered an unusual value.

15. preliminary values: n = 16 $\Sigma x = 4154$ $\Sigma x^2 = 1123116$
    R = 448 − 213 = 235 sec
    $s^2 = [n(\Sigma x^2) − (\Sigma x)^2]/[n(n-1)]$
    $= [16(1123116) − (4154)^2]/[16(15)] = 714140/240 = 2975.6$ sec^2
    $s = \sqrt{2975.6} = 54.5$ sec
    If the highest playing time is omitted, the standard deviation drops to 22.0 sec. The change is substantial. NOTE: Use the formula again. The new Σx and Σx^2 may be found by subtracting from the above totals, and the n is reduced by 1.

17. preliminary values: n = 20 $\Sigma x = 130$ $\Sigma x^2 = 1078.50$
    R = 15 − 4 = 11 yrs
    $s^2 = [n(\Sigma x^2) − (\Sigma x)^2]/[n(n-1)]$
    $= [20(1078.50) − (130)^2]/[20(19)] = 4670/380 = 12.3$ yrs^2
    s = 3.5 yrs
    A value is considered unusual if it differs from the mean by more than two standard deviations. Since 12 years differs from the mean by (12 − 6.5)/3.5 = 1.57 standard deviations, in this context it would not be considered an unusual value.

19. preliminary values: n = 12 $\Sigma x = 1320$ $\Sigma x^2 = 236580$
   R = 371 – 14 = 357 bankruptcies
   $s^2 = [n(\Sigma x^2) - (\Sigma x)^2]/[n(n-1)]$
   $ = [12(236580) - (1320)^2]/[12(11)] = 1096560/132 = 8307.3$ bankruptcies^2
   s = 91.1 bankruptcies
   A value is considered unusual if it differs from the mean by more than two standard deviations.
   Since 371 bankruptcies differs from the mean by (371 – 110.0)/91.1 = 2.86 standard
   deviations, in this context it would be considered an unusual value. None of the other given
   values meets that criterion.

NOTE: In exercises 21-24, CV = $s/\overline{x}$ is found by storing the value of $\overline{x}$ and dividing the
unrounded s by the unrounded stored value of $\overline{x}$. Using the rounded values of $\overline{x}$ and s may
produce slight different values as indicated. The CV is expressed as a percent.

21. 30 Days
   n = 7, $\Sigma x = 1908$, $\Sigma x^2 = 523336$
   $\overline{x} = (\Sigma x)/n = 1908/7 = \$272.6$
   $s^2 = [n(\Sigma x^2) - (\Sigma x)^2]/[n(n-1)]$
   $ = [7(523336) - (1908)^2]/[7(6)]$
   $ = 22888/42 = 545.0$ dollars^2
   s = \$23.3
   CV = $s/\overline{x}$ = (\$23.3)/(\$272.6) = 8.6%
   [8.5% using rounded values for $\overline{x}$ and s]

   One Day
   n = 7, $\Sigma x = 4832$, $\Sigma x^2 = 3661094$
   $\overline{x} = (\Sigma x)/n = 4832/7 = \$690.3$
   $s^2 = [n(\Sigma x^2) - (\Sigma x)^2]/[n(n-1)]$
   $ = [7(3661094) - (4832)^2]/[7(6)]$
   $ = 2279434/42 = 54272.2$ dollars^2
   s = \$233.0
   CV = $s/\overline{x}$ = (\$233.0)/(\$690.3) = 33.7%
   [33.8% using rounded values for $\overline{x}$ and s]

   There is considerably more variation among the costs for tickets purchased one day in
   advance.

23. Non-filtered
   n = 25, $\Sigma x = 31.4$, $\Sigma x^2 = 40.74$
   $\overline{x} = (\Sigma x)/n = 31.4/25 = 1.26$ mg
   $s^2 = [n(\Sigma x^2) - (\Sigma x)^2]/[n(n-1)]$
   $ = [25(40.74) - (31.4)^2]/[25(24)]$
   $ = 32.54/600 = 0.05$ mg^2
   s = 0.23 mg
   CV = $s/\overline{x}$ = (0.23 mg)/(1.26 mg) = 18.5%
   [18.3% using rounded values for $\overline{x}$ and s]

   Filtered
   n = 25, $\Sigma x = 22.9$, $\Sigma x^2 = 22.45$
   $\overline{x} = (\Sigma x)/n = 22.9/25 = 0.92$ mg
   $s^2 = [n(\Sigma x^2) - (\Sigma x)^2]/[n(n-1)]$
   $ = [25(22.45) - (22.9)^2]/[25(24)]$
   $ = 36.84/600 = 0.06$ mg^2
   s = 0.25 mg
   CV = $s/\overline{x}$ = (0.25 mg)/(0.92 mg) = 27.1%
   [27.2% using rounded values for $\overline{x}$ and s]

   The nicotine values for the filtered cigarettes appear to exhibit slightly more variation. This is
   reasonable, since the filter is an added element: the variability in filters adds more variation to
   the natural variation already present in the tobaccos.

25. The Minitab DESCRIBE results are as follows.

Variable	N	Mean	SE Mean	StDev	Minimum	Q1	Median	Q3	Maximum
BodyTemp	106	98.200	0.0605	0.623	96.500	97.800	98.400	98.600	99.600

   R = 99.6 – 96.5 = 3.1 °F; $s^2 = 0.623^2 = 0.39$ (°F)^2 ; s = 0.62 °F

27. The Minitab DESCRIBE results are as follows.

Variable	N	Mean	SE Mean	StDev	Minimum	Q1	Median	Q3	Maximum
home	40	123.66	0.0380	0.240	123.30	123.43	123.70	123.90	124.20
generator	40	124.66	0.0457	0.289	124.00	124.50	124.70	124.80	125.20
UPS	40	123.59	0.0487	0.308	122.90	123.53	123.70	123.80	123.90

   home: R = 124.2 – 123.3 = 0.9 volts; $s^2 = 0.240^2 = 0.06$ volts^2; s = 0.24 volts
   generator: R = 125.2 – 124.0 = 1.2 volts; $s^2 = 0.289^2 = 0.08$ volts^2; s = 0.29 volts
   UPS: R = 123.9 – 122.9 = 1.0 volts; $s^2 = 0.308^2 = 0.09$ volts^2; s = 0.31 volts
   The three different sources appear to have about the same amount of variation.

29. The x values below are the class midpoints from the given frequency table, the class limits of which indicate the original data were integers.

x	f	f·x	f·x^2
11.5	1	11.5	132.25
15.5	0	0	0
19.5	15	292.5	5703.75
23.5	7	164.5	3865.75
27.5	2	55.0	1512.50
	25	523.5	11214.25

$s^2 = \{n[\Sigma(f \cdot x^2)] - [\Sigma(f \cdot x)]^2\}/[n(n-1)]$
$= \{25[11214.25] - [523.5]^2\}/[25(24)]$
$= 6304/600$
$= 10.507$
$s = 3.2$ mg

This is the same as the true value of 3.2 mg.

31. The range rule of thumb suggests s ≈ range/4. In this case, the range is 24 – 18 = 6, and so we estimate s ≈ 6/4 = 1.5 years.

33. a. The range from 154 to 168 is $\bar{x} \pm 1s$. The empirical rule suggests that about 68% of the data values should fall within those limits.
    b. The range from 147 to 175 is $\bar{x} \pm 2s$. The empirical rule suggests that about 95% of the data values should fall within those limits.

35. Chebyshev's Theorem states that for any set of data there must be at least $1 - 1/2^2 = 1 - 1/4 = 3/4$ (i.e., 75%) of the data values within 2 standard deviations of the mean. In this context, the limits are $161 \pm 2(7)$: there must be at least 75% of the heights between 147 cm and 175 cm.

37. For greater accuracy and understanding, we use 3 decimal places and formula 3-4.
    a. the original population

x	x-μ	(x-μ)^2
1	-5	25
3	-3	9
14	8	64
18	0	98

$\mu = (\Sigma x)/N = 18/3 = 6$

$\sigma^2 = \Sigma(x-\mu)^2/N = 98/3 = 32.667$
$[\sigma = 5.715]$

b. the nine samples: using $s^2 = \Sigma(x-\bar{x})^2/(n-1)$ [for each sample, n=2]

sample	$\bar{X}$	s^2	s
1,1	1.0	0	0
1,3	2.0	2.0	1.414
1,14	7.5	84.5	9.192
3,1	2.0	2.0	1.414
3,3	3.0	0	0
3,14	8.5	60.5	7.778
14,1	7.5	84.5	9.192
14,3	8.5	60.5	7.778
14,14	14.0	0	0
	54.0	294.0	36.768

mean of the 9 s^2 values

$(\Sigma s^2)/9 = 294.0/9 = 32.667$

c. the nine samples: using $\sigma^2 = \Sigma(x-\mu)^2/N$ [for each sample, N=2]

sample	μ	σ^2
1,1	1.0	0
1,3	2.0	1.00
1,14	7.5	42.25
3,1	2.0	1.00
3,3	3.0	0
3,14	8.5	30.25
14,1	7.5	42.25
14,3	8.5	30.25
14,14	14.0	0
	54.0	147.00

mean of the 9 σ^2 values

$(\Sigma\sigma^2)/9 = 147.00/9 = 16.333$

d. The approach in (b) of dividing by n-1 when calculating the sample variance gives a better estimate of the population variance. On the average, the approach in (b) gave the correct population variance of 32.667. The approach in (c) of dividing by n underestimated the correct population variance. When computing sample variances, divide by n-1 and not by n.

e. No. An unbiased estimator is one that gives the correct answer on the average. Since the average value of s^2 in part (b) was 6, which was the correct value calculated for σ^2 in part (a), s^2 is an unbiased estimator of σ^2. Since the average value of s in part (b) is $(\Sigma\ s)/9 =$ 36.768/9 = 4.085, which is not the correct value of 5.715 calculated for σ in part (a), s is not an unbiased estimator of σ.

NOTE: Since the average value of $\overline{X}$ in part (b) is $(\Sigma\overline{X})/9 = 54.0/9 = 6.0$, which is the correct value calculated for μ in part (a), $\overline{X}$ is an unbiased estimator of μ.

3-4 Measures of Relative Standing and Boxplots

1. For a z score of -0.61, the negative sign indicates that her age is below the mean and the numerical portion indicates that her age is 0.61 standard deviations away from the mean.

3. The values shown in the boxplot constitute the 5-number summary as follows.
0 hours = the minimum value, the length of the shortest flight
166 hours = the first quartile Q_1, the length below which the shortest 25% of the flights occur
215 hours = the second quartile Q_2, the median length of the flights
269 hours = the third quartile Q_3, the length above which the longest 25% of the flights occur
423 hours = the maximum value, the length of the longest flight

5. a. $|x - \overline{x}| = |61 - 35.8| = |25.2| = 25.2$ years
   b. 25.2/11.3 = 2.23
   c. $z = (x - \overline{x})/s = (61-35.8)/11.3 = 2.23$
   d. Since 2.23 > 2.00, Helen Mirren's age is considered unusual in this context.

7. a. $|x - \overline{x}| = |110 - 245.0| = |-135.0| = 135.0$ seconds
   b. 135.0/36.4 = 3.71
   c. $z = (x - \overline{x})/s = (110 - 245.0)/36.4 = -3.71$
   d. Since -3.71 < -2.00, a duration time of 110 seconds is considered unusual in this context.

9. a. $z = (x - \overline{x})/s = (101.00 - 98.20)/0.62 = 4.52$; unusual, since 4.52 > 2.00
   b. $z = (x - \overline{x})/s = (96.90 - 98.20)/0.62 = -2.10$; unusual, since -2.10 < -2.00
   c. $z = (x - \overline{x})/s = (96.98 - 98.20)/0.62 = -1.97$; usual, since -2.00 < -1.97 < 2.00

11. $z = (x - \overline{x})/s = (308 - 268)/15 = 40/15 = 2.67$
    Yes, since 2.67 > 2.00, a pregnancy of 308 days is considered unusual. "Unusual" is just what the word implies – out of the ordinary. Other exercises in this section note that there are unusual ages and heights, and so there can be unusual pregnancies.

13. SAT: $z = (x - \overline{x})/s = (1840 - 1518)/325 = 322/325 = 0.99$
    ACT: $z = (x - \overline{x})/s = (26.0 - 21.1)/4.8 = 4.9/4.8 = 1.02$
    Since 1.02 > 0.99, the ACT score of 26.0 is the relatively better score.

For exercises 15-27, refer to the list of ordered scores at the right. The units are total points.

#	points
1	36
2	37
3	37
4	39
5	39
6	41
7	43
8	44
9	44
10	47
11	50
12	53
13	54
14	55
15	56
16	56
17	57
18	59
19	61
20	61
21	65
22	69
23	69
24	75

15. Let b = # of scores below x; n = total number of scores
    In general, the percentile of score x is (b/n)·100.
    The percentile score of 47 is (9/24)·100 = 38.

17. Let b = # of scores below x; n = total number of scores
    In general, the percentile of score x is (b/n)·100.
    The percentile score of 54 is (12/24)·100 = 50.

19. To find P_{20}, L = (20/100)·24 = 4.8, rounded up to 5.
    Since the 5th score is 39, P_{20} = 39.

21. To find $Q_3 = P_{75}$, L = (75/100)·24 = 18, a whole number.
    The mean of the 18th and 19th scores, Q_3 (59+61)/2 = 60.

23. To find P_{50}, L = (50/100)·24 = 12, a whole number.
    The mean of the 12th and 13th scores, P_{50} = (53+54)/2 = 53.5.

25. To find P_{25}, L = (25/100)·24 = 6, a whole number.
    The mean of the 6th and 7th scores, P_{25} = (41+43)/2 = 42.

27. Arrange the 24 values in order. The 5-number summary and
    boxplot are given below. All measurements are in total points.
    min = x_1 = 36
    $Q_1 = (x_6+x_7)/2 = (41+43)/2 = 42$
    median = $(x_{12}+x_{13})/2 = (53+54)/2 = 53.5$
    $Q_3 = (x_{18}+x_{19})/2 = (59+61)/2 = 60$
    max = x_{24} = 75

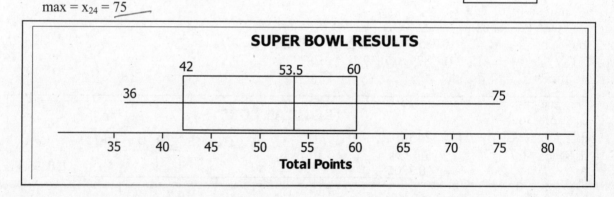

SUPER BOWL RESULTS

GENERAL NOTE ABOUT BOXPLOTS: There is no universally accepted format for boxplots, and
the final appearance is often determined by the software used to generate the figure. The examples
in the text generated by STATDISK, for example, include a vertical "barrier" at the minimum and
maximum values even though such a feature is not included in the text's definition of a boxplot. If
the boxplot includes an adequate horizontal scale, the 5-number summary values can be inferred
from the scale and do not need to clutter up the figure – in fact, the purpose of boxplots and other
figures is to give a visual image of the data without such clutter. While this manual includes the
5-number summary values in the figure, it is more for student understanding than for good display.
In addition, some formats extend the line connecting the minimum and maximum values through
the box and others do not. In general, follow the format preferred by the instructor.

29. Arrange the 12 values in order. The 5-number summary and boxplot are given below. All measurements are in FICO rating units.

$$\text{min} = x_1 = 664$$
$$Q_1 = (x_3+x_4)/2 = (698+714)/2 = 706$$
$$\text{median} = (x_6+x_7)/2 = (753+779)/2 = 766$$
$$Q_3 = (x_9+x_{10})/2 = (802+818)/2 = 810$$
$$\text{max} = x_{12} = 836$$

31. Arrange the 36 values in order. The 5-number summaries and boxplots are given below.
Regular Coke

$$\text{min} = x_1 = 0.7901 \text{ lbs}$$
$$Q_1 = (x_9+x_{10})/2 = (0.8143+0.8150)/2 = 0.81465 \text{ lbs}$$
$$\text{median} = (x_{18}+x_{19})/2 = (0.8170+0.8172)/2 = 0.8171 \text{ lbs}$$
$$Q_3 = (x_{27}+x_{28})/2 = (0.8207+0.8211)/2 = 0.8209 \text{ lbs}$$
$$\text{max} = x_{36} = 0.8295 \text{ lbs}$$

Diet Coke

$$\text{min} = x_1 = 0.7758 \text{ lbs}$$
$$Q_1 = (x_9+x_{10})/2 = (0.7822+0.7822)/2 = 0.7822 \text{ lbs}$$
$$\text{median} = (x_{18}+x_{19})/2 = (0.7852+0.7852)/2 = 0.7852 \text{ lbs}$$
$$Q_3 = (x_{27}+x_{28})/2 = (0.7879+0.7879)/2 = 0.7879 \text{ lbs}$$
$$\text{max} = x_{36} = 0.7923 \text{ lbs}$$

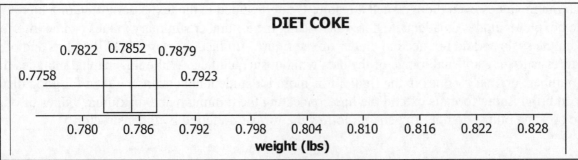

The weights of the diet Coke appear to be substantially less than those of the regular Coke.

33. Arrange the 40 values in order. The 5-number summaries and boxplots are given below.
   Pre-1964 Quarters
   $\quad$ min = x_1 = 6.0002 grams
   $\qquad$ Q_1 = $(x_{10}+x_{11})/2$ = (6.1309+6.1352)/2 = 6.13305 grams
   $\quad$ median = $(x_{20}+x_{21})/2$ = (6.1940+6.1947)/2 = 6.19435 grams
   $\qquad$ Q_3 = $(x_{30}+x_{31})/2$ = (6.2647+6.2674)/2 = 6.26605 grams
   $\qquad$ max = x_{40} = 6.3669 grams
   Post-1964 Quarters
   $\quad$ min = x_1 = 5.5361 grams
   $\qquad$ Q_1 = $(x_{10}+x_{11})/2$ = (5.5928+5.5941)/2 = 5.59345 grams
   $\quad$ median = $(x_{20}+x_{21})/2$ = (5.6274+5.6449)/2 = 5.63615 grams
   $\qquad$ Q_3 = $(x_{30}+x_{31})/2$ = (5.6841+5.6848)/2 = 5.68445 grams
   $\qquad$ max = x_{40} = 5.7790 grams

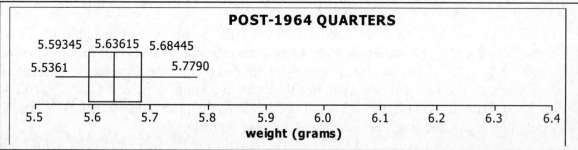

The weights of the pre-1964 quarters are substantially greater than the weights of the post-1964 quarters.

35. Arrange the 40 values in order. The 5-number summary and boxplot are given below. All measurements are in centimeters
   $\quad$ min = x_1 = 27.0
   $\qquad$ Q_1 = $(x_{10}+x_{11})/2$
   $\qquad\qquad$ = (37.5+38.0)/2 = 37.75
   $\quad$ median = $(x_{20}+x_{21})/2$
   $\qquad\qquad$ = (39.0+39.1)/2 = 39.05
   $\qquad$ Q_3 = $(x_{30}+x_{31})/2$
   $\qquad\qquad$ = (41.0+41.0)/2 = 41.0
   $\qquad$ max = x_{40} = 48.6

Outliers are identified as follows.
   $\qquad$ IQR = $Q_3 - Q_1$
   $\qquad\qquad$ = 41.0 – 37.75 = 3.25
   $\quad$ 1.5(IQR) = 1.5(3.25) = 4.875
   lower limit = $Q_1 - 1.5(IQR)$
   $\qquad\qquad$ = 37.75 – 4.875 = 32.875
   upper limit = $Q_3 + 1.5(IQR)$
   $\qquad\qquad$ = 41.0 + 4.875 = 45.875
4 outliers: 27.0, 31.1, 32.1, 48.6

37. There are 35 scores. To find $Q_1 = P_{25}$, $L = (25/100) \cdot 35 = 8.75$. Since $x_8 = 30$ and $x_9 = 35$, we estimate $x_{8.75}$ to be $30 + (0.75)(35-30) = 30 + 3.75 = \$33.75$ million. This close to the result $x_9 = \$35$ million found using Figure 3-5 without interpolation.

Statistical Literacy and Critical Thinking

1. No, the correct mean amount of Coke in each can is not sufficient to guarantee that the production process is proceeding as it should. If the standard deviation of the amounts is large, then unacceptable numbers of under-filled and over-filled cans are being produced. Quality control engineers need to monitor both the mean and the standard deviation (i.e., both the central tendency and the variation) of the process under consideration. Even if the standard deviation is small, approximately half of the cans will be underweight – not good for customer relations. The mean should be over 12 oz to ensure that most of the cans contain at least the 12 ounces stated on the can.

2. The mean and standard deviation of the 11 zip codes have no mathematical meaning and convey no useful information. Zip codes are merely categories at the nominal level of measurement, and they do not indicate quantities of anything. Because assigning other appropriate zip code numbers would produce a different mean and standard deviation, but would not change the nature of the situation, the mean and standard deviation are not meaningful in this context.

3. The mean is based on all the values and will increase noticeably but not dramatically – in a sample of size n=25, one outlier four times larger than the other values will increase the mean by about 3/25 = 12%. The median, which is determined by the middle value, will essentially be unaffected. The standard deviation, like the mean in a sample of size n=25, is based on all the values and will increase noticeably but not dramatically. The range, which is determined by only the largest and smallest values, will increase dramatically.

4. When the data are presented sorted by magnitude, any trends or patterns over time are lost.

Chapter Quick Quiz

1. $\bar{x} = (\Sigma x)/n = 20/5 = 4.0$ cm

2. Arranged in order, the values are 2,2,3,5,8. The median is the middle value, 3 cm.

3. The mode is the most frequently occurring value, 2 cm.

4. The variance is the square of the standard deviation. $s^2 = 5.0^2 = 25.0$ ft^2

5. $z = (x - \bar{x})/s = (4.0 - 10.0)/2.0 = -3.00$

6. The range, standard deviation, and variance are all measures of variation.

7. In general, sample statistics are indicated by English letters and population parameters are indicated by the corresponding Greek letter. The symbol for the standard deviation of a sample is s, and the symbol for the standard deviation of a population is σ.

8. The symbol for the mean of a sample is $\bar{x}$, and the symbol for the mean of a population is μ.

9. The 25th percentile separates the lowest 25% from the highest 75%. Approximately 75 percent of the values in a sample are greater than or equal to the 25th percentile.

10. True. Both the median and the 50th percentile are defined as the middle value – the value with at least 50% of the values less than or equal to it, and with at least 50% of the values greater than or equal to it.

Review Exercises

1. Arranged in order, the values are: 17 18 18 19 19 20 20 20 21 21
   summary statistics: n = 10 $\Sigma x = 193$ $\Sigma x^2 = 3741$
   a. $\bar{x} = (\Sigma x)/n = 193/10 = 19.3$ oz
   b. $\tilde{x} = (x_5 + x_6)/2 = (19+20)/2 = 19.5$ oz
   c. M = 20 oz
   d. m.r. $= (x_1 + x_n)/2 = (17+21)/2 = 19.0$ oz
   e. $R = x_n - x_1 = 21 - 17 = 4$ oz
   f. s = 1.3 oz [the square root of the answer given in part g]
   g. $s^2 = [n(\Sigma x^2) - (\Sigma x)^2]/[n(n-1)] = [10(3741) - (193)^2]/[10(9)] = 161/90 = 1.8$ oz^2
   h. $Q_1 = x_3 = 18$ oz [L = (25/100)(10) = 2.5, rounded up to 3]
   i. $Q_3 = x_8 = 20$ oz [L = (75/100)(10) = 7.5, rounded up to 8]

2. Arrange the 10 values in order. The 5-number summary and boxplot are given below.
         min $= x_1 = 17$ oz
          $Q_1 = x_3 = 18$ oz
   median $= (x_5 + x_6)/2 = (19+20)/2 = 19.5$ oz
          $Q_3 = x_8 = 20$ oz
         max $= x_{10} = 21$ oz

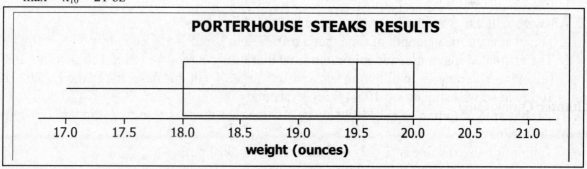

3. Arranged in order, the values are: 878 880 923 924 928 930 934 936 936
   summary statistics: n = 9 $\Sigma x = 8269$ $\Sigma x^2 = 7601621$
   a. $\bar{x} = (\Sigma x)/n = 8269/9 = 918.8$ mm
   b. $\tilde{x} = x_5 = 928$ mm
   c. M = 936 mm
   d. m.r. $= (x_1 + x_n)/2 = (878+936)/2 = 907$ mm
   e. $R = x_n - x_1 = 936 - 878 = 58$ mm
   f. s = 23.0 mm [the square root of the answer given in part g]
   g. $s^2 = [n(\Sigma x^2) - (\Sigma x)^2]/[n(n-1)] = [9(7601621) - (8269)^2]/[9(8)] = 38228/72 = 530.9$ mm^2
   h. $Q_1 = x_3 = 923$ mm [L = (25/100)(9) = 2.25, rounded up to 3]
   i. $Q_3 = x_7 = 934$ mm [L = (75/100)(9) = 6.75, rounded up to 7]

4. $z = (x - \bar{x})/s = (878 - 918.8)/23.0 = -1.77$
   No. Since $-2 < -1.77 < 2$, the sitting height of 878 mm is not unususal.

5. Arrange the 10 values in order. The 5-number summary and boxplot are given below.
   $$\text{min} = x_1 = 878 \text{ mm}$$
   $$Q_1 = x_3 = 923 \text{ mm}$$
   $$\text{median} = x_5 = 928 \text{ mm}$$
   $$Q_3 = x_7 = 934 \text{ mm}$$
   $$\text{max} = x_9 = 936 \text{ mm}$$

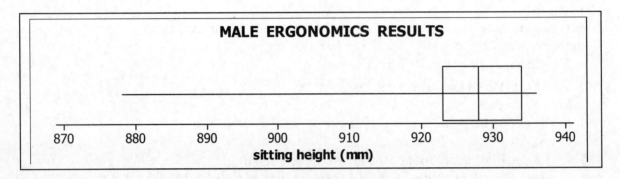

6. SAT: $z = (x - \bar{x})/s = (1030 - 1518)/325 = -488/325 = -1.50$
   ACT: $z = (x - \bar{x})/s = (14.0 - 21.1)/4.8 = -7.1/4.8 = -1.48$
   Since $-1.48 > -1.50$, the ACT score of 14.0 is the relatively better score.

7. Answers will vary, but the following are reasonable responses.
   a. The estimated mean age of cars driven by students is 6.0 years.
   b. The estimated maximum and minimum age for cars driven by students is 12 years and 0 years. The range rule of thumb suggests $s \approx \text{range}/4$. In this case, the range is $12 - 0 = 12$, and so we estimate $s \approx 12/4 = 3.0$ years.

8. Answers will vary, but the following are reasonable responses.
   a. The estimated mean length of time that a traffic light is red is 45 seconds.
   b. The estimated maximum and minimum time that traffic lights are red is 120 seconds and 15 seconds. The range rule of thumb suggests $s \approx \text{range}/4$. In this case, the range is $120 - 15 = 105$, and so we estimate $s \approx 105/4 = 26.25$ seconds.

9. The range rule of thumb suggests that "usual" values are those within two standard deviations of the mean.
   minimum usual value $= \bar{x} - 2s = 1212 - 2(51) = 1110$ mm
   maximum usual value $= \bar{x} + 2s = 1212 + 2(51) = 1314$ mm
   Since accessibility to controls is the issue, the minimum usual value of 1110 mm is the more relevant value. No controls should be placed beyond the reach of the person with the smallest likely grip reach.

10. The range rule of thumb suggests that "usual" values are those within two standard deviations of the mean.
    minimum usual value $= \bar{x} - 2s = 97.5 - 2(6.9) = 83.7$ cm
    maximum usual value $= \bar{x} + 2s = 97.5 + 2(6.9) = 111.3$ cm
    No; since $83.7 < 87.8 < 111.3$, a height of 87.8 cm is not considered unusual. On this basis, the physician should not be concerned.

Cumulative Review Exercises

1. a. Since the values can be any number on a continuum, the sitting heights are continuous data.
   b. Since differences between the values are meaningful and there is a meaningful zero, the sitting heights are data at the ratio level of measurement.

2. The frequency distribution is given below.

sitting height (mm)	frequency
870 – 879	1
880 – 889	1
890 – 899	0
900 – 909	0
910 – 919	0
920 – 929	3
930 – 939	4
	9

3. The histogram is given at the right. The distribution appears to be skewed.
   NOTE: The true boundaries for the bars are as given in the figure, but for visual simplicity it is permissible to use the labels 870, 880, etc. In general, visual images are intended to convey a concise summary of the data and not technical detail.

4. The dotplot is given below.

5. The stemplot is given below.
   sitting height (mm)

```
87 | 8
88 | 0
89 |
90 |
91 |
92 | 348
93 | 0466
```

6. a. The mode is the only appropriate measure of central tendency for data at the nominal level of measurement. All the other measures require that the data have some underlying order, which does not exist for data at the nominal level.

   b. This is a convenience sample, because the sample was obtained without detailed planning from those people conveniently at hand. Had the botanist divided the phone book into groups of 250, selected one of those groups at random, and happened to get the first such group, then he would have a cluster sample.

   c. This is a cluster sample, since the population was divided into groups, groups were selected at random, and everyone in each selected group was sampled.

   d. The problem is that there appears to be too much variation. Of the measures listed, only the standard deviation is a measure of variation. The is the statistic that should be monitored, with the hope lowering its value.

7. No, the responses cannot be considered representative of the population of the United States for at least two reasons: the responders were a voluntary response sample (i.e., persons who tend to feel strongly about the issue under consideration), and the sample was limited to AOL Internet subscribers (who may not be representative in education, incomes, etc.).

8. A simple random sample of n subjects is selected in such a way that every possible combination of n subjects in the population has the same chance of being selected. A voluntary response sample is one in which the subjects themselves decide whether to be included in the study. A simple random sample is generally a better sample because it is more likely to be representative of the population. Voluntary response samples tend to include mostly those with strong feelings on and/or a vested interest in the issue under consideration.

9. An observational study involves the measurement of some response or characteristic not under the influence of the researcher. An experiment involves the measurement of some response or characteristic in the presence of an influence (usually called a "treatment") from the researcher.

10. By not starting the vertical axis at zero, the figure exaggerates the differences between the categories. A bar that is twice the height of another, for example, gives the faulty visual impression of twice the frequency count – when the actual difference is less than that.

Chapter 4

Probability

4-2 Basic Concepts of Probability

1. To say that the probability of being injured while using recreation equipment in 1/500 means that approximately one injury occurs for every 500 times that recreation equipment is used. Note that this is an aggregate result, and not necessarily true for any one particular piece of equipment – since some pieces of equipment are more dangerous than others. Because the probability $1/500 = .002$ is small, such an injury is considered unusual.

3. If A denotes the fact that some event occurs, then $\overline{A}$ denotes the fact that the event does not occur. If $P(A) = 0.995$, then $P(\overline{A}) = 1 - 0.995 = 0.005$. If $P(A) = 0.995$, then A is very likely to occur and it would be unusual for $\overline{A}$ to occur – i.e., it would be unusual for A not to occur.

5. "4 in 21" $= 4/21 = 0.190$

7. "50−50 chance" $= 50\% = 50/100 = 0.50$

9. $6/36 = 0.167$

11. "impossible" $= 0\% = 0/100 = 0$

13. All probability must be between 0 and 1 inclusive.
The value $3:1 = 3/1 = 3$ cannot be a probability because $3 > 1$.
The value $5/2 = 2.5$ cannot be a probability because $2.5 > 1$.
The value -0.5 cannot be a probability because $-0.5 < 0$.
The value $321/123 = 2.610$ cannot be a probability, because $2.610 > 1$.

15. a. Let having exactly one girl be the set with 3 simple events A = {bbg,bgb,gbb}.
$P(A) = 3/8 = 0.375$
b. Let having exactly two girls be the set with 3 simple events B = {bgg,gbg,ggb}.
$P(B) = 3/8 = 0.375$
c. Let having all three girls be the set with 1 simple event C = {ggg}.
$P(C) = 1/8 = 0.125$

17. a. The total number of responses summarized in the table is $15 + 42 + 32 + 9 = 98$.
b. The number of negative responses (the test says the person did not lie) is $32 + 9 = 41$.
c. Let N = selecting one of the negative test responses. $P(N) = 41/98$.
d. $41/98 = 0.418$

19. Let F = selecting a false positive response. $P(F) = 15/98 = 0.153$.
The probability of this type of error is high enough that such an occurrence would not be considered unusual; the test is not highly accurate.

21. There are $84 + 16 = 100$ total senators.
Let W = selecting a woman. $P(W) = 16/100 = 0.16$.
No; this probability is too far below 0.50 to agree with the claim that men and women have equal opportunities to become a senator.

23. Let L = getting struck by lightning. By the relative frequency approximation of probability, the estimate is P(L) = 281/290,789,000 = 0.000000966. No; engaging in at risk behavior increases the probability of experiencing negative consequences.

25. a. Let B = the birthdate is correctly identified. P(B) = 1/365 = 0.003.
    b. Yes. Since 0.003 ≤ 0.05, it would be unusual to guess correctly on the first try.
    c. Most people would probably believe that Mike's correct answer was the result of having inside information – and not making a lucky guess.
    d. Fifteen years is a big error. If the guess were serious, Kelly would likely think that Mike was not knowledgeable and/or be personally insulted – and a second date is unlikely. If the guess were perceived as being made in jest, Kelly might well appreciate his sense of humor and/or handling of a potentially awkward situation – and a second date is likely.

27. Let F = a pacemaker malfunction is caused by firmware.
    Based on the given results, P(F) = 504/8834 = 0.0571.
    No. Since 0.0571 > 0.05, a firmware malfunction is not unusual among pacemaker malfunctions.

29. There were 795 + 10 = 805 total persons executed.
    Let W = a randomly selected execution was that of a woman. P(W) = 10/805 = 0.0124.
    Yes. Since 0.0124 ≤ 0.05, it is unusual for an executed person to be a woman. This is due to the fact that more crimes worthy of the death penalty are committed by men than by women.

31. There were 211+ 288 + 366 + 144 + 89 = 1098 total persons in the survey.
    Let F = a household has 4 or more cellphones in use.
    Based on these results, P(F) = 89/1098 = 0.0811.
    No. Since 0.0811 > 0.05, it is not unusual for a household to have 4 or more cellphones in use.
    NOTE: Technically, the given survey cannot be used to answer the question as posed. A survey made of <u>persons</u> selected at random cannot be used to answer questions about <u>households</u>. If the goal is to make statements about households, then households (and not persons) should be selected at random. Selecting persons at random means that households with more people have a higher probability of being in the survey than households with fewer people. Since households with more people are more likely to have more cellphones, the survey will tend to overestimate the number of cellphones per household.

33. a. Listed by birth order, the sample space contains 4 simple events: bb bg gb gg.
    b. P(gg) = 1/4 = 0.25
    c. P(bg or gb) = 2/4 = 0.50

35. a. Listed with the father's contribution first, the sample space has 4 simple events:
       brown/brown brown/blue blue/brown blue/blue.
    b. P(blue/blue) = 1/4 = 0.25
    c. P(brown eyes) = P(brown/brown or brown/blue or blue/brown) = 3/4 = 0.75

37. The odd against winning are P(not winning)/P(winning) = (423/500)/(77/500) = 423/77, or 423:77 – which is approximately 5.5:1, or 11:2.

39. a. If a $2 bet results in a net return of $14.20, the net profit is 14.20 – 2.00 = $12.20.
    b. If you get back your bet plus $12.20 for every $2 you bet, the payoff odds are 12.2:2 [typically expressed as either 6.1:1 or 61:10], or approximately 6:1.
    c. odds against W = P(not win)/P(W) = (443/500)/(57/500) = 443/57 or 443:57 [typically expressed as 7.77:1], or approximately 8:1.

d. If the payoff odds are 7.77:1, a win gets back your bet plus $7.77 for every $1 bet.
   If you bet $2 and win, you get back 2 + 15.54 = $17.54. The ticket would be worth $17.54
   and your profit would be 17.54 – 2.00 = $15.54.

41. Making a table like the one on the right
    organizes the information and helps to
    understand the concepts involved.
    $p_t = 26/2103 = 0.0124$
    $p_c = 22/1671 = 0.0132$

Treatment		Headache?		
		no	yes	
	Nasonex	2077	26	2103
	control	1649	22	1671
		3276	48	3774

a. The relative risk is the P(risk) for the treatment as a percent of the P(risk) for the control.
   relative risk = p_t/p_c = 0.0124/0.0132 = 0.939 = 93.9%
b. The odds in favor of the risk are P(risk)/P(not risk).
        for the treatment: odds in favor = $p_t/(1-p_t)$ = (26/2103)/(2077/2103) = 26/2077 = 0.012518/1
          for the control: odds in favor = $p_c/(1-p_c)$ = (22/1671)/(1649/1671) = 22/1649 = 0.013341/1
   The odds ratio is the odds in favor of the risk for the treatment as a percent of the odds in
   favor of the risk for the control.
   odds ratio = $[p_t/(1-p_t)]/[p_c/(1-p_c)]$ = 0.012518/0.013314 = 0.938 = 93.8%
Both the relative risk and the odds ratio are less than 1.00, suggesting that the risk of headache
is actually smaller for Nasonex than it is for the control. Nasonex does not appear to pose a
risk of headaches.

43. This difficult problem will be broken into two events and a conclusion. Let L denote the
    length of the stick. Label the midpoint of the stick M, and label the first and second breaking
    points X and Y.
    • Event A: X and Y must be on opposite sides of M. If X and Y are on the same side of M,
      the side without X and Y would be longer than 0.5L and no triangle will be possible.
      Once X is set, P(Y on the same side as X) = P(Y on opposite side from X) = ½.
    • Event B: The distance |XY| must be less than 0.5L. Assuming X and Y are on opposite sides
      of M, let Q denote the end closest to X and R denote the end closest to Y – so that the
      following sketch applies.

```
|--------|------------|----|----------------|
 Q       X          M Y              R
```

      In order to form a triangle, it must be true that |XY| = |XM| + |MY| < 0.5L. This happens
      only when |MY| < |QX|. With all choices at random, there is no reason for |QX| to be larger
      than |MY|, or vice-versa. This means that P(|MY|<|QX|) = ½.
    • Conclusion: For a triangle, we need both events A and B.
      Since P(A) = ½ and P(B occurs assuming A has occurred) = ½, events A and B together will
      happen only ½ of ½ of the time: the final answer would be ½ of ½, or ¼. In the notation of
      section 4-4, P(A and B) = P(A)·P(B|A) = (0.5)(0.5) = 0.25.
NOTE: In situations where the laws of science and mathematics are difficult to apply (like
finding the likelihood that a particular thumbtack will land point up when it is dropped),
probabilities are frequently estimated using Rule #1 of this section. Suppose an instructor
gives 100 sticks to a class of 100 students and says, "Break your stick at random two times so
that you have 3 pieces." Does this mean that only about 25% of the students should be able to
form a triangle with their 3 pieces? In theory, yes. In practice, no. It is likely the students
would break their sticks "conveniently" and not "at random" – e.g., there would likely be
fewer (if any) pieces less than one inch than you would expect by chance, because such pieces
would be difficult to break off by hand.

4-3 Addition Rule

1. Two events in a single trial are disjoint if they are outcomes that cannot happen at the same time.

3. In the context of the addition rule defined in this section, P(A and B) is the probability that both event A and event B occur at the same time in a single trial.

5. Not disjoint, since a physician can be a female surgeon.

7. Disjoint, since a car that is free of defects cannot have a dead battery – which is a defect.

9. Not disjoint, since a subject who believes the evidence for global warming can also be opposed to stem cell research.

11. Not disjoint, since an R-rated movie can receive a four-star review.

13. $P(\overline{M})$ is the probability that a STATDISK user is not a Macintosh computer user
$$\begin{aligned}P(\overline{M}) &= 1 - P(M)\\ &= 1 - 0.05\\ &= 0.95\end{aligned}$$

15. Let M = a selected American believes it morally wrong not to report all income. P(M) = 0.79.
$$\begin{aligned}P(\overline{M}) &= 1 - P(M)\\ &= 1 - 0.79\\ &= 0.21\end{aligned}$$

NOTE: For exercises 17-20, refer to the table at the right and use the following notation.
Let P = the polygraph test indicates a positive
Let L = the person actually lied.

		Subject Lie?		
		no	yes	
Test Result	positive	15	42	57
	negative	32	9	41
		47	51	98

17. $P(P \text{ or } \overline{L}) = P(P) + P(\overline{L}) - P(P \text{ and } \overline{L})$
$$\begin{aligned}&= 57/98 + 47/97 - 15/98\\ &= 89/98 = 0.908\end{aligned}$$

19. $P(\overline{L} \text{ and } \overline{P}) = 32/98 = 0.327$

NOTE: For exercises 21-26, refer to the table at the right and use the following notation.
Let M = the challenge was made by a male.
Let S = the challenge was successful.

		Successful?		
		yes	no	
Gender	male	201	288	489
	female	126	224	350
		327	512	839

21. $P(\overline{S}) = 512/839 = 0.610$

23. $P(M \text{ or } S) = P(M) + P(S) - P(M \text{ and } S)$
$$\begin{aligned}&= 489/839 + 327/839 - 201/839\\ &= 615/839 = 0.733\end{aligned}$$

25. $P(M \text{ or } \overline{S}) = P(M) + P(\overline{S}) - P(M \text{ and } \overline{S})$
$$\begin{aligned}&= 489/839 + 512/839 - 288/839\\ &= 713/839 = 0.850\end{aligned}$$

NOTE: For exercises 27-32, refer to the following table and notation.
For a person selected at random, let the following designations apply.
Y = there is a response
N = there is a refusal
A = the age is 18-21
B = the age is 22-29
C = the age is 30-39
D = the age is 40-49
E = the age is 50-59
F = the age is 60+

		Age						
		18-21	22-29	30-39	40-49	50-59	60+	
Response?	yes	73	255	245	136	138	202	1049
	no	11	20	33	16	27	49	156
		84	275	278	152	165	251	1205

27. P(N) = 156/1205 = 0.129

Yes. Persons who refuse to answer form a subpopulation whose opinions will not be counted, thus preventing the survey from being representative of the entire population.

29. P(Y or A) = P(Y) + P(A) – P(Y and A)
$$= 1049/1205 + 84/1205 - 73/1205$$
$$= 1060/1205 = 0.880$$

31. Use the intuitive approach rather than the formal addition rule.
P(Y or B or C) = P(Y) + P(B and N) + P(C and N)
$$= 1049/1205 + 20/1205 + 33/1205$$
$$= 1102/1205 = 0.915$$

NOTE: For exercises 33-36 refer to the table at the right and use the following notation.
Let P = the drug test indicates a positive
Let M = the person actually uses marijuana

		Use Marijuana?		
		yes	no	
Test Result	positive	119	24	143
	negative	3	154	157
		122	178	300

33. a. 300
    b. 178
    c. $P(\overline{M}) = 178/300 = 0.593$

35. $P(\overline{P}$ or $\overline{M}) = P(\overline{P}) + P(\overline{M}) - P(\overline{P}$ and $\overline{M})$
$$= 157/300 + 178/300 - 154/300$$
$$= 181/300 = 0.603$$

37. P[(false positive) or (false negative)]
$$= P[(P \text{ and } \overline{M}) \text{ or } (\overline{P} \text{ and } M)]$$
$$= P(P \text{ and } \overline{M}) + (\overline{P} \text{ and } M) - P[(P \text{ and } \overline{M}) \text{ and } (\overline{P} \text{ and } M)]$$
$$= 24/300 + 3/300 - 0/300$$
$$= 27/300 = 0.090$$
Using the 0.05 guideline, the occurrence of a false positive or a false negative is not unusual. It appears that the test is not highly accurate.

39. Assuming girls and boys are equally likely, the birth-toss sample space consists of 4 equally likely simple events: GH GT BH BT.
Using the concepts of the formal addition rule,
P(G or H) = P(G) + P(H) – P(G and H)
$$= 2/4 + 2/4 - 1/4$$
$$= 3/4 = 0.75$$
Using the concepts of the intuitive addition rule,
P(G or H) = 3/4 = 0.75

41. If the *exclusive or* is used instead of the *inclusive or*, then the double-counted probability must be completely removed (i.e., it must be subtracted twice) and the formula becomes as follows.

    $$P(A \text{ or } B) = P(A) + P(B) - 2 \cdot P(A \text{ and } B)$$

43. The following two English/logic facts are used in this exercise.
    * not (A or B) = not A and not B   Is your sister either artistic or bright?

        No? Then she is not artistic and she is not bright.
    * not (A and B) = not A or not B   Is your brother artistic and bright?

        No? Then either he is not artistic or he is not bright.

    a. $P(\overline{A \text{ or } B}) = P(\overline{A} \text{ and } \overline{B})$ from the first fact above

    or (from the rule for complementary events)

    $P(\overline{A \text{ or } B}) = 1 - P(A \text{ or } B) = 1 - [P(A) + P(B) - P(A \text{ and } B)] = 1 - P(A) - P(B) + P(A \text{ and } B$

    b. $P(\overline{A} \text{ or } \overline{B}) = 1 - P(A \text{ and } B)$ from the second fact above
    c. They are different: part (a) gives the complement of "A or B"

    part (b) gives the complement of "A and B"

    NOTE: The following example illustrates this concept. Consider the following 5 students and their majors: Rob (art), Steve (art), Tom (art and biology), Uriah (biology), Vern (math). Select one student at random and consider the major.

    $P(A) = 3/5$ $P(B) = 2/5$ $P(A \text{ or } B) = 4/5$ $P(A \text{ and } B) = 1/5$.

    a. $P(\overline{A \text{ or } B}) = 1 - P(A \text{ or } B)$

    $= 1 - 4/5$

    $= 1/5$ [only Vern is "neither an art major nor a biology major"]

    b. $P(\overline{A} \text{ or } \overline{B}) = P(\overline{A}) + P(\overline{B}) - P(\overline{A} \text{ and } \overline{B})$

    $= 2/5 + 3/5 - 1/5$

    $= 4/5$ [everyone but Tom is either "not an art major" or "not a biology major"]

    c. No, $1/5 \neq 4/5$.

4-4 Multiplication Rule: Basics

1. Answers will vary, but the following are possibilities.
   Independent events: Getting an even number when a die is rolled and getting a head when a coin is tossed.
   Dependent events: Selecting aces on two consecutive draws without replacement from an ordinary deck of cards.

3. The events of selecting adults are dependent, because the adults are selected without replacement. On any given pick, the probability of a particular person being selected depends upon who and how many persons have already been selected.

5. Independent, since the population is so large that the 5% guideline applies.

7. Dependent, if there is visual contact during the invitation, since visual impressions affect responses.
   Independent, if there is no visual contact during the invitation, since the invitee would have no knowledge of the inviter's clothing.

9. Dependent, since the two items operate from the same power supply.

11. Independent, since the population is so large that the 5% guideline applies.

13. Let F = a selected person had false positive results.

$P(F_1 \text{ and } F_2) = P(F_1) \cdot P(F_2|F_1)$
$= (15/98)(14/97) = 0.0221$

Yes; since $0.0221 \leq 0.05$, getting two subjects who had false positives would be unusual.
NOTE: Since n=2 is $2/98 = 0.0204 \leq 0.05$ of the population, one could use the 5% guideline and treat the selections as independent to get $(15/98)^2 = 0.0234$. Typically the 5% guideline is invoked only when the calculations for dependent events are extremely tedious.

15. Let P = a selected person tested positive

$P(P_1 \text{ and } P_2 \text{ and } P_3 \text{ and } P_4) = P(P_1) \cdot P(P_2|P_1) \cdot P(P_3|P_1 \text{ and } P_2) \cdot P(P_4|P_1 \text{ and } P_2 \text{ and } P_3)$
$= (57/98)(56/97)(55/96)(54/95) = 0.109$

No; since $0.109 > 0.05$, getting four subjects who had positive results would not be unusual.

NOTE: For exercises 17-20 refer to the table at the right.

17. P(O and Rh+) = 39/100
    a. (39/100)(39/100) = 0.152
    b. (39/100)(38/99) = 0.150

		Blood Group			
	O	A	B	AB	
Rh+	39	35	8	4	86
Rh-	6	5	2	1	14
	45	40	10	5	100

Type

19. P(O and Rh-) = 6/100
    a. (6/100)(6/100)(6/100)(6/100) = 0.0000130
    b. (6/100)(5/99)(4/98)(3/97) = 0.00000383

21. a. Yes; since the correctness of the first response has no effect on the correctness of the second, the events are independent.
    b. $P(C_1 \text{ and } C_2) = P(C_1) \cdot P(C_2|C_1)$
             $= (1/2)(1/4) = 1/8 = 0.125$
    c. No; since $0.125 > 0.05$, getting both answers correct would not be unusual – but it certainly isn't likely enough to recommend such a practice.

23. Let A = a public opinion poll is accurate within its margin of error.
P(A) = 0.95 for each polling organization
P(all 9 are accurate) = $P(A_1 \text{ and } A_2 \text{ and} \ldots \text{and } A_9)$
                      $= P(A_1) \cdot P(A_2) \cdot \ldots \cdot P(A_9)$
                      $= (0.95)^9 = 0.630$

No. With 9 independent polls the probability that at least one of them is not accurate within its margin of error is $1 - P(\text{all accurate}) = 1 - 0.630 = 0.370$.

25. Let G = getting a girl.
P(G) = 1/2 for each birth
$P(G_1 \text{ and } G_2 \text{ and } G_3) = P(G_1) \cdot P(G_2) \cdot P(G_3)$
                      $= (1/2)(1/2)(1/2) = 1/8 = 0.125$

No; since $0.125 > 0.05$, getting 3 girls by chance alone is not an unusual event. The results are not necessarily evidence that the gender-selection method is effective.

27. Let A = the alarm works.
P(A) = 0.9 for each alarm
    a. $P(\overline{A}) = 1 - P(A) = 1 - 0.9 = 0.1$
    b. $P(\overline{A}_1 \text{ and } \overline{A}_2) = P(\overline{A}_1) \cdot P(\overline{A}_2)$
                   $= (0.1)(0.1) = 0.01$

c. P(being awakened) $= P(\overline{\overline{A}_1 \text{ and } \overline{A}_2})$

$$= 1 - P(\overline{A}_1 \text{ and } \overline{A}_2)$$
$$= 1 - 0.01 = 0.99$$

NOTE: The above parts (b) and (c) assume that the alarm clocks work independently of each other. This would not be true if they are both electric alarm clocks.

29. Let G = getting a tire that is good
   P(G) = 4800/5000 = 0.96
   a. $P(G_1 \text{ and } G_2 \text{ and } G_3 \text{ and } G_4) = P(G_1) \cdot P(G_2|G_1) \cdot P(G_3|G_1 \text{ and } G_2) \cdot P(G_4|G_1 \text{ and } G_2 \text{ and } G_3)$
      $$= (4800/5000)(4799/4999)(4798/4998)(4797/4997) = 0.849$$
   b. Since n = 100 represents 100/5000 = 0.02 ≤ 0.05 of the population, use the 5% guideline
      and treat the repeated selections as being independent.
      $P(G_1 \text{ and } G_2 \text{ and} \ldots \text{and } G_{100}) = P(G_1) \cdot P(G_2) \cdot \ldots \cdot P(G_{100})$
      $$= (0.96)(0.96)\ldots(0.96) = (0.96)^{100} = 0.0169$$
      Yes; since 0.0169 ≤ 0.05, getting 100 good tires would be an unusual event and the outlet
      should plan on dealing with returns of defective tires.

31. Let W = the surge protector works correctly
   P(W) = 0.99 for each of p and q
   a. The TV is protected if at least one of the surge protectors works correctly.
      $P(W_p \text{ or } W_q) = P(W_p) + P(W_q) - P(W_p \text{ and } W_q)$
      $$= 0.99 + 0.99 - 0.9801 \quad [.9801 \text{ comes from part (b)}]$$
      $$= 0.9999$$
   b. The TV is protected only if both surge protectors work correctly.
      $P(W_p \text{ and } W_q) = P(W_p) \cdot P(W_q)$ [since they operate independently]
      $$= (.99)(.99) = .9801$$
   c. The series arrangement in part (a) provided the better coverage. The parallel arrangement in
      part (b) actually provides poorer protection than a single protector alone.

33. This problem can be done by two different methods. In either case, let
   A = getting an ace
   S = getting a spade.
   • Consider the sample space
      The first card could be any of 52 cards, and for each first card there are 51 possible
      second cards. This makes a total of 52·51 = 2652 equally likely outcomes in the sample
      space. How many of them are A_1 and S_2?
      The aces of hearts, diamond and clubs can be paired with any of the 13 spades for a
      total of 3·13 = 39 favorable possibilities. The ace of spades can only be paired with any
      of the remaining 12 members of that suit for a total of 12 favorable possibilities. Since
      there are 39 + 12 = 51 total favorable possibilities among the equally likely outcomes,
      $P(A_1 \text{ and } S_2) = 51/2652 = 0.0192$
   • Use the formulas (and express the event as two mutually exclusive possibilities)
      $P(A_1 \text{ and } S_2) = P([\text{spade}A_1 \text{ and } S_2] \text{or } [\text{other}A_1 \text{ and } S_2])$
      $$= P(\text{spade}A_1 \text{ and } S_2) + P(\text{other}A_1 \text{ and } S_2)$$
      $$= P(\text{spade}A_1) \cdot P(S_2|\text{spade}A_1) + P(\text{other}A_1) \cdot P(S_2|\text{other}A_1)$$
      $$= (1/52)(12/51) + (3/52)(13/51)$$
      $$= 12/2652 + 39/2652 = 51/2652 = 0.0192$$

4-5 Multiplication Rule: Complements and Conditional Probability

1. If "at least one" of the 10 pacemakers is defective, the exact number of defective pacemakers must be one of the following: 1,2,3,4,5,6,7,8,9,10.

3. The probability of a particular outcome is equal to one divided by the total number of outcomes only when all the outcomes are equally likely and mutually exclusive. In this case the two outcomes are not equally likely, and so that reasoning cannot be used. The actual survival rate was not considered, but it should have been.

5. If it is not true that at least one of the 15 tests positive, then all 15 of them test negative.

7. If it is not true that none of the 4 has the recessive gene, then at least one of the 4 has the recessive gene.

9. P(at least one girl) = 1 – P(all boys)

$$= 1 - P(B_1 \text{ and } B_2 \text{ and } B_3 \text{ and } B_4 \text{ and } B_5 \text{ and } B_6)$$
$$= 1 - P(B_1) \cdot P(B_2) \cdot P(B_3) \cdot P(B_4) \cdot P(B_5) \cdot P(B_6)$$
$$= 1 - (\tfrac{1}{2})(\tfrac{1}{2})(\tfrac{1}{2})(\tfrac{1}{2})(\tfrac{1}{2})(\tfrac{1}{2})$$
$$= 1 - 1/64 = 63/64 = 0.984$$

Yes, that probability is high enough for the couple to be very confident that they will get at least one girl in six children.

11. P(at least one correct) = 1 – (all wrong)

$$= 1 - P(W_1 \text{ and } W_2 \text{ and } W_3 \text{ and } W_4)$$
$$= 1 - P(W_1) \cdot P(W_2) \cdot P(W_3) \cdot P(W_4)$$
$$= 1 - (4/5)(4/5)(4/5)(4/5)$$
$$= 1 - 256/625 = 269/625 = 0.590$$

Since there is a greater chance of passing than of failing, the expectation is that such a strategy would lead to passing. In that sense, one can reasonably expect to pass by guess. But while the expectation for a single test may be to pass, such a strategy can be expected to lead to failing about 4 times every 10 times it is applied.

13. Since each birth is an independent event, $P(G_4 | G_1 \text{ and } G_2 \text{ and } G_3) = P(G_4) = P(G) = \tfrac{1}{2}$. One can also consider the 16-outcome sample space S for 4 births and use the formula for conditional probability as follows.

S = {BBBG, BBGG, BGBG, GBBG, BGGG, GBGG, GGBG, GGGG,
BBBB, BBGB, BGBB, GBBB, BGGB, GBGB, GGBB, GGGB}

Let F = the first 3 children are girls
Let G_4 = the fourth child is a girl
P(F) = 2/16
P(F and G_4) = 1/16

$$P(G_4 | F) = \frac{P(G_4 \text{ and } F)}{P(F)} = \frac{1/16}{2/16} = 1/2$$

Note: Using the sample space S also verifies directly that $P(G_4) = 8/16 = 1/2 = P(G)$, independent of what occurred during the first three births.
No. This probability is not the same as P(GGGG) = 1/16.

15. Let $P(C) = P(crash) = 0.0480$, and $P(N) = P(no\ crash) = 0.9520$.
    $P(at\ least\ one\ crash) = 1 - P(no\ crashes)$
$$= 1 - P(N_1\ and\ N_2\ and\ N_3\ and\ N_4)$$
$$= 1 - P(N_1) \cdot P(N_2) \cdot P(N_3) \cdot P(N_4)$$
$$= 1 - (0.9520)(0.9520)(0.9520)(9.8520)$$
$$= 1 - 0.821 = 0.179$$

Yes, the above probability does not apply to the given story problem. The above solution assumes that the cars are selected at random – so that, for example, $P(C_2|C_1) = P(C) = 0.0480$. These cars were not selected at random, but they were all taken from the same family. Since the cars were driven by the same people, or at least by people likely to have similar driving habits, knowing that one of the cars has a crash would reasonably make a person think that it was more likely that another of the cars would crash – i.e., the events are not independent and $P(C_2|C_1) > P(C) = 0.0480$.

17. $P(at\ least\ one\ with\ vestigial\ wings) = 1 - P(all\ have\ normal\ wings)$
$$= 1 - P(N_1\ and\ N_2\ and\ N_3\ and\ \ldots\ and\ N_{10})$$
$$= 1 - P(N_1) \cdot P(N_2) \cdot P(N_3) \cdot \ldots \cdot P(N_{10})$$
$$= 1 - (3/4)(3/4)(3/4)\ldots(3/4)$$
$$= 1 - (3/4)^{10} = 1 - 0.0563 = 0.9437,\ \text{rounded to } 0.944$$

Yes, the researchers can be 94.4% certain of getting at least one such offspring.

NOTE: For exercises 19-22, refer to the table at the right and use the following notation.
Let P = the lie detector test indicates a positive
Let Y = the person actually lied

		Actually Lie?		
Test Result		no	yes	
	positive	15	42	57
	negative	32	9	41
		47	51	98

19. $P(P|\overline{Y}) = 15/47 = 0.319$
    This case is problematic because it represents the test indicating incorrectly that a person lied when he was telling the truth.

21. $P(Y|\overline{P}) = 9/41 = 0.220$
    This result (the probability that a person is really lying when the test fails to accuse him) is different from the one in Exercise 20 (the probability that the test fails to accuse a person when he really is lying).

NOTE: For exercises 23-26, refer to the table at the right.

		BB	BG	GB	GG	
type	identical	5	0	0	5	10
	fraternal	5	5	5	5	20
		10	5	5	10	30

Twin Genders

23. a. $P(identical) = 10/30 = 1/3 = 0.333$
    b. $P(identical|BB) = 5/10 = 1/2 = 0.5$

25. $P(\text{"BG or GB"}|fraternal) = 10/20 = 1/2 = 0.5$

27. $P(at\ least\ one\ works) = 1 - P(all\ fail)$
$$= 1 - P(F_1\ and\ F_2\ and\ F_3)$$
$$= 1 - P(F_1) \cdot P(F_2) \cdot P(F_3)$$
$$= 1 - (0.10)(0.10)(0.10)$$
$$= 1 - 0.001 = 0.999$$

Yes, the probability of a working clock rises from 90% with just one clock to 99.9% with 3 clocks. If the alarm clocks run on electricity instead of batteries, then the clocks do not operate independently and the failure of one could be the result of a power failure or interruption and may be related to the failure of another – i.e., $P(F_2|F_1)$ is no longer $P(F) = 0.90$.

29. P(HIV positive result) = P(at least one person is HIV positive)

$$= 1 - \text{P(all persons are HIV negative)}$$
$$= 1 - \text{P}(N_1 \text{ and } N_2 \text{ and } N_3)$$
$$= 1 - \text{P}(N_1) \cdot \text{P}(N_2) \cdot \text{P}(N_3)$$
$$= 1 - (0.9)(0.9)(0.9)$$
$$= 1 - 0.729 = 0.271$$

NOTE: This plan is very efficient. Suppose, for example, there were 3000 people to be tested. Only in 0.271 = 27.1% of the groups would a retest need to be done for each of the three individuals. Those (0.271)(1000) = 271 groups would generate (271)(3) = 813 retests. The total number of tests required would be 1813 (the 1000 original + the 813 retests), only 60% of the 3,000 tests that would have been required to test everyone individually.

31. a. Let D = a birthday is different from any yet selected.

$P(D_1) = 365/365 = 1$ NOTE: With nothing to match, it <u>must</u> be different.

$P(D_2) = P(D_2|D_1) = 364/365$

$P(D_3) = P(D_3|D_1 \text{ and } D_2) = 363/365$

...

$P(D_{25}) = P(D_{25}|D_1 \text{ and } D_2 \text{ and } D_3 \text{ and}\ldots\text{and } D_{24}) = 341/365$

P(all different) = $P(D_1 \text{ and } D_2 \text{ and}\ldots\text{and } D_{25})$

$$= P(D_1) \cdot P(D_2) \cdot \ldots \cdot P(D_{25})$$
$$= (365/365)(364/365)\ldots(341/365) = 0.431$$

NOTE: The above analysis ignores February 29. Including leap years (and counting February 29 equal to the other days) changes the final calculation to $(366/366)(365/366)\ldots(342/366) = 0.432$. An algorithm to perform this calculation can be constructed using a programming language, spreadsheet, or statistical software package. In BASIC, for example, use

```
10  LET P=1                    40  LET P=P*(365-K)/365
15  PRINT "HOW MANY PEOPLE?"   50  NEXT K
20  INPUT D                    60  PRINT "P(ALL DIFFERENT BIRTHDAYS) IS   ";P
30  FOR K=1 TO D-1             70  END
```

b. P(at least 2 share the same birthday) = 1 - P(all different)

$$= 1 - 0.431 = 0.569$$

NOTE: The above analysis ignores February 29. Including leap years (and counting February 29 equal to the other days) changes the final calculation to $1 - 0.432 = 0.568$.

33. Let F = getting a seat in the front row.

$P(F) = 2/24 = 1/12 = 0.0833$

• formulate the problem in terms of "n" rides

P(at least one F in n rides) = 1 - P(no F in n rides)

$$= 1 - (11/12)^n$$

• try different values for n until P(at least one F) ≥ 0.95

P(at least one F in 32 rides) = $1 - (11/12)^{32} = 1 - 0.0616 = 0.9384$

P(at least one F in 33 rides) = $1 - (11/12)^{33} = 1 - 0.0566 = 0.9434$

P(at least one F in 34 rides) = $1 - (11/12)^{34} = 1 - 0.0519 = 0.9481$

P(at least one F in 35 rides) = $1 - (11/12)^{35} = 1 - 0.0476 = 0.9524$

You must ride 35 times in order to have at least a 95% chance of getting a first row seat at least once.

4-6 Probabilities Through Simulations

1. No. Generating random numbers between 2 and 12 simulates a uniform distribution of equally likely values, but the sums obtained by rolling two dice do not follow a uniform distribution.

3. Yes. Each "birthday" has the same chance of being chosen for each of the 25 selections.

NOTE: In exercises #5-8 we suggest using random two-digit numbers from 00 to 99 inclusive. For convenience of association with the natural counting numbers, consider 00 to represent 100 so that the conceptual "order" of the digits is 01,02,03,…98,99,00. Even though other schemes using fewer than 100 possibilities could be used, this has the advantage of being able to read through a printed list of random digits and associating them into pairs – without having to discard any values as being outside the range of desired random numbers.

5. Generate 50 random two-digit integers as suggested in the NOTE before exercise #5.
   Let 01-95 represent an adult who recognizes the McDonald's brand name, and let 96-00 represent an adult who does not recognize the McDonald's brand name.

7. Generate a random three-digit integer from 001 to 000 as suggested in the NOTE before exercise #5.
   Let 001-528 represent a made free throw, and let 529-000 represent a missed free throw.

NOTE: The answers for the remaining exercises will vary. The results shown are those obtained in the preparation of this manual.

9. a. The 50 generated random numbers are:
       56 51 38 41 41 74 83 56 22 47 72 73 38 91 37 68 95 32 33 75 32 48 65 21 09
       86 33 27 46 80 18 37 72 81 45 30 61 08 27 28 66 35 52 71 19 98 65 41 47 06
   Based on the scheme in Exercise 5, the corresponding Y-N brand recognition values are:
       Y Y Y Y Y Y Y Y Y Y Y Y Y Y Y Y Y Y Y Y Y Y Y Y Y
       Y Y Y Y Y Y Y Y Y Y Y Y Y Y Y Y Y Y Y Y N Y Y Y Y
   The simulated proportion of adults who recognize the McDonald name is 49/50 = 0.98.
   This is close to the true value of 0.05
   b. The results of the ten simulations are as follows.
   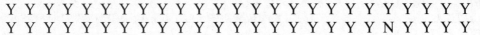
       0.98 0.98 0.98 0.96 0.94 0.90 1.00 0.88 0.94 0.98
   The results are rather variable, from 0.88 to 1.00 – but while they are not narrowly centered at 0.95, their mean value is 0.954. Based on these results, it would be extremely unusual for a sample of size 50 to yield a sample proportion near 0.50.

11. a. The 5 generated random numbers are: 497 424 938 011 655
    Based on the scheme in Exercise 7, the Y-N made free-throw values are: Y Y N Y N
    The simulated proportion of made free-throws is 3/6 = 0.600.
    This is reasonably close to the true value of 0.528.
    b. The proportions of made free-throws in each of the ten simulations are as follows.
        0.600 0.800 0.400 0.600 0.400 0.400 0.600 0.800 0.800 0.600
    Based on these results it appears that it would be unusual for O'Neal to make all five free-throws – at least, based on these results, the chances appear to be less than 1 in 10.

13. Generate 5 random two-digit integers as suggested in the
NOTE before exercise #5. Let 01-50 represent a boy and
let 51-00 represent a girl. Repeat this 20 times. The
results are given in the box at the right.

    The estimated probability of such an occurrence is
$4/20 = 0.25$, and so such a run is not unusual.

NOTE: The sample space for 5 births is the following
32 equally like possibilities. Since 16 of the them include a
run of three of the same gender, the true probability of such
an occurrence is $16/32 = 0.5$ – which is twice the estimate
produced by the simulation. In this instance the simulation
did not perform well.

```
bbbbb bbbbg bbbgb bbbgg bbgbb bbgbg bbggb bbggg
bgbbb bgbbg bgbgb bgbgg bggbb bggbg bgggb bgggg.
gbbbb gbbbg gbbgb gbbgg gbgbb gbgbg gbggb gbggg.
ggbbb ggbbg ggbgb ggbgg gggbb gggbg ggggb ggggg.
```

Exercise 13		
	result	run?
1	bbggb	
2	bbgbg	
3	gbgbg	
4	bgbbg	
5	gbbgg	
6	bbgbb	
7	bbgbb	
8	bggbb	
9	bbbbg	yes
10	bbggb	
11	bbgbb	
12	bbbgg	yes
13	bgbgg	
14	bgbgg	
15	bbggb	
16	ggbgb	
17	gbbbg	yes
18	bgggg	yes
19	ggbgg	
20	gbbgg	

15. Generate 152 random 0's and 1's. Let the 0's represent a girl and let a 1 represent a boy. Sum
the values to determine the number of boys. Repeat this 20 times. The 20 numbers of boys
thus obtained were: 71 91 72 77 72 78 74 77 76 82 83 69 78 79 75 85 76 79 71 72.
It appears that getting 127 boys in 152 would be a very unusual event. It appears that the
YSORT method is effective.

17. The probability of originally selecting the door with the car is 1/3.
    You can either switch or not switch.
    Let W = winning the car.
• If you make a selection and do not switch, $P(W) = 1/3$.
• If you make a selection and switch, there are only 2 possibilities.
    A = you originally had the winning door and switched to a losing door.
    B = you originally had a losing door and switched to a winning door.
  NOTE: You cannot switch from a losing door to another losing door. Monty Hall always
  opens a losing door. If you have a losing door and Monty Hall opens a losing door, then
  the only possible door to switch to is the winning door. Since you win if B occurs,
      $P(W) = P(B) = P(\text{originally had a losing door}) = 2/3$.
• Conclusion: the better strategy is to switch.
While the preceding analysis makes it unnecessary, a simulation of n trails can be done as
follows.
    1. Randomly select door 1,2 or 3 to hold the prize.
    2. Randomly select door 1,2 or 3 as your original choice.
    3. Determine m = the number of times the doors match.
    4. Estimated P(winning if you don't switch) = m/n. [m/n should be about 1/3]
    5. Estimated P(winning if you switch) = (n-m)/n. [(n-m)/n should be about 2/3]
    NOTE: If you switch, you lose m times and win (n-m) times – if you originally selected a

losing door, remember, a switch always wins the car.

The following Minitab commands produce the desired simulation, where column 1 indicates the winning door and column 2 indicates your choice.

```
MTB > RANDOM 100 C1 C2;
SUBC> INTEGER 1 3.
MTB > LET C3= C1-C2
MTB > TABLE C3
```

A zero in C3 indicates a match and that the original choice would win the car. The simulation produced for this manual yielded 37 0's in C3.

Estimated P(winning if you don't switch) = 37/100 = 37%

Estimated P(winning if you switch) = 63/100 = 63%

19. No, his reasoning is not correct. No, the proportion of girls will not increase. Each family would consist of zero to several consecutive girls, followed by a boy. The types of possible families and their associated probabilities would be as follows.

$$P(B) = \frac{1}{2} = 1/2 \qquad\qquad =16/32$$
$$P(GB) = (\frac{1}{2})(\frac{1}{2}) = 1/4 \qquad = 8/32$$
$$P(GGB) = (\frac{1}{2})(\frac{1}{2})(\frac{1}{2}) = 1/8 \qquad = 4/32$$
$$P(GGGB) = (\frac{1}{2})(\frac{1}{2})(\frac{1}{2})(\frac{1}{2}) = 1/16 \qquad = 2/32$$
$$P(GGGGB) = (\frac{1}{2})(\frac{1}{2})(\frac{1}{2})(\frac{1}{2})(\frac{1}{2}) = 1/32 = 1/32$$

etc.

Each collection of 32 families would be expected to look like this, where * represents one family with 5 or more girls and a boy.

B	B	B	B	B	B	B	B
B	B	B	B	B	B	B	B
GB	GB	GB	GB	GB	GB	GB	GB
GGB	GGB	GGB	GGB	GGGB	GGGB	GGGGB	*

A gender count reveals 32 boys and 31 or more girls, an approximately equal distribution of genders. In practice, however, families would likely stop after some maximum number of children – whether they had a boy or not. If that number were 5 for all families, then * = GGGGG and the expected gender count for the 32 families would be an exactly equal distribution of 31 boys and 31 girls.

NOTE: Since the "stopping point" would likely vary from family to family, the problem is actually more complicated. Suppose, for example, that 25% of the families decided in advance to stop at two children regardless of gender. We expect those 1/4 = 8/32 of the families to be distributed approximately evenly among the four rows in the above listing – 2 in each row. The 6 such families in the first 3 rows would be stopped by the king's decree and remain unchanged, but the 2 such families in the last row would both become GG families. Counting the * as GGGGG, this reduces the expected number of boys by 2·7/8 = 1.75 and the expected number of girls by 2·7/8 = 1.75 – and there would still be an equal number of each gender.

4-7 Counting

1. The permutations rule applies when different orderings of the same collection of objects represent a different solution. The combinations rule applies when all orderings of the same collection of objects represent the same solution.

3. Permutations. Because order makes a difference, and a different ordering of the top 3 finishers represents a different race outcome, the trifecta involves permutations and not combinations.

5. $5! = 5 \cdot 4 \cdot 3 \cdot 2 \cdot 1 = 120$

7. $_{52}C_2 = \dfrac{52!}{50!2!} = \dfrac{52 \cdot 51}{2!} = 1326$

   NOTE: This technique of reducing the problem by "canceling out" the 50! from both the numerator and denominator can be used in most combinations and permutations problems. In general, a smaller factorial in the denominator can be completely divided into a larger factorial in the numerator to leave only the "excess" factors not appearing in the denominator. Furthermore, $_nC_r$ and $_nP_r$ will always be integers. More generally, a non-integer answer to <u>any counting problem</u> (but not a probability problem) indicates that an error has been made.

9. $_9P_5 = 9!/4! = 9 \cdot 8 \cdot 7 \cdot 6 \cdot 5 = 15{,}120$

11. $_{42}C_6 = \dfrac{42!}{36!6!} = \dfrac{42 \cdot 41 \cdot 40 \cdot 39 \cdot 38 \cdot 37}{6!} = 5{,}245{,}786$

13. Let W = winning the described lottery with a single selection.
    The total number of possible combinations is
    $$_{54}C_6 = \dfrac{54!}{48!6!} = \dfrac{54 \cdot 53 \cdot 52 \cdot 51 \cdot 50 \cdot 49}{6!} = 25{,}827{,}165.$$
    Since only one combination wins, P(W) = 1/25,827,165.

15. Let W = winning the described lottery with a single selection.
    The total number of possible combinations is
    $$_{36}C_5 = \dfrac{36!}{31!5!} = \dfrac{36 \cdot 35 \cdot 34 \cdot 33 \cdot 32}{5!} = 376{,}992$$
    Since only one combination wins, P(W) = 1/376,992.

17. a. Let G = generating a given social security number in a single trial.
       The total number of possible sequences is
       $10 \cdot 10 \cdot 10 \cdot 10 \cdot 10 \cdot 10 \cdot 10 \cdot 10 \cdot 10 = 10^9 = 1{,}000{,}000{,}000$
       Since only one sequence is correct, P(G) = 1/1,000,000,000.
    b. Let F = generating the first 5 digits of a given social security number in a single trial.
       The total number of possible sequences is
       $10 \cdot 10 \cdot 10 \cdot 10 \cdot 10 = 10^5 = 100{,}000$
       Since only one sequence is correct, P(F) = 1/100,000. Since this probability is so small, one need not worry about the given scenario.

19. Since order makes a difference, use permutations.
    $_{20}P_6 = 20!/14! = 20 \cdot 19 \cdot 18 \cdot 17 \cdot 16 \cdot 15 = 27{,}907{,}200$

21. Since the order of the two wires being tested is irrelevant, use combinations.
    $_5C_2 = \dfrac{5!}{3!2!} = \dfrac{5 \cdot 4}{2!} = 10$

23. There are 8 tasks to perform, and each task can be performed in either of 2 ways.
    The total number of possible sequences is $2 \cdot 2 \cdot 2 \cdot 2 \cdot 2 \cdot 2 \cdot 2 \cdot 2 = 2^8 = 256$.
    Yes. The typical keyboard has approximately 50 keys, each with a lower an upper case possibility, representing about 100 commonly used characters.

25. The number of possible sequences of n different objects is n!, and $5! = 5 \cdot 4 \cdot 3 \cdot 2 \cdot 1 = 120$
    The unscrambled word is JUMBO. Since there is only one correct sequence, the probability of finding it with one random arrangement is 1/120.

27. a. Since order makes a difference, as there are 4 different offices, use permutations.

$_{11}P_4 = 11!/7! = 11 \cdot 10 \cdot 9 \cdot 8 = 7920$

   b. Since the order in which the 4 are picked makes no difference, use combinations.

$_{11}C_4 = \dfrac{11!}{7!4!} = \dfrac{11 \cdot 10 \cdot 9 \cdot 8}{4!} = 330$

29. a. There are 14 tasks to perform, and each task can be performed in either of 2 ways. The total number of possible sequences is $2 \cdot 2 \cdot 2 \cdot 2 \cdot 2 \cdot 2 \cdot 2 \cdot 2 \cdot 2 \cdot 2 \cdot 2 \cdot 2 \cdot 2 \cdot 2 = 2^{14} = 16{,}384$.

   b. The number of possible sequences of n objects is when some are alike is

$\dfrac{n!}{n_1!n_2! \cdots n_k!}$, and $\dfrac{14!}{13!1!} = 14$

   c. P(13G,1B) = (# of ways to get 13G,1B)/(total number of ways to get 14 babies)

        = 14/16,384 = 0.000854

   d. Yes. Since P(13G,1B) is so small, and since 13G,1B so far (only the 14G,0B result is more extreme) from the expected 7G,7B result, the gender-selection method appears to yield results significantly different from those of chance alone.

31. Since order makes no difference when one is choosing groups, use combinations.

The number of possible treatment groups is $_{10}C_5 = \dfrac{10!}{5!5!} = \dfrac{10 \cdot 9 \cdot 8 \cdot 7 \cdot 6}{5!} = 252$.

The number of ways to select a group of 5 men from the 5 males is $_5C_5 = 1$.
The number of ways to select a group of 5 women from the 5 females is $_5C_5 = 1$.
Since the number of possible same sex groups is $1 + 1 = 2$, the probability that random selection determines a treatment group of all the same sex is 2/252 = 0.00794.
Yes. Having a treatment group all of the same sex would create confounding – the researcher could not say whether any difference in the treatment group was a treatment effect or a gender effect.

33. Let A be selecting the correct 5 numbers from 1 to 55. The number of possible selections is

$_{55}C_5 = \dfrac{55!}{50!5!} = \dfrac{55 \cdot 54 \cdot 53 \cdot 52 \cdot 51}{5!} = 3{,}478{,}761.$

Since there is only one winning selection, P(A) = 1/3,478,761.
Let B be selecting the correct winning number from 1 to 42. There are 42 possible selections.
Since there is only one winning selection, P(B) = 1/42.
P(winning Powerball)
    = P(A and B) = P(A)·P(B)
    = (1/3,478,761)(1/42) = 1/146,107,962 = 0.00000000684

35. The first digit could be any of the 8 numbers 2,3,4,5,6,7,8,9.
The second digit could be either of the 2 numbers 0,1.
The third digit could be any of 9 numbers as follows.
     1,2,3,4,5,6,7,8,9 if the second digit is 0.
     0,2,3,4,5,6,7,8,9 if the second digit is 1.
By the fundamental counting rule, the number of possible 3-digit area codes is $8 \cdot 2 \cdot 9 = 144$.

37. There are 26 possible first characters, and 36 possible characters for the other positions. Find the number of possible names using 1,2,3,…,8 characters and then add to get the total.

characters	possible names		
1	26	=	26
2	26·36	=	936
3	26·36·36	=	33,696
4	26·36·36·36	=	1,213,056
5	26·36·36·36·36	=	43,670,016
6	26·36·36·36·36·36	=	1,572,120,576
7	26·36·36·36·36·36·36	=	56,596,340,736
8	26·36·36·36·36·36·36·36	=	2,037,468,266,496

total = 2,095,681,645,538

39. a. The calculator factorial key gives 50! = 3.04140932 x 10^{64}

Using the approximation, K = (50.5)·log(50) + 0.39908993 – 0.43429448(50)

$$= 85.79798522 + 0.39908993 – 21.71472400$$
$$= 64.48235225$$
$$\text{and then } 50! = 10^K$$
$$= 10^{64.48235225}$$
$$= 3.036345215 \text{ x } 10^{64}$$

NOTE: The two answers differ by 0.0051 x 10^{64} – i.e., 5.1 x 10^{61}, or 51 followed by 60 zeros. While such an error of "zillions and zillions" may seem quite large, it is only an error of (5.1 x 10^{61})/(3.04 x 10^{64}) = 1.7%.

b. The number of possible routes is 300!

Using the approximation, K = (300.5)·log(300) + 0.39908993 – 0.43429448(300)

$$= 744.374937 + 0.39908993 – 130.288344$$
$$= 614.485683$$
$$\text{and then } 300! = 10^K$$
$$= 10^{614.485683}$$

Since the number of digits in 10^x is the next whole number above x, 300! has 615 digits.

41. This problem cannot be solved directly using permutations, combinations or other techniques presented in this chapter. It is best solved by listing all the possible solutions in an orderly fashion and then counting the number of solutions. Often this is the most reasonable approach to a counting problem.

While you are encouraged to develop your own systematic approach to the problem, the following table represents one way to organize the solution. The table is organized by rows, according to the numbers of pennies in each way that change can be made. The numbers in each row give the numbers of the other coins as explained in the footnotes below the table.

The three numbers in the 80 row, for example, indicate that there are three ways to make change using 80 pennies. The 4 in the "only 5¢" column represents one way (80 pennies, 4 nickels). The 2 in the 10¢ column of the "nickels and one other coin" columns represents another way (80 pennies, 2 dimes). The 1 in the 10¢ column of the "nickels and one other coin" columns represents a third way (80 pennies, 1 dime, and [by default] 2 nickels).

The bold-face **3** in the 20 row and the 10¢/25¢ column of the "nickels and two other coins" columns represents 20 pennies, 3 dimes, one quarter, and [by default] 5 nickels.

Using the following table, or the system of your own design, you should be able to
(1) take a table entry and determine what way to make change it represents and
(2) take a known way to make change and find its representation in the table.

1¢	5¢ (only)	nickels and one other coin[a] 10¢	25¢	50¢	nickels and two other coins[b] 10¢/25¢	10¢/50¢	25¢/50¢	all[c] 10¢/25¢/50¢
100	0	.			.			.
95	1	.			.			.
90	2	1			.			.
85	3	1			.			.
80	4	2,1	.		.			.
75	5	2,1	1	.	.			.
70	6	3,2,1	1	.	.			.
65	7	3,2,1	1	.	1			.
60	8	4,3,2,1	1	.	1			.
55	9	4,3,2,1	1	.	2,1	.		.
50	10	5,4,3,2,1	2,1	1	2,1	.		.
45	11	5,4,3,2,1	2,1	1	3,2,1	.		.
40	12	6,5,4,3,2,1	2,1	1	3,2,1	1	.	.
35	13	6,5,4,3,2,1	2,1	1	4,3,2,1	1	.	.
30	14	7,6,5,4,3,2,1	2,1	1	4,3,2,1	2,1	.	.
25	15	7,6,5,4,3,2,1	3,2,1	1	5,4,3,2,1	2,1	1	.
20	16	8,7,6,5,4,3,2,1	3,2,1	1	5,4,**3**,2,1	3,2,1	1	
15	17	8,7,6,5,4,3,2,1	3,2,1	1	6,5,4,3,2,1	3,2,1	1	1
10	18	9,8,7,6,5,4,3,2,1	3,2,1	1	6,5,4,3,2,1	4,3,2,1	1	1
5	19	9,8,7,6,5,4,3,2,1	3,2,1	1	7,6,5,4,3,2,1	4,3,2,1	1	2,1
0	20	10,9,8,7,6,5,4,3,2,1	4,3,2,1	2,1	7,6,5,4,3,2,1	5,4,3,2,1	2,1	2,1
ways 21		100	34	12	56	25[d]	7	6[e]

[a]The possible numbers of the non-nickel coin are given. The numbers of nickels are found by default.

[b]This is for a single occurrence of the second coin listed. The possible numbers of the leading non-nickel coin are given. The numbers of nickels are found by default.

[c]This is for a single occurrence of the second and third coin listed. The possible numbers of the leading non-nickel coin are given. The numbers of nickels are found by default.

[d]There are another 25 ways when the 50¢ is in the form of 2 quarters.

[e]There are another 6 ways when the 50¢ is in the form of 2 quarters.

The total number of ways identified is 21+100+34+12+56+25+25+7+6+6 = 292.

If a one-dollar coin is also considered as change for a dollar, there are 293 ways.

The one dollar bill itself should not be counted.

Statistical Literacy and Critical Thinking

1. An event that has probability 0.004 is unlikely to occur by chance. It would occur by chance only about 4 times in 1000 opportunities.

2. No. The detectors would be related in the sense that they would all be inoperable whenever the home loses its electricity.

3. No, her reasoning is incorrect for two reasons. (1) Since she is a new recruit she has no reason to assume that the historical rate of 12.7% will apply to her. (2) Even if the 12.7% rate applies, the (0.127)(0.127) = 0.0161 calculation applies only to burglaries that are selected at random and/or handled independently. Since the cases were handled by the same person, knowing that one of the cases closes with an arrest would reasonably make a person think that it was more likely the other case would close with an arrest – i.e., the events are not independent and $P(C_2|C_1) > P(C) = 0.127$.

4. No. Lottery numbers are selected so that past results have no effect on future drawings. Each drawing is statistically independent of the all the others.

Chapter Quick Quiz

1. $P(\overline{C}) = 1 - P(C)$
   $= 1 - 0.70 = 0.30$

2. $P(C_1 \text{ and } C_2) = P(C_1){\cdot}P(C_2)$
   $= (0.70)(0.70) = 0.49$

3. Answers will vary, but such an occurrence is not very common. A reasonable guess for such an interruption might be $P(I) = 0.005$.

4. No. A result with probability 0.342 could easily have occurred by chance.

5. $P(\overline{A}) = 1 - P(A) = 1 - 0.4 = 0.6$

NOTE: For exercises 6-10, refer to the table at the right and use A,B,P,F to identify in the natural way the groups and qualifying exam results.

		Result		
		passed	failed	
Group	A	10	14	24
	B	417	145	562
		427	159	586

6. $P(P) = 427/586 = 0.729$

7. $P(B \text{ or } P) = P(B) + P(P) - P(B \text{ and } P)$
   $= 562/586 + 427/586 - 417/586$
   $= 572/586 = 0.976$

8. $P(A_1 \text{ and } A_2) = P(A_1){\cdot}P(A_2|A_1)$
   $= (24/586)(23/585)$
   $= 0.00161$

9. This problem may be worked either of two ways.
   directly from chart: $P(A \text{ and } P) = 10/586 = 0.0171$
   by formula: $P(A \text{ and } P) = P(A){\cdot}P(P|A)$
   $= (24/586)(10/24) = 10/586 = 0.0171$

10. $P(P|A) = 10/24 = 0.417$

Review Exercises

NOTE: For exercises 1-10, refer to the table at the right and use the following notation.
Let H = the rider wore a helmet.
Let I = the rider experienced a head injury.

		Head Injury?		
		yes	no	
Wore Helmet?	yes	96	656	752
	no	480	2330	2810
		576	2986	3562

1. $P(I) = 576/3562 = 0.162$

2. $P(H) = 752/3562 = 0.211$

3. $P(I \text{ or } H) = P(I) + P(H) - P(I \text{ and } H)$
   $= 576/3562 + 752/3562 - 96/3562$
   $= 1232/3562 = 0.346$

4. $P(\overline{H} \text{ or } \overline{I}) = P(\overline{H}) + P(\overline{I}) - P(\overline{H} \text{ and } \overline{I})$
$$= 2810/3562 + 2986/3562 - 2330/3562$$
$$= 3466/3562 = 0.973$$

5. This problem may be worked either of two ways.
directly from chart: $P(H \text{ and } I) = 96/3562 = 0.0270$
by formula: $P(H \text{ and } I) = P(H) \cdot P(I|H)$
$$= (752/3562)(96/752) = 96/3562 = 0.0270$$

6. This problem may be worked either of two ways.
directly from chart: $P(\overline{H} \text{ and } \overline{I}) = 2330/3562 = 0.654$
by formula: $P(\overline{H} \text{ and } \overline{I}) = P(\overline{H}) \cdot P(\overline{I}|\overline{H})$
$$= (2810/3562)(2330/2810) = 2330/3562 = 0.654$$

7. $P(H_1 \text{ and } H_2) = P(H_1) \cdot P(H_2|H_1)$
$$= (752/3562)(751/3561) = 0.0445$$

8. $P(I_1 \text{ and } I_2) = P(I_1) \cdot P(I_2|I_1)$
$$= (576/3562)575/3561) = 0.0261$$

9. $P(\overline{H}|I) = 480/576 = 0.833$

10. $P(\overline{I}|H) = 656/752 = 0.872$

11. Answers will vary. Black cars are not popular, but they are not rare enough to be considered unusual. $P(B) = 0.08$ seems like a reasonable guess.

12. a. $P(\overline{B}) = 1 - P(B)$
$$= 1 - 0.35 = 0.65$$
   b. $P(B_1 \text{ and } B_2 \text{ and } B_3 \text{ and } B_4) = P(B_1) \cdot P(B_2) \cdot P(B_3) \cdot P(B_4)$
$$= (0.35)(0.35)(0.35)(0.35) = (0.35)^4 = 0.0150$$
   c. Yes. Since $0.015 \leq 0.05$, selecting four people at random and finding they all have blue eyes would be considered an unusual event.

13. a. $P(O18) = 1/365 = 0.00274$
   b. $P(O) = 31/365 = 0.0849$
   c. This would be a very rare event, probably occurring less than one time in a million tries. A reasonable guess would be $P(E) = 0.0000001$
   d. Yes, based on the answer in (c). Since $0.0000001 \leq 0.05$, randomly selecting an adult American and fifing someone who knows that date would be considered an unusual event.

14. a. $P(D) = 15.2/100,000 = 0.000152$
   b. $P(D_1 \text{ and } D_2) = P(D_1) \cdot P(D_2)$
$$= (0.000152)(0.000152) = (0.000152)^2 = 0.0000000231$$
   c. $P(\overline{D}) = 1 - P(D) = 1 - 0.000152 = 0.999848$
$P(\overline{D}_1 \text{ and } \overline{D}_2) = P(\overline{D}_1) \cdot P(\overline{D}_2)$
$$= (0.999848)(0.999848) = (0.999848)^2 = 0.999696$$

15. a. $P(\overline{A}) = 1 - P(A)$
$$= 1 - 0.40 = 0.60$$
   b. No, for two reasons. First, the sample was limited to America OnLine subscribers – who are not necessarily representative of the general population. Second, the poll was based on a voluntary response sample – and not on people selected at random. People who respond in a

voluntary response sample are typically those with strong feelings one way or the other (i.e., either people who really like Sudoku, or those who can't stand it) and are not necessarily representative of the general population.

16. For each person, P(negative) = 1 – P(positive) = 1 – 0.00320 = 0.99680.
    P(positive group result) = P(at least one person is positive)
    $$= 1 - \text{P(all persons are negative)}$$
    $$= 1 - \text{P}(N_1 \text{ and } N_2 \text{ and...and } N_{10})$$
    $$= 1 - \text{P}(N_1) \cdot \text{P}(N_2) \cdot ... \cdot \text{P}(N_{10})$$
    $$= 1 - (0.99680)(0.99680)...(0.99680)$$
    $$= 1 - (0.99680)^{10} = 1 - 0.9685 = 0.0315$$
    No. Since $0.0315 \leq 0.05$, a positive group result is not considered likely.
    NOTE: This plan is very efficient. Suppose, for example, there were 3000 people to be tested. Only in 0.0315 = 3.15% of the 300 groups would a retest need to be done for each of the ten individuals. Those $(0.0315)(300) \approx 9$ groups would generate $(9)(10) = 90$ retests. The total number of tests required would be 390 (the 300 original + the 90 retests), only about 13% of the 3,000 tests that would have been required to test everyone individually.

17. Let D = getting a Democrat.
    P(D) = 0.85 for each selection
    $$\text{P}(D_1 \text{ and } D_2 \text{ and...and } D_{30}) = \text{P}(D_1) \cdot \text{P}(D_2) \cdot ... \cdot \text{P}(D_{30})$$
    $$= (0.85)(0.85)...(0.85) = (0.85)^{30} = 0.00763$$
    Yes; since $0.00763 \leq 0.05$, getting 30 Democrats by chance alone is an unusual event. Since the probability 0.00763 is so small, the results are evidence that the pollster is not telling the truth.

18. Let $\overline{M}$ = a male in that age bracket not surviving. $\text{P}(\overline{M}) = 114.4/100,000 = 0.001144$
    Let $\overline{F}$ = a female in that age bracket not surviving. $\text{P}(\overline{F}) = 44/100,000 = 0.00044$
    a. $\text{P}(M) = 1 - \text{P}(\overline{M})$
       $$= 1 - 0.001144 = 0.998856$$
    b. $\text{P}(M_1 \text{ and } M_2) = \text{P}(M_1) \cdot \text{P}(M_2)$
       $$= (.998856)(0.998856) = (0.998856)^2 = 0.998$$
    c. $\text{P}(F) = 1 - \text{P}(\overline{F}) = 1 - 0.00044 = 0.99956$
       $\text{P}(F_1 \text{ and } F_2) = \text{P}(F_1) \cdot \text{P}(F_2)$
       $$= (.99956)(0.99956) = (0.99956)^2 = 0.999$$
    d. Males are more likely than females to be involved in situations where death is a possible result (e.g., military combat, violent crimes, hazardous occupations, etc.).

19. The number of possible selections is $_{38}C_5 = \dfrac{38!}{33!5!} = \dfrac{38 \cdot 37 \cdot 36 \cdot 35 \cdot 34}{5!} = 501,942$.
    Since there is only one winning selection, P(W) = 1/501,942 = 0.00000199.
    Since $0.00000199 \leq 0.05$, winning the jackpot with a single selection is an unusual event. But since hundreds of thousands of tickets are purchased, it is not unusual for there to be a winner.

20 There are 10 possibilities for each digit: 0,1,2,3,4,5,6,7,8,9.
    By the fundamental counting rule, the number of possibilities is now
    $$10 \cdot 10 \cdot ... \cdot 10 = 10^{13} = 10,000,000,000,000 = 10 \text{ trillion}$$

Cumulative Review Exercises

1. values in order are: 17 18 18 18 18 19 19 19 19 19 20 20 20 20 20 20 20 20 21 21
   The summary statistics are: n = 20 $\Sigma x = 386$ $\Sigma x^2 = 7472$

   a. $\bar{x} = (\Sigma x)/n = 386/20 = 19.3$ oz

   b. $\tilde{x} = (19+20)/2 = 19.5$ oz

   c. $s^2 = [n(\Sigma x^2) - (\Sigma x)^2]/[n(n-1)]$
      $= [20(7472) - (386)^2]/[20(19)]$
      $= 444/380 = 1.168$
      s = 1.081, rounded to 1.1 oz

   d. $s^2 = 1.168$, rounded to 1.2 oz^2 [from part (c) above]

   e. No. The mean and the median are below the supposed weight of 21 oz, as are 18/20 = 90%
      of the individual values.

2. The total number of responses is 3042 + 2184 = 5226
   a. 3042/5226 = 58.2%
   b. P(Y) = 3042/5226 = 0.582
   c. This is a voluntary response sample. If the intended population is all people, this is also a
      convenience sample (of those who could be easily reached by AOL since they were AOL
      subscribers). Neither voluntary response samples nor convenience samples are suitable for
      making statements about the general population.
   d. A simple random sample is one for which every sample of size n (in this case, n = 5226)
      from the population has an equal chance of being selected. A simple random sample would
      be better than a voluntary response sample or a convenience sample because it is more likely
      to be representative of the general population.

3. Organize the data as follows.
   values in order
        regular: 370 370 371 372 372 374
           diet: 352 353 356 357 357 358
   summary statistics
        regular: n = 6 $\Sigma x = 2229$ $\Sigma x^2 = 828085$
           diet: n = 6 $\Sigma x = 2133$ $\Sigma x^2 = 758311$

   a. $\bar{x} = (\Sigma x)/n$
        regular: 2229/6 = 371.5 g
           diet: 2133/6 = 355.5 g
      The mean weight for the diet Coke appears to be significantly smaller.

   b. $\tilde{x} = (x_3+x_4)/2$
        regular: (371+372)/2 = 371.5 g
           diet: (356+357)/2 = 356.5 g
      The median weight for the diet Coke appears to be significantly smaller.

   c. $s^2 = [n(\Sigma x^2) - (\Sigma x)^2]/[n(n-1)]$
        regular: $s^2 = [6(828085) - (2229)^2]/[6(5)] = 69/30 = 2.3$; s = 1.5 g
           diet: $s^2 = [6(758311) - (2133)^2]/[6(5)] = 177/30 = 5.9$; s = 2.4 g
      The diet Coke weights appear to have significantly more variability.

d. The answers for the variance are taken from part (c) above.
regular: $s^2 = 2.3 \text{ g}^2$
diet: $s^2 = 5.9 \text{ g}^2$

e. No. The weights for the diet Coke appear to be significantly smaller and have more variability.

4. a. According to one criterion, unusual values are those that are more than two standard deviations from the mean. A score of 38 is unusual because it is $(38-67.4)/11.6 = -2.53$ standard deviations from the mean.
   b. According to one criterion, unusual events are those for which $P(E) \le 0.05$. Getting all ten T-F questions correct by guessing is unusual because $P(E) = (1/2)10 = 1/1024 = 0.000977$.

5. a. This is a convenience sample because the data was gathered from those who happened to be at hand.
   b. If eye color is related to ethnicity and the college is not ethnically representative of the United States, then this particular sample might have a sample bias preventing it from being representative of the general population.
   c. P(Brown or Blue) = P(Brown) + P(Blue) – P(Brown and Blue)
$$= 0.40 + 0.35 - 0$$
$$= 0.75$$
   d. $P(Brown_1 \text{ and } Brown_2) = P(Brown_1) \cdot P(Brown_2|Brown_1) = (0.40)(0.40) = 0.16$
   P(at least one Brown) = $P(Brown_1 \text{ or } Brown_2)$
$$= P(Brown_1) + P(Brown_2) - P(Brown_1 \text{ and } Brown_2)$$
$$= 0.40 + 0.40 - 0.16$$
$$= 0.64$$

6. The first note is given by *. There are 3 possibilities for each of the next 15 notes.
By the fundamental counting rule, there are
$$3 \cdot 3 \cdot 3 \cdot 3 \cdot 3 \cdot 3 \cdot 3 \cdot 3 \cdot 3 \cdot 3 \cdot 3 \cdot 3 \cdot 3 \cdot 3 \cdot 3 = 3^{15} = 14,348,907 \text{ possible sequences.}$$
NOTE: This assumes that each song has at least 16 notes, and it does not guarantee that two different melodies will not have the same representation – if one goes up two steps every time the other goes up one step, for example, they both will show a U in that position.

Chapter 5

Discrete Probability Distributions

5-2 Random Variables

1. As defined in the text, a random variable is a variable that takes on a single numerical value, determined by chance, for each outcome of a procedure. In this exercise, the random variable is the number of winning lottery tickets obtained over a 52 week period. The possible values for this random variable are 0,1,2,…,52.

3. The probability distribution gives each possible non-overlapping outcome O_i of a procedure and the probability $P(O_i)$ associated with each of those outcomes.
   Since it is a certainty that one of the outcomes will occur,
   $P(O_1 \text{ or } O_2 \text{ or } O_3 \text{ or}…) = 1$.
   Since the outcomes are mutually exclusive,
   $P(O_1 \text{ or } O_2 \text{ or } O_3 \text{ or}…) = P(O_1) + P(O_2) + P(O_3) +…$.
   Therefore,
   $P(O_1) + P(O_2) + P(O_3) +… = \Sigma P(O_i) = 1$.

5. a. Discrete, since such a number is limited to certain values – viz., the non-negative integers.
   b. Continuous, since weight can be any value on a specified continuum.
   c. Continuous, since height can be any value on a specified continuum.
   d. Discrete, since such a number is limited to certain values – viz., the non-negative integers.
   e. Continuous, since time can be any value on a specified continuum.

NOTE: If one of the conditions for a probability distribution does not hold, the formulas do not apply – and they produce numbers that have no meaning. When working with probability distributions and formulas in the exercises that follow, always keep the following important facts in mind.
- $\Sigma P(x)$ must always equal 1.000
- $\Sigma[x \cdot P(x)]$ gives the mean of the x values and must be a number between the highest and lowest x values.
- $\Sigma[x^2 \cdot P(x)]$ gives the mean of the x^2 values and must be a number between the highest and lowest x^2 values.
- Σx and Σx^2 have no meaning and should not be calculated
- The quantity "$\Sigma[x^2 \cdot P(x)] - \mu^2$" cannot possibly be negative – if it is, there is a mistake.
- Always be careful to use the unrounded mean in the calculation of the variance, and to take the square root of the unrounded variance to find the standard deviation.

7. The given table is a probability distribution since $0 \le P(x) \le 1$ for each x and $\Sigma P(x)=1$.

x	P(x)	x·P(x)	x^2	x^2·P(x)	$\mu = \Sigma[x \cdot P(x)]$
0	0.125	0	0	0	= 1.500, rounded to 1.5 children
1	0.375	0.375	1	0.375	$\sigma^2 = \Sigma[x^2 \cdot P(x)] - \mu^2$
2	0.375	0.750	4	1.500	$= 3.000 - (1.500)^2$
3	0.125	0.375	9	1.125	= 0.750
	1.000	1.500		3.000	$\sigma = 0.866$, rounded to 0.9 children

9. The given table is not a probability distribution since $\Sigma P(x) = 0.984 \neq 1$.

11. The given table is a probability distribution since $0 \leq P(x) \leq 1$ for each x and $\Sigma P(x)=1$.

x	P(x)	x·P(x)	x^2	x^2·P(x)	
0	0.02	0	0	0	$\mu = \Sigma[x \cdot P(x)]$
1	0.15	0.15	1	0.15	$\quad$ = 2.75, rounded to 2.8 TV sets
2	0.29	0.58	4	1.16	$\sigma^2 = \Sigma[x^2 \cdot P(x)] - \mu^2$
3	0.26	0.78	9	2.34	$\quad = 9.21 - (2.75)^2$
4	0.16	0.64	16	2.56	$\quad = 1.6475$
5	0.12	0.60	25	3.00	$\sigma = 1.284$, rounded to 1.3 TV sets
	1.00	2.75		9.21	

13. The given table is a probability distribution since $0 \leq P(x) \leq 1$ for each x and $\Sigma P(x)=1$.

x	P(x)	x·P(x)	x^2	x^2·P(x)	
0	0+	0	0	0	$\mu = \Sigma[x \cdot P(x)]$
1	0+	0	1	0	$\quad$ = 6.000, rounded to 6.0 peas
2	0.004	0.008	4	0.016	$\sigma^2 = \Sigma[x^2 \cdot P(x)] - \mu^2$
3	0.023	0.069	9	0.207	$\quad = 37.484 - (6.000)^2$
4	0.087	0.348	16	1.382	$\quad = 1.484$
5	0.208	1.040	25	5.200	$\sigma = 1.218$, rounded to 1.2 peas
6	0.311	1.866	36	11.196	
7	0.267	1.869	49	13.083	
8	0.100	0.800	64	6.400	
	1.000	6.000		37.484	

15. a. $P(x=7) = 0.267$

$\quad$ b. $P(x \geq 7) = P(x=7 \text{ or } x=8)$
$\qquad\qquad\quad = P(x=7) + P(x=8)$
$\qquad\qquad\quad = 0.267 + 0.100$
$\qquad\qquad\quad = 0.367$

$\quad$ c. The answer in part (b) is the one which determines whether 7 is an unusually high result. When there are several different possible outcomes, the probability of getting any one of them exactly (even the more likely ones near the middle of the distribution) may be small. An outcome is unusually high if it is in the upper tail of the distribution – i.e., if the probability of getting it or a higher value is small.

$\quad$ d. No; since $P(x \geq 7) = 0.367 > 0.05$, 7 is not an unusually high number of peas with green pods.

17. a. The given table is a probability distribution since $0 \leq P(x) \leq 1$ for each x and $\Sigma P(x)=1$.

x	P(x)	x·P(x)	x^2	x^2·P(x)	
4	0.1919	0.7676	16	3.0704	$\mu = \Sigma[x \cdot P(x)]$
5	0.2121	1.0605	25	5.3025	$\quad$ = 5.7772, rounded to 5.8
6	0.2222	1.3332	36	7.9992	$\sigma^2 = \Sigma[x^2 \cdot P(x)] - \mu^2$
7	0.3737	2.6159	49	18.3113	$\quad = 34.6834 - (5.7772)^2$
	0.9999	5.7772		34.6834	$\quad = 1.3074$
					$\sigma = 1.1434$, rounded to 1.1

$\quad$ b. $\mu = 5.8$ games and $\sigma = 1.1$ games

$\quad$ c. No; since $P(x=4) = 0.1919 > 0.05$, winning in 4 games is not an unusual event.

19. a. The given table is a probability distribution since $0 \le P(x) \le 1$ for each x and $\Sigma P(x)=1$.

x	P(x)	x·P(x)	x^2	x^2·P(x)	
0	0.824	0	0	0	$\mu = \Sigma[x \cdot P(x)]$
1	0.083	0.083	1	0.083	$= 0.435$, rounded to 0.4
2	0.039	0.078	4	0.156	$\sigma^2 = \Sigma[x^2 \cdot P(x)] - \mu^2$
3	0.014	0.042	9	0.126	$= 1.821 - (0.435)^2$
4	0.012	0.048	16	0.192	$= 1.635$
5	0.008	0.040	25	0.200	$\sigma = 1.279$, rounded to 1.3
6	0.008	0.048	36	0.288	
7	0.004	0.028	49	0.196	
8	0.004	0.032	64	0.256	
9	0.004	0.036	81	0.324	
	1.000	0.435		1.821	

b. $\mu = 0.4$ bumper stickers and $\sigma = 1.3$ bumper stickers

c. The range rule of thumb suggests that "usual" values are those within two standard deviations of the mean.

minimum usual value = $\mu - 2\sigma = 0.4 - 2(1.3) = -2.2$, truncated to 0
maximum usual value = $\mu + 2\sigma = 0.4 + 2(1.3) = 3.0$

The range of values for usual numbers of bumper stickers is from 0 to 3.0.

d. No; since $0 < 1 < 3.0$, it is not unusual to have more than one bumper sticker.

21. There are eight equally likely possible outcomes: GGG GGB GBG BGG GBB BGB BBG BBB. The following table describes the situation, where x is the number of girls per family of 3.

x	P(x)	x·P(x)	x^2	x^2·P(x)	
0	0.125	0	0	0	$\mu = \Sigma[x \cdot P(x)]$
1	0.375	0.375	1	0.375	$= 1.500$, rounded to 1.5 girls
2	0.375	0.750	4	1.500	$\sigma^2 = \Sigma[x^2 \cdot P(x)] - \mu^2$
3	0.125	0.375	9	1.125	$= 3.000 - (1.500)^2$
	1.000	1.500		3.000	$= 0.75$
					$\sigma = 0.8660$, rounded to 0.9 girls

No. Since $P(x=3) = 0.125 > 0.05$, it is not unusual for a family of 3 children to have all girls.

23. The following table describes the situation.

x	P(x)	x·P(x)	x^2	x^2·P(x)	
0	0.1	0	0	0	$\mu = \Sigma[x \cdot P(x)]$
1	0.1	0.1	1	0.1	$= 4.5$
2	0.1	0.2	4	0.4	$\sigma^2 = \Sigma[x^2 \cdot P(x)] - \mu^2$
3	0.1	0.3	9	0.9	$= 28.5 - (4.5)^2$
4	0.1	0.4	16	1.6	$= 8.25$
5	0.1	0.5	25	2.5	$\sigma = 2.8723$, rounded to 2.9
6	0.1	0.6	36	3.6	
7	0.1	0.7	49	4.9	
8	0.1	0.8	64	6.4	
9	0.1	0.9	81	8.1	
	1.0	4.5		28.5	

The probability histogram for this distribution would be flat, with each bar having the same height.

25. a. Since each of the 3 positions could be filled (with replacement) by any of the 10 digits 0,1,2,3,4,5,6,7,8,9, there are $10 \cdot 10 \cdot 10 = 1000$ possible different selections.

b. Let W = winning (i.e., matching the selection drawn).
   Since only one of the possible 1000 selections is a inner, $P(W) = 1/1000 = 0.001$

c. The net profit is the payoff minus the original bet, in this case $250.00 − $0.50 = $249.50.

d. The following table describes the situation.

x	P(x)	x·P(x)
-0.50	0.999	-0.4995
249.50	0.001	0.2495
	1.000	-0.2500

E = Σ[x·P(x)] = -0.2500 [i.e., a loss of 25¢]

The expected value is -25¢.

e. Since both games have the same expected value, neither bet is better than the other.

27. a. The following table describes the situation.

x	P(x)	x·P(x)
-5	33/38	-165/38
30	5/38	150/38
	38/38	-15/38

E = Σ[x·P(x)] = -15/38 = -.3947, rounded to -39.4¢

b. Since -26 > -39.4, wagering $5 on the number 13 is the better bet.

29. a. From the 30-year-old male's perspective, the two possible outcome values are
   -$161, if he lives
   100,000 – 161 = $99,839, if he dies

b. The following table describes the situation.

x	P(x)	x·P(x)
-161	0.9986	-160.7746
99,839	0.0014	139.7746
	1.0000	-21.0000

E = Σ[x·P(x)] = -21.0000, rounded to -$21.0

c. Yes; the insurance company can expect to make an average of $21.0 per policy.

31. For every $1000, Bond A gives a profit of (0.06)($1000) = $60 with probability 0.99.
The following table describes the situation.

x	P(x)	x·P(x)
-1000	0.01	-10.00
60	0.99	59.40
	1.00	49.40

E = Σ[x·P(x)] = 49.40, rounded to 49.4 [i.e., a profit of $49.4]

For every $1000, Bond B gives a profit of (0.08)($1000) = $80 with probability 0.95.
The following table describes the situation.

x	P(x)	x·P(x)
-1000	0.05	-50.00
80	0.95	76.40
	1.00	26.00

E = Σ[x·P(x)] = 26.00, rounded to 26.0 [i.e., a profit of $26.0]

Bond A is the better bond since it has the higher expected value – i.e., 49.4 > 26.0. Since both bonds have positive expectations, either one would be a reasonable selection. Although Bond A has a higher expectation, a person willing to assume more risk in hope of a higher payoff might opt for Bond B.

33. Since each die has 6 faces, there will be 6·6 = 36 possible outcomes. If each of the 12 sums 1,2,3,…,12 is to be equally likely, each sum must occur exactly 3 times. As a starting point, suppose one of the dice is normal. If one die contains the usual digits 1,2,3,4,5,6, the other die
   (1) must have three 0's to pair with the 1 to get three sums of 1.
   (2) must have three 6's to pair with the 6 to get three sums of 12.

The 36 possible outcomes generated by such dice are Given in the box at the right. Inspection indicates that each of the sums 1,2,3,…,12 appears exactly 3 times so that P(x) = 3/36 = 1/12 for x = 1,2,3,…,12. A solution, therefore, is to mark one die normally 1,2,3,4,5,6 and mark the other die 0,0,0,6,6,6.

1-0	2-0	3-0	4-0	5-0	6-0
1-0	2-0	3-0	4-0	5-0	6-0
1-0	2-0	3-0	4-0	5-0	6-0
1-6	2-6	3-6	4-6	5-6	6-6
1-6	2-6	3-6	4-6	5-6	6-6
1-6	2-6	3-6	4-6	5-6	6-6

5-3 Binomial Probability Distributions

1. Getting two people with blue eyes followed by three people with eyes that are not blue is only one way of getting exactly two people with blue eyes in a sample of five randomly selected people. All the ways of getting exactly two people as described must be considered.

3. Yes. Since 30/1236 = 0.024 ≤ 0.05, the 5% rule can be applied to determine that the repeated selections can be treated as if they were independent.

5. Yes. All four requirements are met.

7. No. Requirement (3) is not met. There are more than two possible outcomes for the ages of the parents.

9. No. Requirements (2) and (4) are not met. Since 20/100 = 0.20 > 0.05, and the selections are done without replacement, the trials cannot be considered independent. The value p of obtaining a success changes from trial to trial as each selection without replacement changes the population from which the next selection is made.

11. Yes. All four requirements are met. Since 500/2,800,000 = 0.00018 ≤ 0.05, the selections can considered to be independent – even though the they are made without replacement.

13. a. P(WWC) = P(W)·P(W)·P(C)
$$= (4/5)(4/5)(1/5)$$
$$= 16/125 = 0.128$$
   b. There are three possible arrangements: WWC, WCW, CWW
$$P(WWC) = P(W) \cdot P(W) \cdot P(C) = (4/5)(4/5)(1/5) = 16/125$$
$$P(WCW) = P(W) \cdot P(C) \cdot P(W) = (4/5)(1/5)(4/5) = 16/125$$
$$P(CWW) = P(C) \cdot P(W) \cdot P(W) = (1/5)(4/5)(4/5) = 16/125$$
   c. P(exactly one correct) = P(WWC or WCW or CWW)
$$= P(WWC) + P(WCW) + P(CWW)$$
$$= 16/125 + 16/125 + 16/125$$
$$= 48/125 = 0.384$$

15. From Table A-1 in the .30 column and the 2-1 row, P(x=1) = 0.420.

17. From Table A-1 in the .99 column and the 15-11 row, P(x=11) = 0+.

19. From Table A-1 in the .05 column and the 10-2 row, P(x=2) = 0.075.

21. $P(x) = \dfrac{n!}{(n-x)!x!}p^x q^{n-x}$
$$P(x=10) = [12!/(2!10!)](0.75)^{10}(0.25)^2$$
$$= [66](0.0563)(0.0625)$$
$$= 0.232$$

NOTE: The intermediate values 66, 0.0563, and 0.0625 are given in Exercise 21 to help those with an incorrect answer to identify the portion of the problem in which the mistake was made. In the future, only the value n!/[(n-x)!x!] will be given separately. In practice, all calculations can be done in one step on a calculator. You may choose (or be asked) to write down intermediate values for your own (or the instructor's) benefit, but…
  • Never round off in the middle of a problem.
  • Do not write the values down on paper and then re-enter them in the calculator.
  • Use the memory to let the calculator remember with complete accuracy any intermediate values
      that will be used in subsequent calculations.
In addition, always make certain that n!/[(n-x)!x!] is a whole number and that the final answer is between 0 and 1.

23. $P(x) = \dfrac{n!}{(n-x)!x!}p^x q^{n-x}$

   $P(x=4) = [20!/(16!4!)](0.15)^4(0.85)^{16}$
   $\qquad\quad = [4845](0.15)^4(0.85)^{16}$
   $\qquad\quad = 0.182$

25. $P(x \geq 1) = 1 - P(x=0)$
   $\qquad\qquad = 1 - 0.050328$
   $\qquad\qquad = 0.949672$, rounded to 0.0950
   Yes, it is reasonable to expect that at least one group O donor will be obtained.

27. $P(x=5) = 0.018453$, rounded to 0.018
   Yes. Since $0.018 \leq 0.05$, getting all 5 donors from group O would be considered unusual.

29. Let x = the number of consumers who recognize the Mrs. Fields brand name.
   binomial problem: n = 10 and p = 0.90, use Table A-1
   a. $P(x=9) = 0.387$
   b. $P(x\neq9) = 1 - P(x=9)$
   $\qquad\qquad = 1 - 0.387$
   $\qquad\qquad = 0.613$

31. Let x = the number of people with brown eyes.
   binomial problem: n = 14 and p = 0.40, use Table A-1
   $P(x \geq 12) = P(x=12) + P(x=13) + P(x=14)$
   $\qquad\qquad = 0.001 + 0^+ + 0^+$
   $\qquad\qquad = 0.001$
   Yes. Since $0.001 \leq 0.05$, getting at least 12 persons with brown eyes would be unusual.

32. Let x = the number of delinquencies.
   binomial problem: n = 12 and p = 0.01, use Table A-1
   $P(x \geq 1) = 1 - P(x=0)$
   $\qquad\qquad = 1 - 0.886$
   $\qquad\qquad = 0.114$
   Yes. Since $0.114 > 0.05$, it would not be unusual for at least one of the people to become delinquent. The bank should make plans for dealing with a delinquency.

33. Let x = the number of offspring peas with green pods.
    binomial problem: n = 10 and p = 0.75, use the binomial formula

    $$P(x) = \frac{n!}{(n-x)!x!}p^x q^{n-x}$$

    $P(x \geq 9) = P(x=9) + P(x=10)$
    $= [10!/(1!9!)](0.75)^9(0.25)^1 + [10!/(0!10!)](0.75)^{10}(0.25)^0$
    $= [10](0.75)^9(0.25)^1 + [1](0.75)^{10}(0.25)^0$
    $= 0.1877 + 0.0563$
    $= 0.2440$, rounded to 0.244

    No. Since 0.244 > 0.05, getting at least 9 peas with green pods is not unusual.

35. Let x = the number of special program students who graduated.
    binomial problem: n = 10 and p = 0.94, use the binomial formula

    $$P(x) = \frac{n!}{(n-x)!x!}p^x q^{n-x}$$

    a. $P(x \geq 9) = P(x=9) + P(x=10)$
    $= [10!/(1!9!)](0.94)^9(0.06)^1 + [10!/(0!10!)](0.94)^{10}(0.06)^0$
    $= [10](0.94)^9(0.06)^1 + [1](0.94)^{10}(0.06)^0$
    $= 0.3438 + 0.5386$
    $= 0.8824$, rounded to 0.882

    b. $P(x=8) = [10!/(2!8!)](0.94)^8(0.06)^2$
    $= [45](0.94)^8(0.06)^2$
    $= 0.0988$

    $P(x \geq 8) = P(x=8) + P(x=9) + P(x=10)$
    $= 0.0988 + 0.3438 + 0.5386$
    $= 0.9812$

    $P(x \leq 7) = 1 - P(x \geq 8)$
    $= 1 - 0.9812$
    $= 0.0188$

    Yes. Since $P(x \leq 7) = 0.0188 \leq 0.05$, getting only 7 that graduated would be unusual.
    NOTE: Remember that in situations involving multiple ordered outcomes, the unusualness of
    a particular outcome is generally determined by the probability of getting that outcome or a
    more extreme outcome.

37. Let x = the number of households tuned to the NBC game if the stated share is correct.
    binomial problem: n = 20 and p = 0.22, use the binomial formula

    $$P(x) = \frac{n!}{(n-x)!x!}p^x q^{n-x}$$

    a. $P(x=0) = [20!/(20!0!)](0.22)^0(0.78)^{20}$
    $= [1](0.22)^0(0.78)^{20}$
    $= 0.00695$

    b. $P(x \geq 1) = 1 - P(x=0)$
    $= 1 - 0.00695$
    $= 0.99305$, rounded to 0.993

    c. $P(x=1) = [20!/(19!1!)](0.22)^1(0.78)^{19}$
    $= [20](0.22)^1(0.78)^{19}$
    $= 0.0392$

$$P(\text{at most } 1) = P(x \leq 1)$$
$$= P(x=0) + P(x=1)$$
$$= 0.0069 + 0.0392$$
$$= 0.0461$$

d. Yes. Since $0.0461 \leq 0.05$, getting at most one household tuned to the NBC game would be unusual. Either an unusual event has occurred or the 22% share value is incorrect.

39. Let x = the number who stayed less than one year.
binomial problem: n = 15 and p = 0.36, use the binomial formula

$$P(x) = \frac{n!}{(n-x)!x!} p^x q^{n-x}$$

a. $P(x=5) = [15!/(10!5!)](0.36)^5(0.64)^{10}$
$$= [3003](0.36)^5(0.64)^{10}$$
$$= 0.209$$

b. The binomial distribution requires that the repeated selections be independent. Since these persons are selected from the original group of 320 without replacement, the repeated selections are not independent and the binomial distribution should not be used. In part (a), however, the sample size is 15/320 = 4.6% ≤ 5% of the population and the repeated samples may be treated as though they are independent. If the sample size is increased to 20, the sample is 20/320 = 6.25% > 5% of the population and the criteria for using independence to get an approximate probability is no longer met.

41. Let x = the number of aspirin tablets that do not meet the specifications.
binomial problem: n = 40 and p = 0.03, use the binomial formula

$$P(x) = \frac{n!}{(n-x)!x!} p^x q^{n-x}$$

$P(\text{accept shipment}) = P(x=0) + P(x=1)$
$$= [40!/(40!0!)](0.03)^0(0.97)^{40} + [40!/(39!1!)](0.03)^1(0.97)^{39}$$
$$= [1](0.03)^0(0.97)^{40} + [40](0.03)^1(0.97)^{39}$$
$$= 0.2957 + 0.3658$$
$$= 0.6615, \text{ rounded to } 0.662$$
Approximately 2/3 of such shipments will be accepted.

43. Let x = the number of females hired if there is no discrimination.
binomial problem: n = 24 and p = 0.50, use the binomial formula

$$P(x) = \frac{n!}{(n-x)!x!} p^x q^{n-x}$$

$P(x \leq 3) = P(x=0) + P(x=1) + P(x=2) + P(x=3)$
$$= [24!/(24!0!)](0.5)^0(0.5)^{24} + [24!/(23!1!)](0.5)^1(0.5)^{23} + [24!/(22!2!)](0.5)^2(0.05)^{22}$$
$$+ [24!/(21!3!)](0.5)^3(0.5)^{21}$$
$$= [1](0.5)^0(0.5)^{24} + [24](0.5)^1(0.5)^{23} + [276](0.5)^2(0.5)^{22} + [2024](0.5)^3(0.05)^{21}$$
$$= 0.00000006 + 0.00000143 + 0.00001645 + 0.00012063$$
$$= 0.00013857, \text{ rounded to } 0.000139$$
Yes. Since $0.000139 \nleq 0.05$, such an event would be unusual. Either a very unusual event has occurred or there is gender discrimination.

$$\frac{24!\ (0.5)^0}{24!0!}$$

45. Let x = the number of peas with green pods
    binomial problem: n = 580 and p = 0.75, use the binomial formula

    $$P(x) = \frac{n!}{(n-x)!x!} p^x q^{n-x}$$

    $P(x=428) = [580!/(152!428!)](0.75)^{428}(0.25)^{152}$
    $\qquad = 0.0301231$, rounded to 0.0301

    Since $0.0301 \leq 0.05$, this particular result might be considered unusual – but so would any of the 581 possible individual outcomes. In situations involving multiple ordered outcomes, the unusualness of a particular outcome is generally determined by the probability of getting that outcome or a more extreme outcome. In this instance the appropriate probability is $P(x \leq 428) = 0.264956$, rounded to 0.265 Since $0.265 > 0.05$, the obtained result is not considered unusual – and there is no suggestion that Mendel's probability value of 0.75 is wrong.
    NOTE: The large factorials required are beyond the capabilities of most calculators. The above answers [$P(x=428) = 0.0301231$, and $P(x \leq 428) = 0.264956$] were obtained using the Minitab software package.

47. Let x = the number of winners selected.
    hypergeometric problem: A=6 (winners), B=53 (losers) and n=6 (selections)

    $$P(x) = \frac{A!}{(A-x)!x!} \cdot \frac{B!}{(B-n+x)!(n-x)!} \div \frac{(A+B)!}{(A+B-n)!n!}$$

    a. $P(x=6) = [6!/(0!6!)] \cdot [53!/(53!0!)] \div [59!/(53!6!)]$
    $\qquad = [1] \cdot [1] \div [45,057,474] = 0.0000000222$
    b. $P(x=5) = [6!/(1!5!)] \cdot [53!/(52!1!)] \div [59!/(53!6!)]$
    $\qquad = [6] \cdot [53] \div [45,057,474] = 0.00000706$
    c. $P(x=3) = [6!/(3!3!)] \cdot [53!/(50!3!)] \div [59!/(53!6!)]$
    $\qquad = [20] \cdot [23,426] \div [45,057,474] = 0.0104$
    d. $P(x=0) = [6!/(6!0!)] \cdot [53!/(47!6!)] \div [59!/(53!6!)]$
    $\qquad = [1] \cdot [22,957,480] \div [45,057,474] = 0.510$
    NOTE: This formula is actually $[_AC_x] \cdot [_BC_{n-x}] \div [_{A+B}C_n]$ and follows directly from the methods of section 4-7.

5-4 Mean, Variance, and Standard Deviation for the Binomial Distribution

1. Yes. By definition $q = 1-p$. The two expressions represent the same numerical quantity, and they may be used interchangeably

3. Using the exact formula, $\sigma^2 = npq = 1236(0.05)(0.95) = 58.71$, rounded to 58.7 people^2.
   Using the rounded value given for σ, $\sigma^2 = (\sigma)^2 = (7.7)^2 = 59.29$, rounded to 59.3 people^2.

5. $\mu = np = (50)(0.2) = 10.0$
   $\sigma^2 = npq = (50)(0.2)(0.8) = 8$; $\sigma = 2.828$, rounded to 2.8
   minimum usual value = $\mu - 2\sigma = 10 - 2(2.828) = 4.3$
   maximum usual value = $\mu + 2\sigma = 10 + 2(2.828) = 15.7$

7. $\mu = np = (300)(0.48) = 144.0$
   $\sigma^2 = npq = (300)(0.48)(0.52) = 74.88$; $\sigma = 8.653$, rounded to 8.7
   minimum usual value = $\mu - 2\sigma = 144.0 - 2(8.653) = 126.7$
   maximum usual value = $\mu + 2\sigma = 144.0 + 2(8.653) = 161.3$

9. Let x = the number of correct answers.
   binomial problem: n = 75 and p = 0.5
   a. μ = np = (75)(0.5) = 37.5
   σ^2 = npq = (75)(0.5)(0.5) = 18.75; σ = 4.330, rounded to 4.3
   b. Unusual values are those outside $\mu \pm 2\sigma$
   $$37.5 \pm 2(4.330)$$
   $$37.5 \pm 8.7$$
   $$28.8 \text{ to } 46.2 \text{ [using rounded values, 28.9 to 46.1]}$$
   No. Since 45 is within the above limits, it would not be unusual for a student to pass by getting at least 45 correct answers.

11. Let x = the number of green M&M's.
   binomial problem: n = 100 and p = 0.16
   a. μ = np = (100)(0.16) = 16.0
   σ^2 = npq = (100)(0.16)(0.84) = 13.44; σ = 3.666, rounded to 3.7
   b. Unusual values are those outside $\mu \pm 2\sigma$
   $$16.0 \pm 2(3.666)$$
   $$16.0 \pm 7.3$$
   $$8.7 \text{ to } 23.3 \text{ [using rounded values, 8.6 to 23.4]}$$
   No. Since 19 is within the above limits, it would not be unusual for 100 M&M's to include 19 green ones. This is not evidence that the claimed rate of 16% is wrong.

13. Let x = the number of baby girls.
   binomial problem: n = 574 and p = 0.5
   a. μ = np = (574)(0.5) = 287.0
   σ^2 = npq = (574)(0.5)(0.5) = 143.5; σ = 11.979, rounded to 12.0
   b. Unusual values are those outside $\mu \pm 2\sigma$
   $$287.0 \pm 2(11.979)$$
   $$287.0 \pm 24.0$$
   $$263.0 \text{ to } 311.0$$
   Yes. Since 525 is not within the above limits, it would be unusual for 574 births to include 525 girls. The results suggest that the gender selection method is effective.

15. Let x = the number who stay at their first job less than two years.
   binomial problem: n = 320 and p = 0.5
   a. μ = np = (320)(0.5) = 160.0
   σ^2 = npq = (320)(0.5)(0.5) = 80.0; σ = 8.944, rounded to 8.9
   b. Unusual values are those outside $\mu \pm 2\sigma$
   $$160.0 \pm 2(8.944)$$
   $$160.0 \pm 17.9$$
   $$142.1 \text{ to } 177.9 \text{ [using rounded values, 142.2 to 177.8]}$$
   c. x = (0.78)(320) = 250 [actually 248,249,250,251 all round to 78%].
   Since 250 is not within the above limits, it would be unusual for 320 graduates to include 250 persons who stayed at their first job less than two years if the true proportion were 50%. Since 250 is greater than the above limits, the true proportion is most likely greater than 50%. The result suggests that the headline is justified.
   d. The statement suggests that the 320 participants were a voluntary response sample, and so the results might not be representative of the target population.

17. Let x = the number of persons who voted.
    binomial problem: n = 1002 and p = 0.61
    a. $\mu = np = (1002)(0.61) = 611.22$
       $\sigma^2 = npq = (1002)(0.61)(0.39) = 238.3758$; $\sigma = 15.439$, rounded to 15.4
    b. Unusual values are those outside $\mu \pm 2\sigma$
       $$611.22 \pm 2(15.439)$$
       $$611.22 \pm 30.88$$
       $$580.3 \text{ to } 642.1 \ [\text{using rounded values, } 580.4 \text{ to } 642.0]$$
    No. Since 701 is not within the above limits, it would be unusual for 1002 potential voters to include 701 who actually voted. The results are not consistent with the actual voter turnout and suggest that those polled may not be telling the truth.
    c. No. It appears that asking persons how they acted may not yield accurate results.

19. Let x = the number of persons receiving the drug who experience nausea.
    binomial problem: n = 821 and p = 0.0124
    a. $\mu = np = (821)(0.0124) = 10.18$
       $\sigma^2 = npq = (821)(0.0124)(0.9876) = 10.0542$; $\sigma = 3.171$, rounded to 3.2
    b. Unusual values are those outside $\mu \pm 2\sigma$
       $$10.18 \pm 2(3.171)$$
       $$10.18 \pm 6.34$$
       $$3.8 \text{ to } 16.5 \ [\text{using rounded values, } 3.8 \text{ to } 16.6]$$
    Yes. Since 30 is not within the above limits, it is an unusual result.
    c. Perhaps. The results suggest that Chantix does increase the likelihood of experiencing nausea, but at 30/821 = 3.65% that probability is still relatively low.

21. Refer to the Exercise 47 in Section 5-3 and the NOTE at the end of the problem.
    Let x = the number of females selected.
    $P(x) = [_{10}C_x][_{30}C_{12-x}]/[_{40}C_{12}]$
    $P(0) = [_{10}C_0][_{30}C_{12}]/[_{40}C_{12}] = \quad 1 \cdot 86493225/5586853480 = \quad 86{,}493{,}225/5{,}586{,}853{,}480$
    $P(1) = [_{10}C_1][_{30}C_{11}]/[_{40}C_{12}] = \quad 10 \cdot 54627300/5586853480 = \quad 546{,}273{,}000/5{,}586{,}853{,}480$
    $P(2) = [_{10}C_2][_{30}C_{10}]/[_{40}C_{12}] = \quad 45 \cdot 30045015/5586853480 = 1{,}352{,}025{,}675/5{,}586{,}853{,}480$
    $P(3) = [_{10}C_3][_{30}C_9]/[_{40}C_{12}] = 120 \cdot 14307150/5586853480 = 1{,}716{,}858{,}000/5{,}586{,}853{,}480$
    $P(4) = [_{10}C_4][_{30}C_8]/[_{40}C_{12}] = 210 \cdot 5852925/5586853480 = 1{,}229{,}114{,}250/5{,}586{,}853{,}480$
    $P(5) = [_{10}C_5][_{30}C_7]/[_{40}C_{12}] = 252 \cdot 2035800/5586853480 = \quad 513{,}021{,}600/5{,}586{,}853{,}480$
    $P(6) = [_{10}C_6][_{30}C_6]/[_{40}C_{12}] = 210 \cdot 593775/5586853480 = \quad 124{,}692{,}750/5{,}586{,}853{,}480$
    $P(7) = [_{10}C_7][_{30}C_5]/[_{40}C_{12}] = 120 \cdot 142506/5586853480 = \quad 17{,}100{,}720/5{,}586{,}853{,}480$
    $P(8) = [_{10}C_8][_{30}C_4]/[_{40}C_{12}] = \quad 45 \cdot 27405/5586853480 = \quad 1{,}233{,}225/5{,}586{,}853{,}480$
    $P(9) = [_{10}C_9][_{30}C_3]/[_{40}C_{12}] = \quad 10 \cdot 4060/5586853480 = \quad 40{,}600/5{,}586{,}853{,}480$
    $P(10) = [_{10}C_{10}][_{30}C_2]/[_{40}C_{12}] = \quad 1 \cdot 435/5586853480 = \quad \underline{\quad 435/5{,}586{,}853{,}480}$
    $$5{,}586{,}853{,}440/5{,}586{,}853{,}480$$

There are $_{40}C_{12}$ = 5,586,853,440 total ways that a group of 12 persons can be selected from a class of 40. The number of females in each group can range from x=0 to x=10. The probability of each of those x values is given above – in fraction form, to indicate the exact numbers of ways involved for each type of group.

The following table, with the probabilities converted to decimal form, follows the methods of this chapter to calculate the mean and standard deviation of the probability distribution. The given table is a probability distribution since $0 \le P(x) \le 1$ for each x and $\Sigma P(x)=1$.

x	P(x)	x·P(x)	x^2	x^2·P(x)	
0	0.0154815	0	0	0	$\mu = \Sigma[x \cdot P(x)]$
1	0.0977782	0.0977783	1	0.0977783	$= 3$
2	0.2420012	0.4840026	4	0.9680051	$\sigma^2 = \Sigma[x^2 \cdot P(x)] - \mu^2$
3	0.3073032	0.9219096	9	2.7657289	$= 10.6153846 - 3^2$
4	0.2200011	0.8800046	16	3.5200186	$= 1.6153846$
5	0.0918266	0.4591329	25	2.2956643	$\sigma = 1.2709$, rounded to 1.3
6	0.0223190	0.1339137	36	0.8034825	
7	0.0030608	0.0214262	49	0.1499834	
8	0.0002207	0.0017659	64	0.0141272	
9	0.0000073	0.0000654	81	0.0005886	
10	0.0000001	0.0000008	100	0.0000078	
	1.0000000	3.0000000		10.6153846	

5-5 Poisson Probability Distributions

1. The Poisson distribution describes the numbers of occurrences, over some defined interval, of independent random events that are uniformly distributed over that interval.

3. The probability 4.82×10^{-64} is so small that for all practical purposes it is equal to zero.

5. $P(x) = \mu^x e^{-\mu}/x!$ with $\mu = 2$
$P(x=3) = (2)^3(e^{-2})/3!$
$= (8)(0.1353)/6$
$= 0.180$

NOTE: The intermediate values 8, 0.1353 and 6 are given in exercise #5 to help students who are unable to get the correct answer and need to identify where they are going wrong. In practice, the answer can be obtained on a calculator in one step without writing down intermediate values. If you write down intermediate values (for your sake or the instructor's), DO NOT USE ROUNDED INTERMEDIATE VALUES IN SUBSEQUENT CALCULATIONS. IF NECESSARY, USE THE MEMORY TO STORE INTERMEDIATE VALUES WITH COMPLETE ACCURACY. Intermediate values given in subsequent Poisson problems are for instructional purposes only.

7. $P(x) = \mu^x e^{-\mu}/x!$ with $\mu = 0.75$
$P(x=3) = (0.75)^3(e^{-0.75})/3!$
$= (0.4219)(0.4624)/6$
$= 0.0332$

IMPORTANT NOTE: In the problems that follow, store the underlined value for μ for use in all Poisson calculations.

9. Let x = the number of motor vehicle deaths per day.
   Poisson problem: use $P(x) = \mu^x e^{-\mu}/x!$
   a. Ignoring leap years, $\mu = 35.4/365 = 0.0970$
   b. $P(x=0) = (0.0970)^0(e^{-0.0970})/0! = (1)(0.9076)/1 = 0.907568$
      $P(x=1) = (0.0970)^1(e^{-0.0970})/1! = (0.0970)(0.9076)/1 = 0.088022$
      $P(x=2) = (0.0970)^2(e^{-0.0970})/2! = (0.0094)(0.9076)/2 = 0.004268$
      $P(x \leq 2) = P(x=0) + P(x=1) + P(x=2)$
      $\qquad\quad = 0.907568 + 0.088022 + 0.004268$
      $\qquad\quad = 0.999859$
      $P(x>2) = 1 - P(x \leq 2)$
      $\qquad\quad = 1 - 0.999859$
      $\qquad\quad = 0.000141$
   c. Yes. Since $0.000141 \leq 0.05$, having more than 2 motor vehicle deaths on the same day is unusual.

11. Let x = the number of atoms lost per day.
    Poisson problem: use $P(x) = \mu^x e^{-\mu}/x!$
    a. The number of atoms lost is $1,000,000 - 997,287 = 22,713$.
       $\mu = 22,713/365 = 62.2$
    b. $P(x=50) = (62.2)^{50}(e^{-62.2})/50!$
       $\qquad\quad = 0.0155$

NOTE: Exercises 12 and 13 compare actual results to predictions by the Poisson probability distribution. The comparison can be made either by changing the actual frequencies to relative frequencies (by dividing by $n = \Sigma f$) and comparing them to the predicted Poisson probabilities, or by changing the Poisson probabilities to predicted frequencies (by multiplying by $n = \Sigma f$) and comparing them to the actual frequencies. We use the first approach, but the choice is arbitrary.

13. Let x = the number of homicides per day.
    Poisson problem: $\mu = 116/365 = 0.3178$, use $P(x) = \mu^x e^{-\mu}/x!$
    a. $P(x=0) = (0.3178)^0(e^{-0.3178})/0! = \quad (1)(0.7277)/1 \ = 0.7277$
    b. $P(x=1) = (0.3178)^1(e^{-0.3178})/1! = (0.3178)(0.7277)/1 \ = 0.2313$
    c. $P(x=2) = (0.3178)^2(e^{-0.3178})/2! = (0.1010)(0.7277)/2 \ = 0.0368$
    d. $P(x=3) = (0.3178)^3(e^{-0.3178})/3! = (0.0321)(0.7277)/6 \ = 0.00389$
    e. $P(x=4) = (0.3178)^4(e^{-0.3178})/4! = (0.0102)(0.7277)/24 = 0.000309$
    The following table compares the actual relative frequencies to the Poisson probabilities.

x	f	r.f.	P(x)	
0	268	0.7342	0.7277	NOTE: r.f. = f/Σf = f/365
1	79	0.2164	0.2313	
2	17	0.0466	0.0368	
3	1	0.0027	0.0038	
4 or more	0	0.0000	0.0004	(by subtraction)
	365	1.0000	1.0000	

The agreement is very good.

15. Let x = the number of such women who die during the year.
    binomial problem: $n = 700$ and $p = 1 - 0.9968 = 0.0032$
    a. $\mu = np = (700)(0.0032) = 2.24$

b. Since n = 700 ≥ 100 and np = 2.24 ≤ 10,
   use the Poisson approximation to the binomial: $\mu = 2.24$ and $P(x) = \mu^x e^{-\mu}/x!$
   $P(x=0) = (2.24)^0 (e^{-2.24})/0! = $ $(1)(0.1065)/1 = 0.10645$
   $P(x=1) = (2.24)^1 (e^{-2.24})/1! = (2.24)(0.1065)/1 = 0.23847$
   $P(x=2) = (2.24)^2 (e^{-2.24})/2! = (2.24)^2(0.1065)/2 = 0.26708$
   $P(x=3) = (2.24)^3 (e^{-2.24})/3! = (2.24)^3(0.1065)/6 = 0.19942$
   $P(x<4) = P(x=0) + P(x=1) + P(x=2) + P(x=3)$
   $\qquad = 0.10645 + 0.23847 + 0.26708 + 0.19942$
   $\qquad = 0.811$

No. The probability of making a profit is relatively high, but the company is not "almost sure" to make a profit.

NOTE: Part (b) can be answered using the exact binomial distribution as follows.

$$P(x) = \frac{n!}{(n-x)!x!} p^x q^{n-x}$$

$P(x=0) = [700!/(700!0!)](0.0032)^0(0.9968)^{700} = $ $[1](0.0032)^0(0.9968)^{700} = 0.10608$
$P(x=1) = [700!/(699!1!)](0.0032)^1(0.9968)^{699} = $ $[700](0.0032)^1(0.9968)^{699} = 0.23837$
$P(x=2) = [700!/(698!2!)](0.0032)^2(0.9968)^{698} = [244650](0.0032)^2(0.9968)^{698} = 0.26745$
$P(x=3) = [700!/(697!3!)](0.0032)^3(0.9968)^{697} = [56921900](0.0032)^3(0.9968)^{697} = 0.19977$
$P(x<4) = P(x=0) + P(x=1) + P(x=2) + P(x=3)$
$\qquad = 0.10608 + 0.23837 + 0.26745 + 0.19977$
$\qquad = 0.812$

The answers are very close. The Poison is a good approximation to the binomial.

17. As illustrated by the table at the right, the Poisson approximations are not acceptable.

The binomial probabilities [taken from Table A-1 using n=10 and p=0.5] are symmetric around x = 5.

The Poisson probabilities [calculated from $P(x) = \mu^x e^{-\mu}/x!$ using $\mu = np = (10)(0.5) = 5$] are positively skewed around x = 4.5.

For x = 4, for example, the relative error is approximately $(0.205 - 0.1755)/0.205 = 14\%$. In addition, the Poisson shows that the impossible $x \geq 11$ has probability $1 - 0.9863 = 0.0137$.

x	binomial P(x)	Poisson P(x)
0	0.001	0.0067
1	0.010	0.0337
2	0.044	0.0842
3	0.117	0.1404
4	0.205	0.1755
5	0.246	0.1755
6	0.205	0.1462
7	0.117	0.1044
8	0.044	0.0653
9	0.010	0.0363
10	0.001	0.0181
	1.000	0.9863

Statistical Literacy and Critical Thinking

1. A random variable is a variable that has a single value, determined by chance, for each outcome of a procedure. Yes, a discrete random variable can have an infinite number of possible values – e.g., the number of coin tosses required to get the first head can be 1,2,3,4,…

2. A discrete random variable has a countable number of possible values – i.e., its possible values are either finite, or they can by placed in a well-defined 1-to-1 correspondence with the positive integers. A continuous random variable has an infinite number of possible values that are associated with measurements on a continuous scale – i.e., a scale in which there are generally no gaps or interruptions between the possible values.

3. In a binomial probability distribution, there are exactly two possible outcomes: p is the probability of one of the outcomes (usually called "success"), and q is the probability of the other outcome (usually called "failure"). Since the two outcomes are mutually exclusive and exhaustive, p+q = 1. This means that q = 1–p and that p = 1–q.

4. No. Other named discrete probability distributions identified in the exercises are the geometric distribution (Section 5-3, Exercise 46) and the hypergeometric distribution (Section 5-3, Exercise 47). There are also general unnamed discrete probability distributions (e.g., Section 5-2, Exercise 11.

Chapter Quick Quiz

1. No. Since P(x=0)+P(x=1) = 0.8+0.8 = 1.6 violates the requirement that $\Sigma P(x) = 1$, the given scenario does not define a probability distribution.

x	P(x)	x·P(x)
0	0.3	0.0
1	0.7	0.7
	1.0	0.7

2. As shown by the table at the right, $\mu = \Sigma[x \cdot P(x)] = 0.7$.

3. The scenario describes a binomial probability distribution with n = 400 and p = 0.5.
   $\mu = np = (400)(0.5) = 200$

4. The scenario describes a binomial probability distribution with n = 400 and p = 0.5.
   $\sigma^2 = npq = (400)(0.5)(0.5) = 100$; $\sigma = 10$

5. Yes. Any value more than two standard deviations from the mean is generally considered unusual, and 35 is (35-20.0)/4.0 = 3.75 standard deviations from the mean.

6. Yes. The scenario describes a binomial probability distribution with n = 5 and p = 0.2
   A check of the given P(x) values indicates they were generated by the binomial formula
   $$P(x) = \frac{n!}{(n-x)!x!}p^x q^{n-x}.$$

7. $P(x \geq 1) = 1 - P(x=0)$
   $= 1 - 0.4096$
   $= 0.5904$

8. $P(x \leq 4) = P(x=0) + P(x=1) + P(x=2) + P(x=3) + P(x=4)$
   $= 0.4096 + 0.4096 + 0.1536 + 0.0256 + 0.0016$
   $= 0.9984$
   $P(x=5) = 1 - P(x \leq 4)$
   $= 1 - 0.9984$
   $= 0.0016$

9. $P(x=2 \text{ or } x=3) = P(x=2) + P(x=3)$
   $= 0.1536 + 0.0256$
   $= 0.1792$

10. Yes. Since P(x=5) = 0.0016 ≤ 0.05, getting all five answers correct by random guessing would be an unusual event.

Review Exercises

1. Let x = the number of deaths in the week before Thanksgiving.
   The scenario describes a binomial probability distribution
        with n = 8 and p = 0.5.
   The table at the right was constructed using Table A-1.

   It could also be constructed using the binomial formula $P(x) = \dfrac{n!}{(n-x)!x!}p^x q^{n-x}$.

x	P(x)
0	0.004
1	0.031
2	0.109
3	0.219
4	0.273
5	0.219
6	0.109
7	0.031
8	0.004
	1.000

2. Since the random variable follows a binomial probability distribution.
   $\mu = np = (8)(0.5) = 4.0$
   $\sigma^2 = npq = (8)(0.5)(0.5) = 2$; $\sigma = 1.414$, rounded to 1.4
   Unusual values are those outside $\mu \pm 2\sigma$
   $$4.0 \pm 2(1.414)$$
   $$4.0 \pm 2.8$$
   $$1.2 \text{ to } 6.8$$
   Yes. Since 8 is outside the above limits, it would be unusual for all 8 of the deaths to occur
   during the week before Thanksgiving. Also, $P(8) = 0.004 < 0.05$ implies an unusual result.
   NOTE: All 8 deaths occurring in the week <u>before</u> Thanksgiving does not support the theory
   about postponing death. That theory suggests that there should be a larger than usual number
   of deaths during the week <u>after</u> the holiday.

3. Let x = the number of deaths in the week before Thanksgiving.
   The scenario describes a binomial probability distribution with n = 20 and p = 0.5.
   a. $P(x) = \dfrac{n!}{(n-x)!x!}p^x q^{n-x}$.
   $$P(x=14) = [20!/(6!14!)](0.5)^{14}(0.5)^6$$
   $$= [38760](0.5)^{14}(0.5)^6$$
   $$= 0.0370$$
   b. Yes. Since $0.0370 \leq 0.05$, having exactly 14 of the deaths during the week before
      Thanksgiving would be considered unusual.
   c. No. In situations involving multiple ordered outcomes, the unusualness of a particular
      outcome is generally determined by the probability of getting that outcome or a more
      extreme outcome. In this instance the appropriate probability is $P(x \geq 14)$ – or, alternatively,
      one could use the $\mu \pm 2\sigma$ guideline to determine whether 14 was outside the range of usual
      values.

4. The following table describes the situation.
   Based on expectations, since 315,075 ≳ 193,000, she should continue.

x	P(x)	x·P(x)
75	0.2	15.0
300	0.2	60.0
75,000	0.2	15,000.0
500,000	0.2	100,000.0
1,000,000	0.2	200,000.0
	1.0	315,075.0

$E = \Sigma[x \cdot P(x)] = \$315,075.0$

5. The table at the right describes the situation.
   a. $E = \Sigma[x \cdot P(x)] = 0.012474746$
   $$= 1.25¢$$
   b. expected loss = price of stamp − 1.25¢
   $$= 42¢ - 1.25¢$$
   $$= 40.75¢ \text{ [as of fall 2008]}$$

x	P(x)	x·P(x)
1,000,000	0.000000011	0.011111111
100,000	0.000000009	0.000909091
25,000	0.000000009	0.000227273
5,000	0.000000027	0.000136362
2,500	0.000000036	0.000090909
0	0.999999908	0
	1.000000000	0.012474746

No. From a purely financial point of view, it is not worth entering this contest.
NOTE: The biggest winner in the sweepstakes appears to be the USPS. Based on the probabilities of winning, it appears that the company expects about 100,000,000 entries. At 42¢ per entry, the sweepstakes would generate 42 million dollars of entry postage alone!

6. No. Since $P(x=0)+P(x=1)+P(x=2)+P(x=3)+P(x=4) = 0.0016+0.0250+0.1432+0.3892+0.4096 = 0.9686$ violates the requirement that $\Sigma P(x) = 1$, the given scenario does not describe a probability distribution.

7. a. There are 10 possible values [0,1,2,3,4,5,6,7,8,9] for each digit.
   There are $10 \cdot 10 \cdot 10 \cdot 10 = 10,000$ ways to select the 4 digits.
   Since only one selection wins, $P(W) = 1/10,000 = 0.0001$
   b. Refer to the table in part (e). The first two columns give the desired probability distribution.
   c. The expected number of wins in 365 tries is $365(.0001) = 0.0365$.
   d. The scenario describes a binomial probability distribution with $n = 365$ and $p = 0.0001$.

   $$P(x) = \frac{n!}{(n-x)!x!}p^x q^{n-x}.$$

   $P(x=1) = [365!/(364!1!)](0.0001)^1(0.9999)^{364}$
   $$= [365](0.0001)^1(0.9999)^{364}$$
   $$= 0.0352$$

x	P(x)	x·P(x)
−1	0.9999	−0.9999
4999	0.0001	0.4999
	1.0000	−0.5000

   e. Refer to the table. $E = \Sigma[x \cdot P(x)] = -0.5000 = -50¢$.

8. a. Let x = the number fired for not getting along with others.
   binomial problem: $n = 5$ and $p = 0.17$, use the binomial formula

   $$P(x) = \frac{n!}{(n-x)!x!}p^x q^{n-x}$$

   $P(x \geq 4) = P(x=4) + P(x=5)$
   $$= [5!/(1!4!)](0.17)^4(0.83)^1 + [5!/(0!5!)](0.17)^5(0.83)^0$$
   $$= [5](0.17)^4(0.83)^1 + [1](0.17)^5(0.83)^0$$
   $$= 0.003466 + 0.000142$$
   $$= 0.003608, \text{ rounded to } 0.00361$$
   b. Yes. Since $0.00361 \leq 0.05$, such an event would be unusual for a typical company operating with $p = 0.17$. Either a rare event has occurred or, more likely, this company is different.

9. Let x = the number of checks with leading digit 1.
   binomial problem: $n = 784$ and $p = 0.301$, use the binomial formula
   a. The expected number is the mean, $\mu = np = (784)(0.301) = 236.0$
   b. For the binomial distribution,
   $$\mu = np = (784)(0.301) = 236.0$$
   $$\sigma^2 = npq = (784)(0.301)(0.699) = 164.9528; \sigma = 12.843, \text{ rounded to } 12.8$$

c. Usual values are those within $\mu \pm 2\sigma$

$$236.0 \pm 2(12.843)$$
$$236.0 \pm 25.7$$
$$210.3 \text{ to } 261.7 \text{ [using rounded values, 210.4 to 261.6]}$$

d. Yes. Since 0 is so far outside the above limits, that would be very strong evidence that the checks are different (and likely fraudulent).

10. Let x = the number of bombs per region
Poison problem: use $P(x) = \mu^x e^{-\mu}/x!$
a. $\mu = 535/576 = 0.9288$
b. $P(x=0) = (0.9288)^0 e^{-0.9288}/0!$
$$= (1)(0.3950)/1$$
$$= 0.395$$
c. We expect $(576)(0.395) = 227.53$, or about 228, regions not to be hit.
d. The actual result of 229 is very close to the result predicted by the Poison formula.

Cumulative Review Exercises

1. Arranged in order, the values are: 115.00   134.83   142.94   188.00   217.60
summary statistics: $n = 5$ $\Sigma x = 798.37$ $\Sigma x^2 = 134529.7325$
a. $\bar{x} = (\Sigma x)/n = 798.37/5 = \$159.674$
b. $\tilde{x} = x_3 = \$142.94$
c. $R = x_n - x_1 = 217.60 - 115.00 = \$102.60$
d. $s = \$41.985$ [the square root of the answer given in part e]
e. $s^2 = [n(\Sigma x^2) - (\Sigma x)^2]/[n(n-1)]$
$$= [5(134529.7325) - (798.37)^2]/[5(4)]$$
$$= 35254.0056/20$$
$$= 1762.700 \text{ dollars}^2$$
f. Usual values are those within $\bar{x} \pm 2s$
$$159.674 \pm 2(41.9845)$$
$$159.674 \pm 83.969$$
$$75.705 \text{ to } 243.643 \text{ [using rounded values, 75.704 to 243.644]}$$
g. No. Since all of the sample values are within the above limits, none of them is unusual.
h. Ratio, since differences are meaningful and there is a meaningful zero.
i. Discrete, since checks must be written in whole numbers of cents.
j. Convenience, since they were selected because they happened to be close at hand.
k. Since the sample was a random sample, the sample mean of the n items observed should be a good estimate of the population mean of all N items. The estimate for the population total is $N\bar{x} = (134)(159.674) = \$21,396.316$.

2. Let x = the number of companies that test employees for substance abuse.
binomial problem: $n = 10$ and $p = 0.80$, use Table A-1
a. $P(x=5) = 0.026$
b. $P(x \geq 5) = P(x=5) + P(x=6) + P(x=7) + P(x=8) + P(x=9) + P(x=10)$
$$= 0.026 + 0.088 + 0.201 + 0.302 + 0.268 + 0.107$$
$$= 0.992$$

c. For the binomial distribution,
$\mu = np = (10)(0.80) = 8.0$
$\sigma^2 = npq = (10)(0.80)(0.20) = 1.60$; $\sigma = 1.2649$, rounded to 1.3
d. Usual values are those within $\mu \pm 2\sigma$
$$8.0 \pm 2(1.2649)$$
$$8.0 \pm 2.5$$
$$5.5 \text{ to } 10.5 \text{ [using rounded values, 5.4 to 10.6]}$$

3. Let x = the number of HIV cases.
binomial problem: n = 150 and p = 0.10, use the binomial formula
$$P(x) = \frac{n!}{(n-x)!x!} p^x q^{n-x}.$$
a. $\mu = np = (150)(0.10) = 15.0$
$\sigma^2 = npq = (150)(0.10)(0.90) = 13.5$; $\sigma = 3.674$, rounded to 3.7
b. Usual values are those within $\mu \pm 2\sigma$
$$15.0 \pm 2(3.674)$$
$$15.0 \pm 7.3$$
$$7.7 \text{ to } 22.3 \text{ [using rounded values, 7.6 to 22.4]}$$
No. Since 12 is within the above limits, it is not an unusually low result. There is not sufficient evidence to suggest that the program is effective in lowering the 10% rate.

4. Let S = the selected passenger was one who survived.
a. $P(S) = 706/2223 = 0.318$
b. $P(S_1 \text{ and } S_2) = P(S_1) \cdot P(S_2|S_1)$
$$= (706/2223)(705/2222)$$
$$= 0.101$$
c. $P(\overline{S}) = 1 - P(S) = 1 - 706/2223 = 1517/2223$
$P(\overline{S}_1 \text{ and } \overline{S}_2) = P(\overline{S}_1) \cdot P(\overline{S}_2|\overline{S}_1)$
$$= (1517/2223)(1516/2222)$$
$$= 0.466$$

5. No. The correct value is the weighted mean of the 50 state means, with the population of each state as the weight. If the means are not weighted, then small population states count equal with large population states – and a small population state with unusually high or low per capita consumption has undue influence and incorrectly affects the national mean as much as a large population state.

Chapter 6

Normal Probability Distributions

6-2 The Standard Normal Distribution

1. The word "normal" as used when referring to a normal distribution does carry with it some of the meaning the word has in ordinary language. Normal distributions occur in nature and describe the normal, or natural, state of many common phenomena. But in statistics the term "normal" has a specific and well-defined meaning in addition to its generic connotations of being "typical" – it refers to a specific bell-shaped distribution generated by a particular mathematical formula.

3. A normal distribution can be centered about any value and have any level of spread. A *standard* normal distribution has a center (as measured by the mean) of 0 and has a spread (as measured by the standard deviation) of 1.

5. The height of the rectangle is 0.5. Probability corresponds to area, and the area of a rectangle is (width)·(height).
   $P(x>124.0)$ = (width)·(height)
   = (125.0−124.0)(0.5)
   = (1.0)(0.5)
   = 0.50

7. The height of the rectangle is 0.5. Probability corresponds to area, and the area of a rectangle is (width)·(height).
   $P(123.2<x<124.7)$ = (width)·(height)
   = (124.7−123.2)(0.5)
   = (1.7)(0.5)
   = 0.75

NOTE: For problems 9-16, the answers are re-expressed (when necessary) in terms of items that can be read directly from Table A-2. In general, this step is omitted in subsequent exercises and the reader is referred to the accompanying sketches to se how the indicated probabilities and z scores relate to Table A-2. "A" is used to denote the tabled value of the area to the left of the given z score. As a crude check, always verify that

    A>0.5000 corresponds to a positive z score and z>0 corresponds to an A >0.5000
    A<0.5000 corresponds to a negative z score and z<0 corresponds to an A < 0.5000

9. P(z<0.75) = 0.7734

11. P(-0.60<z<1.20) = P(z<1.20) – P(z<-0.60)
$$= 0.8849 – 0.2743$$
$$= 0.6106$$

13. For A = 0.9798, z = 2.05.

15. If the area to the right of z is 0.1075, A = 1 – 0.1075 = 0.8925.
For A = 0.8925, z = 1.24.

NOTE: The sketch is the key to Exercises 17-36. It tells whether to subtract two Table A-2 probabilities, to subtract a Table A-2 probability from 1, etc. For the remainder of chapter 6, THE ACCOMPANYING SKETCHES ARE NOT TO SCALE and are intended only as aids to help the reader understand how to use the table values to answer the questions. In addition, the probability of any single point in a continuous distribution is zero – i.e., P(x=a) = 0 for any single point a. For normal distributions, therefore, this manual ignores P(x=a) and uses P(x>a) = 1 – P(x<a).

17. P(z<-1.50) = 0.0668

19. P(z<1.23) = 0.8907

21. P(z>2.22)
$$= 1 – 0.9868$$
$$= 0.0132$$

23. P(z>-1.75)
$$= 1 – 0.0401$$
$$= 0.9599$$

25. P(0.50<z<1.00)
    = 0.8413 − 0.6915
    = 0.1498

27. P(-3.00<z<-1.00)
    = 0.1587 − 0.0013
    = 0.1574

29. P(-1.20<z<1.95)
    = 0.9744 − 0.1151
    = 0.8593

31. P(-2.50<z<5.00)
    = 0.9999 − 0.0062
    = 0.9937

33. P(z<3.55) = 0.9999

35. P(z>0.00)
    = 1 − 0.5000
    = 0.5000

37. $P(-1<z<1) = 0.8413 - 0.1587 = 0.6826$
    about 68%

39. $P(-3<z<3) = 0.9987 - 0.0013 = 0.9974$
    About 99.7%

41. For $z_{0.05}$, the cumulative area is 0.9500.
    The closest entry is 0.9500,
        for which $z = 1.645$

43. For $z_{0.10}$, the cumulative area is 0.9000.
    The closest entry is 0.8997,
        for which $z = 1.28$.

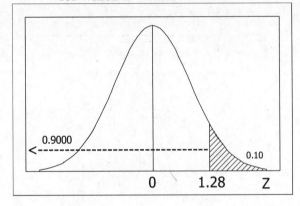

45. $P(-1.96<z<1.96)$
        $= 0.9750 - 0.0250$
        $= 0.9500$

47. $P(z<-2.575 \text{ or } z>2.575)$
        $= P(z<-2.575) + P(z>2.575)$
        $= 0.0050 + (1 - 0.9950)$
        $= 0.0050 + 0.0050$
        $= 0.0100$

49. For P_{95}, the cumulative area is 0.9500.
    The closest entry is 0.9500,
       for which z = 1.645

51. For the lowest 2.5%, the cumulative
    area is 0.0250 (exact entry in the table),
    indicated by z = -1.96.
    By symmetry, the highest 2.5%
    [= 1–.9750] are above z = 1.96.

53. Rewrite each statement in terms of z, recalling that z is the number of standard deviations a
    score is from the mean.

a. P(-2<z<2)
    = 0.9772 – 0.0228
    = 0.9544 or 95.44%

b. P(z<-1 or z >1)
    = P(z<-1) + P(z>1)
    = 0.1587 + (1 – 0.8413)
    = 0.1597 + 0.1587 = 0.3174 or 31.74%

c. P(z<-1.96 or z > 1.96)
    = P(z<-1.96) + P(z>1.96)
    = 0.0250 + (1 – 0.9750)
    = 0.250 + 0.250 = 0.0500 or 5.00%

d. P(-3<z<3)
    = 0.9987 – 0.0013
    = 0.9974 or 99.74%

e. P(z<-3 or z>3)
   = P(z<-3) + P(z>3)
   = 0.0013 + (1 − 0.9987)
   = 0.0013 + 0.0013 = 0.0026 = 0.26%

55. The sketches are the key. They tell what
    probability (i.e., cumulative area) to look
    up when reading Table A-2 "backwards."
    They also provide a check against gross
    errors by indicating whether a score is
    above or below zero.

    a. P(z<a) = 0.9599
            a = 1.75

       (see the sketch at the right)

b. P(z>b) = 0.9722
   P(z<b) = 1- 0.9972
          = 0.0028
          b = -2.00

c. P(z>c) = 0.0668
   P(z<c) = 1 − 0.0668
          = 0.9332
          c = 1.50

d. Since P(z<-d) = P(z>d) by symmetry
   and ΣP(z) = 1,
   $$P(z<-d) + P(-d<z<d) + P(z>d) = 1$$
   $$P(z<-d) + 0.5878 + P(z<-d) = 1$$
   $$2 \cdot P(z<-d) + 0.5878 = 1$$
   $$2 \cdot P(z<-d) = 0.4122$$
   $$P(z<-d) = 0.2061$$
   $$-d = -0.82$$
   $$d = 0.82$$

e. Since P(z<-e) = P(z>e) by symmetry
   and ΣP(z) = 1,
   $$P(z<-e) + P(-e<z<e + P(z>e) = 1$$
   $$P(z<-e) + 0.0956 + P(z<-e) = 1$$
   $$2 \cdot P(z<-e) + 0.0956 = 1$$
   $$2 \cdot P(z<-e) = 0.9044$$
   $$P(z<-e) = 0.4522$$
   $$-e = -0.12$$
   $$e = 0.12$$

Observe that 0.7939 – 0.2061 = 0.5878,
as given in the problem.

Observe that 0.5478 – 0.4522 = 0.0956,
as given in the problem.

6-3 Applications of Normal Distributions

1. A normal distribution can have any mean and any positive standard deviation. A standard normal distribution has mean 0 and standard deviation 1 – and it follows a "nice" bell-shaped curve. Non-standard normal distributions can follow bell-shaped curves that are tall and thin, or short and fat.

3. For any distribution, converting to z scores using the formula $z = (x-\mu)/\sigma$ produces a same-shaped distribution with mean 0 and standard deviation 1.

5. P(x<120) = P(z<1.33)
   $$= 0.9082$$

7. P(90<x<115) = P(-0.67<z<1.00)
   $$= P(z<1.00) - P(z<-0.67)$$
   $$= 0.8413 - 0.2514$$
   $$= 0.5899$$

9. The z score with 0.6 below it is z = 0.25 [closest entry 0.5987].
   $$x = \mu + z\sigma$$
   $$= 100 + (0.25)(15)$$
   $$= 100 + 3.75 = 103.75, \text{ rounded to } 103.8$$

11. The z score with 0.95 above is the z score with 0.05 below it; z = -1.645 [bottom of table].
    $$x = \mu + z\sigma$$
    $$= 100 + (-1.645)(15)$$
    $$= 100 - 24.675 = 75.325, \text{ rounded to } 75.3$$

13. normal distribution: $\mu = 100$ and $\sigma = 15$
$P(x<115)$
$= P(z<1.00)$
$= 0.8413$

15. normal distribution: $\mu = 100$ and $\sigma = 15$
$P(90<x<110) = P(-0.67<z<0.67)$
$= 0.7486 - 0.2514$
$= 0.4972$

17. normal distribution: $\mu = 100$ and $\sigma = 15$
For P_{30}, A = 0.3000 [0.3015] and z = -0.52.
and z = -0.52.
$x = \mu + z\sigma$
$= 100 + (-0.52)(15)$
$= 100 - 7.8 = 99.2$

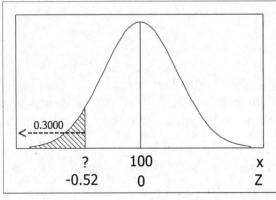

19. normal distribution: $\mu = 100$ and $\sigma = 15$
For Q_3, A = 0.7500 [0.7486]
and z = 0.67.
$x = \mu + z\sigma$
$= 100 + (0.67)(15)$
$= 100 + 10.05 = 110.05$, rounded to 110.1

21. a. normal distribution: $\mu = 69.0$, $\sigma = 2.8$
$P(x<72) = P(z<1.07)$
$= 0.8577$ or 85.77%

b. normal distribution: $\mu = 63.6$, $\sigma = 2.5$
$P(x<72) = P(z<3.36)$
$= 0.9996$ or 99.96%

c. No. It is not adequate in that 14% of the
men need to bend to enter, and they may
be in danger of injuring themselves if they
fail to recognize the necessity to bend.

d. For A = 0.9800 [0.9798],
  and z = 2.05.
  x = μ + zσ
    = 69.0 + (2.05)(2.8)
    = 69.0 + 5.7
    = 74.7 inches
(See the sketch at the right)

23. a. normal distribution: μ = 69.0, σ = 2.8
    P(x>74) = P(z>1.79)
            = 1 − 0.9633
            = 0.0367 or 3.67%

b. normal distribution: μ = 63.6, σ = 2.5
    P(x>70) = P(z>2.56)
            = 1 − 0.9948
            = 0.0052 or 0.52%

c. No. The requirements are not equally fair for men and women, since the percentage of men
that are eligible is so much larger than the percentage of women who are eligible.

25. normal distribution: μ = 63.6 σ = 2.5
    a. P(58<x<80) = P(-2.24<z<6.56)
                  = 0.9999 − 0.0125
                  = 0.9874 or 98.74%
    No. Only 1 − 0.9874 = 0.0126 = 1.26%
    of the women are not eligible because
    of the height requirements.

b. For the shortest 1%, A = 0.0100 [0.0099]
    and z = -2.33.

$x = \mu + z\sigma$
    $= 63.6 + (-2.33)(2.5)$
    $= 63.6 - 5.8$
    $= 57.8$ inches

For the tallest 2%, A = 0.9800 [0.9798]
    and z = 2.05.

$x = \mu + z\sigma$
    $= 63.6 + (2.05)(2.5)$
    $= 63.6 + 5.1$
    $= 68.7$ inches

27. normal distribution: $\mu = 3570$ and $\sigma = 500$
    a. P(x<2700)
        $= P(z<-1.74)$
        $= 0.409$ or 4.09%

b. For the lightest 3%, A = 0.0300 [0.0301]
    and z = -1.88

$x = \mu + z\sigma$
    $= 3570 + (-1.88)(500)$
    $= 3570 - 940 = 2630$ g

c. Not all babies below a certain birth weight require special treatment. The need for special treatment is determined at least as much by developmental considerations as by weight alone. Also, the birth weight identifying the bottom 3% is not a static figure and would have to be updated periodically – perhaps creating unnecessary uncertainty and inconsistency.

29. normal distribution: μ = 98.29 and σ = 0.62

a. P(x≳100.6)

= P(z>3.87)

= 1 – 0.9999

= 0.0001

Yes. The cut-off is appropriate in that there is a small probability of saying that a healthy person has a fever, but many with low grade fevers may erroneously be labeled healthy.

b. For the highest 5%, A = 0.9500

and z = 1.645.

x = μ + zσ

= 98.20 + (1.645)(0.62)

= 98.20 + 1.02

= 99.22 °F

31. normal distribution: μ = 268 and σ = 15

a. P(x>308)

= P(z>2.67)

= 1 – 0.9962

= 0.0038

The result suggests that an unusual event has occurred – but certainly not an impossible one, as about 38 of every 10,000 pregnancies can be expected to last as long.

b. For the lowest 4%, A = 0.0400 [0.0401]

and z = -1.75.

x = μ + zσ

= 268 + (-1.75)(15)

= 268 – 26

= 242 days

33. a. The distribution and summary statistics, obtained using statistical software, are given below. The data appear to have a distribution that is approximately normal.

$n = 40$
$\bar{x} = 118.9$
$s = 10.46$

pressure	frequency
90- 99	1
100-109	4
110-119	17
120-129	12
130-139	5
140-149	0
150-159	1
	40

NOTE: Using rounded $\sigma = 10.5$ in part (b) gives $x_5 = 101.6$ and $x_{95} = 136.2$.

b. normal distribution: $\mu = 118.9$, $\sigma = 10.46$
For P_5, $A = 0.0500$ and $z = -1.645$.
For P_{95}, $A = 0.9500$ and $z = 1.645$.
$x_5 = \mu + z\sigma = 118.9 + (-1.645)(10.46)$
$\qquad = 118.9 - 17.2 = 101.7$ mm
$x_{95} = \mu + z\sigma = 118.9 + (1.645)(10.46)$
$\qquad = 118.9 + 17.2 = 136.1$ mm

35. a. The z scores are always unit free. Because the numerator and the denominator of the fraction $z = (x-\mu)/\sigma$ have the same units, the units will divide out.
b. For a population of size N, $\mu = \Sigma x/N$ and $\sigma^2 = \Sigma(x-\mu)^2/N$.
As shown below, $\mu = 0$ and $\sigma = 1$ will be true for *any* set of z scores – regardless of the shape of the original distribution.
$$\Sigma z = \Sigma(x-\mu)/\sigma = (1/\sigma)[\Sigma(x-\mu)]$$
$$= (1/\sigma)[\Sigma x - \Sigma\mu]$$
$$= (1/\sigma)[N\mu - N\mu] = (1/\sigma)[0] = 0$$
$$\Sigma z^2 = \Sigma[(x-\mu)/\sigma\ ]^2 = (1/\sigma)^2[\Sigma(x-\mu)^2]$$
$$= (1/\sigma^2)[N\sigma^2] = N$$
$$\mu_Z = (\Sigma z)/N = 0/N = 0$$
$$\sigma^2_Z = \Sigma(z-\mu_Z)^2/N = \Sigma(z-0)^2/N = \Sigma z^2/N = N/N = 1 \text{ and } \sigma_Z = 1$$
The re-scaling from x scores to z scores will not affect the basic shape of the distribution – and so in this case the z scores will be normal, as was the original distribution.

37. normal distribution: $\mu = 25$ and $\sigma = 5$
a. For a population of size N, $\mu = \Sigma x/N$ and $\sigma^2 = \Sigma(x-\mu)^2/N$.
As shown below, adding a constant to each score increases the mean by that amount but does not affect the standard deviation. In non-statistical terms, shifting everything by k units does not affect the spread of the scores. This is true for *any* set of scores – regardless of the shape of the original distribution. Let $y = x + k$.

$$\mu_Y = [\Sigma(x+k)]/N$$
$$= [\Sigma x + \Sigma k]/N$$
$$= [\Sigma x + Nk]/N$$
$$= (\Sigma x)/N + (Nk)/N$$
$$= \mu_X + k$$

$$\sigma^2_Y = \Sigma[y - \mu_Y]^2/N$$
$$= \Sigma[(x+k) - (\mu_X + k)]^2/N$$
$$= \Sigma[x - \mu_X]^2/N$$
$$= \sigma^2_X \text{ and so } \sigma_Y = \sigma_X$$

If the teacher adds 50 to each grade,
new mean $= 25 + 50 = 75$ new standard deviation $= 5$ (same as before).

b. No. Curving should consider the variation. Had the test been more appropriately constructed, it is not likely that every student would score exactly 50 points higher. If the typical student score increased by 50, we would expect the better students to increase by more than 50 and the poorer students to increase by less than 50. This would make the scores more spread out and would increase the standard deviation.

c. For the top 10%, A = 0.9000 [0.8997] and z = 1.28.

$$x = \mu + z\sigma$$
$$= 25 + (1.28)(5) = 25 + 6.4 = 31.4$$

For the bottom 70%, A = 0.7000 [0.6985] and z = 0.52.

$$x = \mu + z\sigma$$
$$= 25 + (0.52)(5) = 25 + 2.6 = 27.6$$

For the bottom 30%, A = 0.3000 [0.3015] and z = -0.52.

$$x = \mu + z\sigma$$
$$= 25 + (-0.52)(5) = 25 - 2.6 = 22.4$$

For the bottom 10%, A = 0.1000 [0.1003] and z = -1.28.

$$x = \mu + z\sigma$$
$$= 25 + (-1.28)(5) = 25 - 6.4 = 18.6$$

This produces the grading scheme given at the right.

A: higher than 31.4
B: 27.6 to 31.4
C: 22.4 4o 27.6
D: 18.6 to 22.4
E: less than 18.6

d. The curving scheme in part (c) is fairer because it takes into account the variation as discussed in part (b). Assuming the usual 90-80-70-60 letter grade cut-offs, for example, the percentage of A's under the scheme in part (a) with $\mu = 75$ and $\sigma = 5$ is

$$P(x>90) = 1 - P(x<90)$$
$$= 1 - P(z<3.00)$$
$$= 1 - 0.9987$$
$$= 0.0013 \text{ or } 0.13\%$$

This is considerably less than the 10% A's under the scheme in part (c) and reflects the fact that the variation in part (a) is unrealistically small.

39. Work with z scores to answer the question using a standard normal distribution. This is a four-step process as follows.

Step 1. Find Q_1 and Q_3.

For Q_1, A = 0.2500 [0.2514] and z = -0.67.
For Q_3, A = 0.7500 [0.7486] and z = 0.67.

Step 2. Find the IQR = $Q_3 - Q_1$, and then determine (1.5)(IQR)

$$IQR = Q_3 - Q_1 = 0.67 - (-0.67) = 1.34$$
$$(1.5)(IQR) = (1.5)(1.34) = 2.01$$

Step 3. Find the lower and upper limits L and U beyond which outliers occur.

$$L = Q_1 - (1.5)(IQR) = -0.67 - 2.01 = -2.68$$
$$U = Q_3 + (1.5)(IQR) = 0.67 + 2.01 = 2.68$$

Step 4. Find P(outlier) = P(z<L) + P(z>U).

$$P(outlier) = P(z<-2.68 \text{ or } z>2.68)$$
$$= P(z<-2.68) + P(z>2.68)$$
$$= 0.0037 + (1 - 0.9963)$$
$$= 0.0037 + 0.0037$$
$$= 0.0074$$

6-4 Sampling Distributions and Estimators

1. Given a population distribution of scores, one can take a sample of size n and calculate any of several statistics. A sampling distribution is the distribution of all possible values of such a particular statistic.

3. An unbiased estimator is one whose expected value is the true value of the parameter which it estimates. The sample mean is an unbiased estimator of the population mean because the expected value (or mean) of its sampling distribution is the population mean.

5. No. The students at New York University are not necessarily representative (by race, major, etc.) of the population of all U.S. college students.

7. The sample means will have a distribution that is approximately normal. They will tend to form a symmetric, unimodal and bell-shaped distribution around the value of the population mean.

9. a. The medians of the 9 samples are given in column 2 at the right. The sampling distribution of the median is given in columns 3 and 4 at the right.
   b. The population median is 3. The mean of the sample medians is $\Sigma \tilde{x} \cdot P(\tilde{x}) = 45/9 = 5$.
   c. In general the sample medians do not target the value of the population median. For this reason, sample median is not a good estimator of the population median.

sample	$\tilde{x}$	$\tilde{x}$	$P(\tilde{x})$	$\tilde{x} \cdot P(\tilde{x})$
2,2	2	2	1/9	2/9
2,3	2.5	2.5	2/9	5/9
2,10	6	3	1/9	3/9
3,2	2.5	6	2/9	12/9
3,3	3	6.5	2/9	13/9
3,10	6.5	10	1/9	10/9
10,2	6		9/9	45/9
10,3	6.5			
10,10	10			

NOTE: Section 5-2 defined the mean of a probability distribution of x's as $\mu_x = \Sigma[x \cdot P(x)]$. If the variable is designated by the symbol y, then the mean of a probability distribution of y's is $\mu_y = \Sigma[y \cdot P(y)]$. In this section, the variables are statistics – like $\overline{x}$ and $\hat{p}$. In such cases, the formula for the mean may be adjusted – to $\mu_{\overline{x}} = \Sigma[\overline{x} \cdot P(\overline{x})]$ and $\mu_{\hat{p}} = \Sigma[\hat{p} \cdot P(\hat{p})]$. In a similar manner, the formula for the variance of a probability distribution may also be adjusted to match the variable being considered.

11. a. The variances of the 9 samples are given in column 2 at the right. The sampling distribution of the variance is given in columns 3 and 4 at the right.
    b. Since the values 2,3,10 are considered a population, the population variance is $\sigma^2 = \Sigma(x-\mu)^2/N = (3^2 + 2^2 + 5^2)/3 = 38/3$. The mean of the sample variances is $\Sigma s^2 \cdot P(s^2) = 114/9 = 38/3$.
    c. The sample variance always targets the value of the population variance. For this reason, the sample variance is a good estimator of the population variance.

sample	s^2	s^2	$P(s^2)$	$s^2 \cdot P(s^2)$
2,2	0	0	3/9	0/9
2,3	0.5	0.5	2/9	1/9
2,10	32	24.5	2/9	49/9
3,2	0.5	32	2/9	64/9
3,3	0		9/9	114/9
3,10	24.5			
10,2	32			
10,3	24.5			
10,10	0			

The information below and the box at the right apply to Exercises 13-16.

original population of scores in order: 46 49 56 58
summary statistics: N= 4 $\Sigma x = 209$ $\Sigma x^2 = 11017$
$\mu = \Sigma x/N$
  $= 209/4 = 52.25$
population $\tilde{x} = (x_2+x_3)/2$
           $= (49+56)/2 = 52.5$
population $R = x_n - x_1$
           $= 58 - 46 = 12$
$\sigma^2 = [\Sigma(x-\mu)^2]/N$
  $= [(-6.25)^2 + (-3.25)^2 + (3.75)^2 + (5.75)^2]/4$
  $= [39.0625 + 10.5625 + 14.0625 + 33.0625]/4$
  $= 96.75/4 = 24.1875$

sample	$\bar{x}$	$\tilde{x}$	R	s^2
46,46	46.0	46.0	0	0.0
46,49	47.5	47.5	3	4.5
46,56	51.0	51.0	10	50.0
46,58	52.0	52.0	12	72.0
49,46	47.5	47.5	3	4.5
49,49	49.0	49.0	0	0.0
49,56	52.5	52.5	7	24.5
49,58	53.5	53.5	9	40.5
56,46	51.0	51.0	10	50.0
56,49	52.5	52.5	7	24.5
56,56	56.0	56.0	0	0.0
56,58	57.0	57.0	2	2.0
58,46	52.0	52.0	12	72.0
58,49	53.5	53.5	9	40.5
58,56	57.0	57.0	2	2.0
58,58	58.0	58.0	0	0.0

13. a. The sixteen possible samples are given in the "samples" column of the box preceding this exercise
   b. The sixteen possible means are given in column 2 of the box preceding this exercise. The sampling distribution of the mean is given in the first two columns at the right.
   c. The population mean is 52.25. The mean of the sample means is $\Sigma \bar{x} \cdot P(\bar{x}) = 836.0/16 = 52.25$. They are the same.
   d. Yes. The sample mean always targets the value of the population mean. For this reason, the sample mean is a good estimator of the population mean.

$\bar{x}$	$P(\bar{x})$	$\bar{x} \cdot P(\bar{x})$
46.0	1/16	46.0/16
47.5	2/16	95.0/16
49.0	1/16	49.0/16
51.0	2/16	102.0/16
52.0	2/16	104.0/16
52.5	2/16	105.0/16
53.5	2/16	107.0/16
56.0	1/16	56.0/16
57.0	2/16	114.0/16
58.0	1/16	58.0/16
	16/16	836.0/16

15. a. The sixteen possible samples are given in the "samples" column of the box preceding Exercise 13.
   b. The sixteen possible ranges are given in column 4 of the box preceding Exercise 13. The sampling distribution of the range is given in the first two columns at the right.
   c. The population range is 12. The mean of the sample ranges is $\Sigma R \cdot P(R) = 5.375$. They are not the same.
   d. No. The sample ranges do not always target the value of the population range. For this reason, the sample range is not a good estimator of the population range.

R	P(R)	$R \cdot P(R)$
0	4/16	0/16
2	2/16	4/16
3	2/16	6/16
7	2/16	14/16
9	2/16	18/16
10	2/16	20/16
12	2/16	24/16
	16/16	86/16

17. Let $\hat{p}$ be the symbol for the sample proportion. Refer to the following box.

sample	$\hat{p}$	$\hat{p}$	$P(\hat{p})$	$\hat{p} \cdot P(\hat{p})$
2,2	0.0	0.0	4/9	0.0/9
2,3	0.5	0.5	4/9	2.0/9
2,10	0.0	1.0	1/9	1.0/9
3,2	0.5		9/9	3.0/9
3,3	1.0			
3,10	0.5			
10,2	0.0			
10,3	0.5			
10,10	0.0			

The 9 possible sample proportions are given in column 2 above. The sampling distribution of the proportion is given in columns 3 and 4 above. The population proportion of odd numbers is 1/3. The mean of the sample proportions is $\Sigma \hat{p} \cdot P(\hat{p}) = 3.0/9 = 1/3$. They are the same. The sample proportion always targets the value of the population proportion. For this reason, the sample proportion is a good estimator of the population proportion.

19. Let $\hat{p}$ be the symbol for the sample proportion of females. Refer to the following box.

pair	$\hat{p}$	pair	$\hat{p}$	$\hat{p}$	$P(\hat{p})$	$\hat{p} \cdot P(\hat{p})$
mm	0.0	bm	0.5	0.0	1/16	0.0/16
ma	0.5	ba	1.0	0.5	6/16	3.0/16
mb	0.5	bb	1.0	1.0	9/16	9.0/16
mc	0.5	bc	1.0		16/16	12.0/16
am	0.5	cm	0.5			
aa	1.0	ca	1.0			
ab	1.0	cb	1.0			
ac	1.0	cc	1.0			

a. The sampling distribution of the proportion is given in columns 5 and 6 above.

b. The mean of the sampling distribution is $\Sigma \hat{p} \cdot P(\hat{p}) = 12.0/16 = 0.75$.

c. The population proportion of females is $3/4 = 0.75$. Yes, the two values are the same. The sample proportion always targets the value of the population proportion. For this reason, the sample proportion is a good estimator of the population proportion.

21. Use of the formula for the given values is shown below. The resulting distribution agrees with, and therefore describes, the sampling distribution for the proportion of girls in a family of size 2.

$$P(x) = 1/[2(2-2x)!(2x)!]$$
$$P(x=0) = 1/[2(2)!(0)!] = 1/[2 \cdot 2 \cdot 1] = 1/4$$
$$P(x=0.5) = 1/[2(1)!(1)!] = 1/[2 \cdot 1 \cdot 1] = 1/2$$
$$P(x=1) = 1/[2(0)!(2)!] = 1/[2 \cdot 1 \cdot 2] = 1/4$$

6-5 The Central Limit Theorem

1. The standard error of a statistic is the standard deviation of its sampling distribution. For any given sample size n, the standard error of the mean is the standard deviation of the distribution of all possible sample means of size n. Its symbol and numerical value are $\sigma_{\bar{x}} = \sigma/\sqrt{n}$.

3. The subscript $\bar{x}$ is used to distinguish the mean and standard deviation of the sample means from the mean and standard deviation of the original population. This produces the notation

$$\mu_{\bar{x}} = \mu \qquad \sigma_{\bar{x}} = \sigma/\sqrt{n}$$

5. a. normal distribution
   $\mu = 1518$
   $\sigma = 325$
   $P(x<1500)$
   $\quad = P(z<-0.06)$
   $\quad = 0.4761$

   b. normal distribution,
      since the original distribution is so
   $\mu_{\bar{x}} = \mu = 1518$
   $\sigma_{\bar{x}} = \sigma/\sqrt{n} = 325/\sqrt{100} = 32.5$
   $P(\bar{x}<1500)$
   $\quad = P(z<-0.55)$
   $\quad = 0.2912$

7. a. normal distribution
   $\mu = 1518$
   $\sigma = 325$
   $P(1550<x<1575)$
   $\quad = P(0.10<z<0.18)$
   $\quad = 0.5714 - 0.5398$
   $\quad = 0.0316$

b. normal distribution,
   since the original distribution is so
   $\mu_{\bar{x}} = \mu = 1518$
   $\sigma_{\bar{x}} = \sigma/\sqrt{n} = 325/\sqrt{25} = 65$
   $P(1550 < \bar{x} < 1575)$
       $= P(0.49 < z < 0.88)$
       $= 0.8106 - 0.6879$
       $= 0.1227$

c. Since the original distribution is normal, the Central Limit Theorem can be used in part (b) even though the sample size does not exceed 30

9. a. normal distribution
   $\mu = 172$
   $\sigma = 29$
   $P(x > 180)$
       $= P(z > 0.28)$
       $= 1 - 0.6103$
       $= 0.3897$

b. normal distribution,
   since the original distribution is so
   $\mu_{\bar{x}} = \mu = 172$
   $\sigma_{\bar{x}} = \sigma/\sqrt{n} = 29/\sqrt{20} = 6.48$
   $P(\bar{x} > 180)$
       $= P(z > 1.23)$
       $= 1 - 0.8907$
       $= 0.1093$

c. Yes. A capacity of 20 is not appropriate when the passengers are all adult men, since a 10.93% probability of overloading is too much of a risk.

11. a. normal distribution

$\mu = 172$

$\sigma = 29$

$P(x > 167)$

$= P(z > -0.17)$

$= 1 - 0.4325$

$= 0.5675$

b. normal distribution,

since the original distribution is so

$\mu_{\bar{x}} = \mu = 172$

$\sigma_{\bar{x}} = \sigma/\sqrt{n} = 29/\sqrt{12} = 8.372$

$P(\bar{x} > 167)$

$= P(z > -0.60)$

$= 1 - 0.2743$

$= 0.7257$

c. No. It appears that the 12 person capacity could easily exceed the 2004 lbs – especially when the weight of clothes and equipment is considered. On the other hand, skiers may be lighter than the general population – as the skiing may not be an activity that attracts heavier persons.

13. a. normal distribution

$\mu = 114.8$

$\sigma = 13.1$

$P(x > 140)$

$= P(z > 1.92)$

$= 1 - 0.9726$

$= 0.0274$

b. normal distribution,

since the original distribution is so

$\mu_{\bar{x}} = \mu = 114.8$

$\sigma_{\bar{x}} = \sigma/\sqrt{n} = 13.1/\sqrt{4} = 6.55$

$P(\bar{x} > 140)$

$= P(z > 3.85)$

$= 1 - 0.9999$

$= 0.0001$

c. Since the original distribution is normal, the Central Limit Theorem can be used in part (b) even though the sample size does not exceed 30.

d. No. The mean can be less than 140 when one or more of the values is greater than 140.

15. a. normal distribution

$\mu = 69.0$

$\sigma = 2.8$

$P(x<72)$

$\quad = P(z<1.07)$

$\quad = 0.8577$

b. normal distribution,

　　since the original distribution is so

$\mu_{\bar{x}} = \mu = 69.0$

$\sigma_{\bar{x}} = \sigma/\sqrt{n} = 2.8/\sqrt{100} = 0.28$

$P(\bar{x}<72)$

$\quad = P(z<10.17)$

$\quad = 0.9999$

c. The probability in part (a) is more relevant. Part (a) deals with individual passengers, and these are the persons whose safety and comfort need to be considered. Part (b) deals with group means – and it is possible for statistics that apply "on the average" to actually describe only a small portion of the population of interest.

d. Women are generally smaller than men. Any design considerations that accommodate larger men will automatically accommodate larger women.

17. a. normal distribution

$\mu = 143$

$\sigma = 29$

$P(140<x<211)$

$\quad = P(-0.10<z<2.34)$

$\quad = 0.9904 - 0.4602$

$\quad = 0.5302$

b. normal distribution,
   since the original distribution is so

$\mu_{\bar{x}} = \mu = 143$

$\sigma_{\bar{x}} = \sigma/\sqrt{n} = 29/\sqrt{36} = 4.833$

$P(140 < \bar{x} < 211)$

$= P(-0.62 < z < 14.07)$

$= 0.9999 - 0.2676$

$= 0.7323$

c. The information from part (a) is more relevant, since the seats will be occupied by one woman at a time.

19. normal distribution,
    by the Central Limit Theorem

$\mu_{\bar{x}} = \mu = 12.00$

$\sigma_{\bar{x}} = \sigma/\sqrt{n} = 0.09/\sqrt{36} = 0.015$

$P(\bar{x} > 12.29)$

$= P(z > 19.33)$

$= 1 - 0.9999$

$= 0.0001$

Yes. The results suggest that the Pepsi cans are being filled with more than 12.00 oz of product. This is undoubtedly done on purpose, to minimize probability of producing cans with less than the stated 12.00 oz of product.

21. a. normal distribution

$\mu = 69.0$

$\sigma = 2.8$

The z score with 0.9500 below it is 1.645.

$x = \mu + z\sigma$

$= 69.0 + (1.645)(2.8)$

$= 69.0 + 4.6$

$= 73.6$ inches

b. normal distribution,
   since the original distribution is so

$\mu_{\bar{x}} = \mu = 69.0$

$\sigma_{\bar{x}} = \sigma/\sqrt{n} = 2.8/\sqrt{100} = 0.28$

The z score with 0.9500 below it is 1.645

$\bar{x} = \mu_{\bar{x}} + z\sigma_{\bar{x}}$

$= 69.0 + (1.645)(0.28)$

$= 69.0 + 0.5$

$= 69.5$ inches

c. The probability in part (a) is more relevant. Part (a) deals with individual passengers, and these are the persons whose safety and comfort need to be considered. Part (b) deals with group means – and it is possible for statistics that apply "on the average" to actually describe only a small portion of the population of interest.

23. a. Yes. Since $n/N = 12/210 = 0.0571 > 0.05$, use the finite population correction factor.
   b. For 12 passengers, the 2100 lb limit implies a mean weight of $2100/12 = 175$ lbs.
   c. normal distribution,
         since the original distribution is normal
   $$\mu_{\bar{x}} = \mu = 163$$
   $$\sigma_{\bar{x}} = \frac{\sigma}{\sqrt{n}}\sqrt{\frac{N-n}{N-1}}$$

   $$= \frac{32}{\sqrt{12}}\sqrt{\frac{210-12}{210-1}} = \frac{32}{\sqrt{12}}\sqrt{\frac{198}{209}} = 8.991$$
   $P(\bar{x}>175)$
   $= P(z>1.33)$
   $= 1 - 0.9082$
   $= 0.0918$

   d. The best approach is systematic trial and error. The above calculations indicate that for
         $n = 12$, $P(\Sigma x<2100) = P(\bar{x}<2100/n) = P(\bar{x}< 175) = 0.9082$.
      Repeat the above calculations for
         $n = 11$, 10, etc. until $P(\Sigma x<2100) = P(\bar{x}<2100/n)$ reaches 0.9990.
      *For $n = 11$, $\bar{x} = 2100/11 = 190.91$
      normal distribution,
         since the original distribution is normal
   $$\mu_{\bar{x}} = \mu = 163$$
   $$\sigma_{\bar{x}} = \frac{\sigma}{\sqrt{n}}\sqrt{\frac{N-n}{N-1}}$$

   $$= \frac{32}{\sqrt{11}}\sqrt{\frac{210-11}{210-1}} = \frac{32}{\sqrt{11}}\sqrt{\frac{199}{209}} = 9.415$$
   $P(\bar{x}<190.91)$
   $= P(z<2.96)$
   $= 0.9985$
      Since $0.9985 < 0.999$, try $n = 10$.
      *For $n = 10$, $\bar{x} = 2100/10 = 210$
      normal distribution,
         since the original distribution is normal
   $$\mu_{\bar{x}} = \mu = 163$$
   $$\sigma_{\bar{x}} = \frac{\sigma}{\sqrt{n}}\sqrt{\frac{N-n}{N-1}}$$

   $$= \frac{32}{\sqrt{10}}\sqrt{\frac{210-10}{210-1}} = \frac{32}{\sqrt{10}}\sqrt{\frac{200}{209}} = 9.899$$
   $P(\bar{x}<210)$
   $= P(z<4.75)$
   $= 0.9999$

Since $0.9999 > 0.999$, set $n = 10$.

NOTE: The z score with 0.999 below it is 3.10. To find n exactly, solve the following equation to get $n = 10.9206$. Since n must be a whole number, round down to set $n = 10$.

$$z = \frac{\overline{x} - \mu_{\overline{x}}}{\sigma_{\overline{x}}}$$

$$3.10 = \frac{2100/n - 163}{\frac{32}{\sqrt{n}} \sqrt{\frac{210-n}{210-1}}}$$

6-6 Normal as Approximation to Binomial

1. Assuming the true proportion of households watching *60 Minutes* remains relative constantly during the two years of sampling, the two years of sample proportions represent samples from the same population. Since the sampling distribution of the proportion approximates a normal distribution, the histogram depicting the two years of sample proportions should be approximately normal – e.g., bell-shaped.

3. No. With n=4 and p=0.5, the requirements that $np \geq 5$ and $nq \geq 5$ are not met.

5. The area to the right of 8.5.   In symbols, $P(x>8) = P_C(x>8.5)$.

7. The area to the left of 4.5.   In symbols, $P(x<5) = P_C(x<4.5)$.

9. The area to the left of 15.5.   In symbols, $P(x \leq 15) = P_C(x<15.5)$.

11. The area between 4.5 and 9.5.   In symbols, $P(4 \leq x \leq 9) = P_C(4.5<x<9.5)$.

13. binomial: n=10 and p=0.5
    a. from Table A-1, $P(x=3) = 0.117$
    b. normal approximation appropriate since
       $np = 10(0.5) = 5.0 \geq 5$
       $nq = 10(0.5) = 5.0 \geq 5$
    $\mu = np = 10(0.5) = 5.0$
    $\sigma = \sqrt{npq} = \sqrt{10(0.5)(0.5)} = 1.581$
    $P(x=3)$
       $= P(2.5<x<3.5)$
       $= P(-1.58<z<-0.95)$
       $= 0.1711 - 0.0571$
       $= 0.1140$

15. binomial: n=8 and p=0.9
    a. from Table A-1, $P(x \geq 6) = 0.149 + 0.383 + 0.430 = 0.962$
    b. normal approximation not appropriate since
       $nq = 8(0.1) = 0.8 < 5$

17. binomial: n=40,000 and p=0.03
    normal approximation appropriate since
    $np = 40,000(0.03) = 1200 \geq 5$
    $nq = 40,000(0.97) = 3880 \geq 5$
    $\mu = np = 40,000(0.03) = 1200$
    $\sigma = \sqrt{npq} = \sqrt{40000(0.03)(0.97)} = 34.117$
    $P(x \geq 1300)$
        $= P(x > 1299.5)$
        $= P(z > 2.92)$
        $= 1 - 0.9982$
        $= 0.0018$

No. It is not likely that the goal of at least 1300 will be reached.

19. binomial: n=574 and p=0.50
    normal approximation appropriate since
    $np = 574(0.50) = 287 \geq 5$
    $nq = 574(0.50) = 287 \geq 5$
    $\mu = np = 574(0.50) = 287$
    $\sigma = \sqrt{npq} = \sqrt{574(0.50)(0.50)} = 11.979$
    $P(x \geq 525)$
        $= P(x > 524.5)$
        $= P(z > 19.83)$
        $= 1 - 0.9999$
        $= 0.0001$

Yes. Since the probability of getting at least 525 girls by chance alone is so small, it appears
that the method is effective and that the genders were not being determined by chance alone.

21. binomial: n=580 and p=0.25
    normal approximation appropriate since
    $np = 580(0.25) = 145 \geq 5$
    $nq = 580(0.75) = 435 \geq 5$
    $\mu = np = 580(0.25) = 145$
    $\sigma = \sqrt{npq} = \sqrt{580(0.25)(0.75)} = 10.428$

a. $P(x = 152)$
    $= P(151.5 < x < 152.5)$
    $= P(0.62 < z < 0.72)$
    $= 0.7642 - 0.7324$
    $= 0.0318$

b. $P(x \geq 152)$
    $= P(x > 151.5)$
    $= P(z > 0.62)$
    $= 1 - 0.7324$
    $= 0.2676$

c. The part (b) answer is the useful probability. In situations involving multiple ordered outcomes, the unusualness of a particular outcome is generally determined by the probability of getting that outcome or a more extreme outcome.

d. No. Since 0,2676 > 0.05, Mendel's result could easily occur by chance alone if the true rate were really 0.25.

23. binomial: n=420,095 and p=0.000340
normal approximation appropriate since
np = 420,095(0.000340) = 142.83 ≥ 5
nq = 420,095(0.999660) = 419952.17 ≥ 5
$\mu = np = 420{,}095(0.000340) = 142.83$
$\sigma = \sqrt{npq} = \sqrt{420095(0.000340)(0.999660)}$
  = 11.949
$P(x \le 135)$
  = $P(x < 135.5)$
  = $P(z < -0.61)$
  = 0.2709

To conclude that cell phones increase the likelihood of experiencing such cancers requires x > 142.83. Since 135 < 142.83, these results definitely do not support the media reports.

25. binomial: n=200 and p=0.06
normal approximation appropriate since
np = 200(0.06) = 12 ≥ 5
nq = 200(0.94) = 188 ≥ 5
$\mu = np = 200(0.06) = 12$
$\sigma = \sqrt{npq} = \sqrt{200(0.06)(0.94)} = 3.359$
$P(x \ge 10)$
  = $P(x > 9.5)$
  = $P(z > -0.74)$
  = 1 – 0.2296
  = 0.7704

Yes. Since there is a 77% chance of getting at least 10 universal donors, a pool of 200 volunteers appears to be sufficient. Considering the importance of the need, and the fact that one can never have too much blood on hand, the hospital may want to use a larger pool to further increase the 77% figure and/or determine how large a pool would be necessary to increase the figure to, say, 95%.

27. binomial: n=100 and p=0.24
normal approximation appropriate since
np = 100(0.24) = 24 ≥ 5
nq = 100(0.76) = 76 ≥ 5
$\mu = np = 100(0.24) = 24$
$\sigma = \sqrt{npq} = \sqrt{100(0.24)(0.76)} = 4.271$
$P(x \ge 27)$
  = $P(x > 26.5)$
  = $P(z > 0.59)$
  = 1 – 0.7224
  = 0.2776

No. Since 0.2776 > 0.05, 27 is not an unusually high number of blue M&M's.

29. binomial: n=863 and p=0.019
    normal approximation appropriate since
        np = 863(0.019) = 16.397 ≥ 5
        nq = 863(0.981) = 846.603 ≥ 5
    μ = np = 863(0.019) = 16.397
    σ = $\sqrt{npq}$ = $\sqrt{863(0.019)(0.981)}$ = 4.011
    P(x≥19)
        = P(x>18.5)
        = P(z>0.52)
        = 1 − 0.6985
        = 0.3015

Since the 0.3015 > 0.05, 19 or more persons experiencing flu symptoms is not an unusual occurrence for a normal population. There is no evidence to suggest that flu symptoms are an adverse reaction to the drug.

31. Let x = the number that show up.
    binomial: n=236 and p=0.9005
    normal approximation appropriate since
        np = 236(0.9005) = 212.518 ≥ 5
        nq = 236(0.0995) = 23.482 ≥ 5
    μ = np = 236(0.9005) = 212.518
    σ = $\sqrt{npq}$ = $\sqrt{236(0.9005)(0.0995)}$ = 4.598
    P(x≥213)
        = P(x>213.5)
        = P(z>0.21)
        = 1 − 0.5832
        = 0.4168

No. Since the 0.4168 > 0.05, overbooking is not an unusual event. It is a real concern for both the airline and the passengers.

33. Marc is placing 200(5) = $1000 in bets. To make a profit his return must be more than $1000.
     In part (a) each win returns the original 5 plus 35(5), for a total of 5 + 175 = $180.
         For a profit, he must win more than 1000/180 = 5.55 times – i.e., he needs at least 6 wins.
     In part (b) each win returns the original 5 plus 1(5), for a total of 5 + 5 = $10.
         For a profit, he must win more than 1000/10 = 100 times – i.e., he needs at least 101 wins.
     a. binomial: n=200 and p=1/38
        normal approximation appropriate since
            np = 200(1/38) = 5.26 ≥ 5
            nq = 200(37/38) = 194.74 ≥ 5
        μ = np = 200(1/38) = 5.263
        σ = $\sqrt{npq}$ = $\sqrt{200(1/38)(37/38)}$ = 2.263
        P(x≥6)
            = P(x>5.5)
            = P(z>0.10)
            = 1 − 0.5398
            = 0.4602

b. binomial: n=200 and p=244/495
   normal approximation appropriate since
     $np = 200(244/495) = 98.59 \geq 5$
     $nq = 200(251/495) = 101.41 \geq 5$
   $\mu = np = 200(244/495) = 98.586$
   $\sigma = \sqrt{npq} = \sqrt{200(244/495)(251/495)}$
           $= 7.070$
   $P(x \geq 101)$
     $= P(x > 100.5)$
     $= P(z > 0.27)$
     $= 1 - 0.6064$
     $= 0.3936$

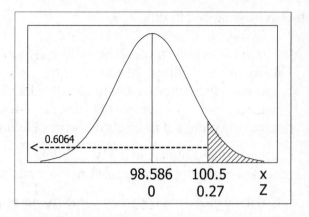

c. Since $0.4602 > 0.3936$, the roulette game is the better "investment" – but since both
   probabilities are less than 0.5000, he would do better not to play at all.

35. a. binomial: n=4 and p=0.350
     $P(x \geq 1) = 1 - P(x=0)$
             $= 1 - [4!/(4!0!)](0.350)^0(0.650)^4$
             $= 1 - 0.1785$
             $= 0.8215$

   b. binomial: n=56(4)=224 and p=0.350
      normal approximation appropriate since
        $np = 224(0.350) = 78.4 \geq 5$
        $nq = 224(0.650) = 145.6 \geq 5$
      $\mu = np = 224(0.350) = 78.4$
      $\sigma = \sqrt{npq} = \sqrt{224(0.350)(0.650)} = 7.139$
      $P(x \geq 56)$
        $= P(x > 55.5)$
        $= P(z > -3.20)$
        $= 1 - 0.0007$
        $= 0.9993$

   c. Let H = getting at least one hit in 4 at bats.
      $P(H) = 0.8215$ [from part (a)]
      For 56 consecutive games, $[P(H)]^{56} = [0.8125]^{56} = 0.0000165$

   d. The solution below employs the techniques and notation of parts (a) and (c).
      for $[P(H)]^{56} > 0.10$
            $P(H) > (0.10)^{1/56}$
            $P(H) > 0.9597$
      for $P(H) = P(x \geq 1) > 0.9597$
              $1 - P(x=0) > 0.9597$
                  $0.0403 > P(x=0)$
                  $0.0403 > [4!/(4!0!)]p^0(1-p)^4$
                  $0.0403 > (1-p)^4$
              $(0.0403)^{1/4} > 1 - p$
                    $p > 1 - (0.0403)^{1/4}$
                    $p > 1 - 0.448$
                    $p > 0.552$

6-7 Assessing Normality

1. A normal quantile plot can be used to determine whether sample data come form a normal distribution. In theory, it compares the z scores for the sample data with the z scores for normally distributed data with the same cumulative relative frequencies as the sample data. In practice, it uses the sample data directly – since converting to z scores is a linear transformation that re-labels the scores but does not change their distribution.

3. Because the weights are likely to follow a normal distribution, on expects the points in a normal quantile plot to approximate a straight line.

5. No, the data do not appear to come from a population with a normal distribution. The points are not reasonably close to a straight line.

7. Yes, the data appear to come from a population with a normal distribution. The points are reasonably close to a straight line.

9. Yes, the data appear to come from a population with a normal distribution. The frequency distribution and histogram indicate that the data is approximately bell-shaped.

duration (hours)	frequency
0 – 49	1
50 – 99	8
100 – 149	18
150 – 199	23
200 – 249	25
250 – 299	19
300 – 349	11
350 – 399	8
400 – 449	2
	115

11. No, the data do not appear to come from a population with a normal distribution. The frequency distribution and the histogram indicate there is a concentration of data at the lower end.

degree days	frequency
0 – 499	20
500 – 999	5
1000 – 1499	8
1500 – 1999	9
2000 – 2499	5
2500 – 2599	1
	48

13. Yes. Since the points approximate a straight line, the data appear to come from a population with a normal distribution. The gaps/groupings in the durations may reflect the fact that the times have to reflect whole numbers of orbits or other physical constraints.

15. No. Since the points do not approximate a straight line, the data do not appear to com form a population with a normal distribution.

17. The two histograms are shown below. The heights (on the left) appear to be approximately normally distributed, while the cholesterol levels (on the right) appear to be positively skewed. Many natural phenomena are normally distributed. Height is a natural phenomenon unaffected by human input, but cholesterol levels are humanly influenced (by diet, exercise, medication, etc.) in ways that might alter any naturally occurring distribution.

19. The corresponding z scores in the table below were determined following the five-step "manual construction" procedure of the text.
   (1) Arrange the n scores in order and place them in the x column.
   (2) For each x_i, calculate the cumulative probability using $cp_i = (2i-1)/2n$ for $i = 1,2,\ldots,n$.
   (3) For each cp_i, find the z_i for which
      $P(z<z_i) = cp_i$ for $i = 1,2,\ldots,n$.
   The resulting normal quantile plot indicates that the data appear to come from a population with a normal distribution.

i	x	cp	z
1	127	0.10	-1.28
2	129	0.30	-0.52
3	131	0.50	0
4	136	0.70	0.52
5	146	0.90	1.28

21. a. Yes. Adding two inches to each height shifts the entire distribution to the right, but it does not affect the shape of the distribution.
   b. Yes. Changing to a different unit measure re-labels the horizontal axis, but it does not change relationships between the data points or affect the shape of the distribution.
   c. No. Unlike a linear transformation of the form $f(x) = ax + b$, the log function $f(x) = \log(x)$ does not have the same effect on all segments of the horizontal axis.

Statistical Literacy and Critical Thinking

1. A normal distribution is one that is symmetric and bell-shaped. More technically, it is one that can be described by the following formula, where μ and σ are specified values,

$$f(x) = \frac{e^{-\frac{1}{2\sigma^2}(x-\mu)^2}}{\sigma\sqrt{2\pi}}.$$

A standard normal distribution is a normal distribution that has a mean of 0 and a standard deviation of 1. More technically, it is the distribution that results when $\mu = 0$ and $\sigma = 1$ in the above formula.

2. In statistics, a normal distribution is one that is symmetric and bell-shaped. This statistical use of the term "normal" is not to be confused with the common English word "normal" in the sense of "typical."

3. Assuming there are no trends over time in the lengths of movies, and that there is mean length that remains fairly constant from year to year, the sample means will follow a normal distribution centered around that enduring mean length.

4. Not necessarily. Depending on how the survey was conducted, the sample may be a convenience sample (if AOL simply polled its own customers) or a voluntary response sample (if the responders were permitted decide for themselves whether or not to participate).

Chapter Quick Quiz

1. The symbol $z_{0.03}$ represents the z score with 0.0300 below it.   According to Table A-1 [closest value is 0.0301], this is -1.88.

2. For n=100, the sample means from any distribution with finite mean and standard deviation follow a normal distribution.

3. In a standard normal distribution, $\mu = 0$ and $\sigma = 1$.

4. $P(z>1.00) = 1 - 0.8413$
    $= 0.1587$

5. $P(-1.50<z<2.50) = 0.9938 - 0.0668$
    $= 0.9270$

6. $P(x<115) = P(<1.00)$
    $= 0.8413$

7. $P(x>118) = P(z>1.20)$
    $= 1 - 0.8849$
    $= 0.1151$

8. $P(88<x<112) = P(-0.80<z<0.80)$
    $= 0.7881 - 0.2119$
    $= 0.5762$

9. For n=25, $\mu_{\bar{x}} = \mu = 100$ and $\sigma_{\bar{x}} = \sigma/\sqrt{n} = 15/\sqrt{25} = 3$.
    $P(\bar{x}<103) = P(z<1.00)$
    $= 0.8413$

10. For n=100, $\mu_{\bar{x}} = \mu = 100$ and $\sigma_{\bar{x}} = \sigma/\sqrt{n} = 15/\sqrt{100} = 1.5$.
    $P(\bar{x}>103) = P(z>2.00)$
    $= 1 - 0.9772$
    $= 0.0228$

Review Exercises

1. a. normal distribution: $\mu = 69.0$ and $\sigma = 2.8$
    $P(x>75)$
        $= P(z>2.14)$
        $= 1 - 0.9838$
        $= 0.0162$ or 1.62%

   b. normal distribution: $\mu = 63.6$ and $\sigma = 2.5$
    $P(x>75)$
        $= P(z>4.56)$
        $= 1 - 0.9999$
        $= 0.0001$ or 0.01%

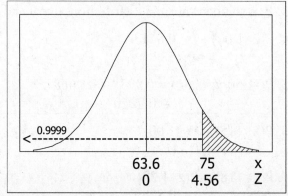

   c. The length of a day bed appears to be adequate to meet the needs of all but the very tallest men and women.

2. For the tallest 5%, $A = 0.9500$ and $z = 1.645$
    $x = \mu + z\sigma$
        $= 69.0 + (1.645)(2.8)$
        $= 69.0 + 4.6$
        $= 73.6$ inches

3. a. normal distribution: $\mu = 69.0$ and $\sigma = 2.8$
    $P(x>78)$
        $= P(z>3.21)$
        $= 1 - 0.9993$
        $= 0.0007$ or 0.07% of the men

   normal distribution: $\mu = 63.6$ and $\sigma = 2.5$
    $P(x>78)$
        $= P(z>5.76)$
        $= 1 - 0.9999$
        $= 0.0001$ or 0.01% of the women

b. For the tallest 1%, A = 0.9900 [0.9901]
            and z = 2.33

    $x = \mu + z\sigma$
      $= 69.0 + (2.33)(2.8)$
      $= 69.0 + 6.5$
      $= 75.5$ inches

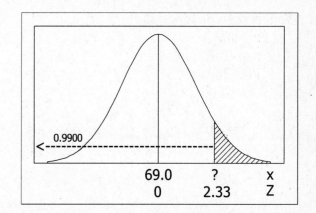

4. normal distribution: $\mu = 63.6$ and $\sigma = 2.5$
    $P(66.5 < x < 71.5)$
        $= P(1.16 < z < 3.16)$
        $= 0.9992 - 0.8770$
        $= 0.1222$ or 12.22%

Yes. The minimum height requirement for Rockettes is greater than th mean height of all women.

5. binomial: n=1064 and p=0.75
    normal approximation appropriate since
        $np = 1064(0.75) = 798 \geq 5$
        $nq = 1064(0.25) = 266 \geq 5$
    $\mu = np = 1064(0.75) = 798$
    $\sigma = \sqrt{npq} = \sqrt{1064(0.75)(0.25)} = 14.124$
    $P(x \leq 787)$
        $= P(x < 787.5)$
        $= P(z < -0.74)$
        $= 0.2296$

Since the 0.2296 > 0.05, obtaining 787 plants with long stems is not an unusual occurrence for a population with p=0.75. The results are consistent with Mendel's claimed proportion.

6. a. True. The sample mean is an unbiased estimator of the population mean.
    See Section 6-4, Exercises 12 and 13.
  b. True. The sample proportion is an unbiased estimator of the population proportion.
    See Section 6-4, Exercises 17 and 18 and 19.
  c. True. The sample variance is an unbiased estimator of the population variance.
    See Section 6-4, Exercises 11 and 16.
  d. Not true. See Section 6-4, Exercises 9 and 14.
  e. Not true. See Section 6-4, Exercise 15.

7. a. normal distribution

$\mu = 178.1$

$\sigma = 40.7$

$P(x>260)$

$= P(z>2.01)$

$= 1 - 0.9778$

$= 0.0222$

b. normal distribution

$\mu = 178.1$

$\sigma = 40.7$

$P(170<x<200)$

$= P(-0.20<z<0.54)$

$= 0.7054 - 0.4207$

$= 0.2847$

c. normal distribution,

since the original distribution is so

$\mu_{\bar{x}} = \mu = 178.1$

$\sigma_{\bar{x}} = \sigma/\sqrt{n} = 40.7/\sqrt{9} = 13.567$

$P(170<\bar{x}<200)$

$= P(-0.60<z<1.61)$

$= 0.9463 - 0.2743$

$= 0.6720$

d. For the top 3%,

$A = 0.9700 [0.9699]$ and $z = 1.88$

$x = \mu + z\sigma$

$= 178.1 + (1.88)(40.7)$

$= 178.1 + 76.5$

$= 254.6$

8. binomial: n=40 and p=0.50

normal approximation appropriate since

$np = 40(0.50) = 20 \geq 5$

$nq = 40(0.50) = 20 \geq 5$

$\mu = np = 40(0.50) = 20$

$\sigma = \sqrt{npq} = \sqrt{40(0.50)(0.50)} = 3.162$

$P(x \leq 15)$

$= P(x<15.5)$

$= P(z<-1.42)$

$= 0.0778$

Since the 0.0778 > 0.05, obtaining 15 or
fewer women is not an unusual occurrence for a population with p=0.75. No. There is not
strong evidence to charge gender discrimination.

9. a. For 0.6700 to the left, A = 0.6700 and z = 0.44.
   b. For 0.9960 to the right, A = 1 – 0.9960 = 0.0040 and z = -2.65.
   c. For 0.025 to the right, A = 1 – 0.0250 = 0.9750 and z = 1.96.

10. a. The sampling distribution of the mean is a normal distribution.

   b. $\mu_{\bar{x}} = \mu = 3420$ grams

   c. $\sigma_{\bar{x}} = \sigma/\sqrt{n} = 495/\sqrt{85} = 53.7$ grams

11. Adding 20 lbs of carry-on baggage for every male will increase the mean weight by 20 but will
   not affect the standard deviation of the
   weights.
   a. normal distribution

   $\mu = 192$

   $\sigma = 29$

   $P(x>195)$

   $= P(z>0.10)$

   $= 1 – 0.5398$

   $= 0.4602$

   b. normal distribution,
   since the original distribution is so

   $\mu_{\bar{x}} = \mu = 192$

   $\sigma_{\bar{x}} = \sigma/\sqrt{n} = 29/\sqrt{213} = 1.987$

   $P(\bar{x}>195)$

   $= P(z>1.51)$

   $= 1 – 0.9345$

   $= 0.0655$

   Yes. Since 0.0655 > 0.05, being
   overloaded under those circumstances
   (i.e., all passengers are male and carrying
   20 lbs of carry-on luggage) would not be unusual.

12. For review purposes, this exercise is worked in detail using a frequency distribution, a frequency histogram, and a normal quantile plot.

a. Using a frequency distribution and a frequency histogram.

weight (grams)	frequency
7.9500 – 7.9999	1
8.0000 – 8.0499	5
8.0500 – 8.0999	8
8.1000 – 8.1444	6
	20

b. Using a normal quantile plot. Using the five-step "manual construction of a normal quantile plot" given in the text, the corresponding z scores in the table below were determined as follows.

(1) Arrange the n=20 scores in order and place them in the x column.

(2) For each x_i, calculate the cumulative probability using $cp_i = (2i-1)/2n$ for $i = 1, 2, \ldots, n$.

(3) For each cp_i, find the z_i for which $P(z<z_i) = cp_i$ for $i = 1, 2, \ldots, n$.

i	x	cp	z
1	7.9817	0.025	-1.96
2	8.0241	0.075	-1.44
3	8.0271	0.125	-1.15
4	8.0307	0.175	-0.93
5	8.0342	0.225	-0.76
6	8.0345	0.275	-0.60
7	8.0510	0.325	-0.45
8	8.0538	0.375	-0.32
9	8.0658	0.425	-0.19
10	8.0719	0.475	-0.06
11	8.0775	0.525	0.06
12	8.0813	0.575	0.19
13	8.0894	0.625	0.32
14	8.0954	0.675	0.45
15	8.1008	0.725	0.60
16	8.1041	0.775	0.76
17	8.1072	0.825	0.93
18	8.1238	0.875	1.15
19	8.1281	0.925	1.44
20	8.1384	0.975	1.96

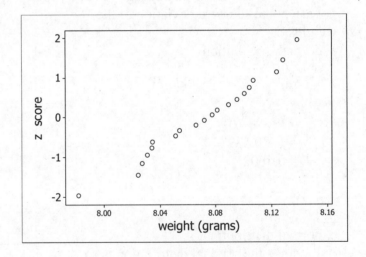

Yes. The weights appear to come from a population that has a normal distribution. The frequency distribution and the histogram suggest a bell-shaped distribution, and the points on the normal quantile plot are reasonably close to a straight line.

Cumulative Review Exercises

1. Arranged in order, the values are: 125   128   138   159   212   235   360   492   530   900
   summary statistics: n = 10 Σx = 3279 Σx^2 = 1639067

   a. $\bar{x} = (\Sigma x)/n$ = 3279/10 = 327.9 or \$327,900

   b. $\tilde{x} = (x_5 + x_6)/2$ = (212 + 235)/2 = 223.5 or \$223,500

   c. s = 250.307 or \$250,307 [the square root of the answer given in part d]

   d. $s^2 = [n(\Sigma x^2) - (\Sigma x)^2]/[n(n-1)]$
      $= [10(1639067) - (3279)^2]/[10(9)]$
      = 5638829/90 = 62653.655556 or 62,653,655,556 dollars^2

   e. z = (235,000 – 327,900)/250,307 = -0.37

   f. Ratio, since differences are meaningful and there is a meaningful zero.

   g. Discrete, since they must be paid in hundredths of dollars (i.e., in whole cents).

2. a. A simple random sample of size n is a sample selected in such a way that every sample of
      size n has the same chance of being selected.

   b. A voluntary response sample is one for which the respondents themselves made the decision
      and effort to be included. Such samples are generally unsuited for statistical purposes
      because they are typically composed of persons with strong feelings on the topic and are not
      representative of the population.

3. a. $P(V_1 \text{ and } V_2) = P(V_1) \cdot P(V_2|V_1)$ = (14/2103)(13/2102) = 0.0000412

   b. binomial: n=5000 and p=14/2103
      normal approximation appropriate since
         np = 5000(14/2103) = 33.29 ≥ 5
         nq = 5000(2089/2103) = 4966.71 ≥ 5
      μ = np = 5000(14/2103) = 33.286
      $\sigma = \sqrt{npq} = \sqrt{5000(14/2103)(2089/2103)}$
              = 5.750
      P(x≥40)
         = P(x>39.5)
         = P(z>1.08)
         = 1 – 0.8599
         = 0.1401

   c. No. Since 0.1401 > 0.05, 40 is not an unusually high number of viral infections.

   d. No. To determine whether viral infections are an adverse reaction to using Nasonex, we
      would need to compare to 14/2103 = 0.00666 rate for Nasonex users to the rate in the
      general population – and that information is not given.

4. No. By not beginning the vertical scale at zero, the graph exaggerates the differences.

5. a. Let L = a person is left-handed.

$P(L) = 0.10$, for each random selection

$P(L_1 \text{ and } L_2 \text{ and } L_3) = P(L_1) \cdot P(L_2) \cdot P(L_3)$
$= (0.10)(0.10)(0.10)$
$= 0.001$

b. Let N = a person is not left-handed.

$P(N) = 0.90$, for each random selection

$P(\text{at least one left-hander}) = 1 - P(\text{no left-handers})$
$= 1 - P(N_1 \text{ and } N_2 \text{ and } N_3)$
$= 1 - P(N_1) \cdot P(N_2) \cdot P(N_3)$
$= 1 - (0.90)(0.90)(0.90)$
$= 1 - 0.729$
$= 0.271$

c. binomial: n=3 and p=0.10

normal approximation not appropriate since
$np = 3(0.10) = 3 < 5$

d. binomial: n=50 and p=0.10

$\mu = np = 50(0.10) = 5$

e. binomial problem: n=50 and p=0.10

$\sigma = \sqrt{npq} = \sqrt{50(0.10)(0.90)} = 2.121$

f. There are two previous approaches that may be used to answer this question.

(1) An unusual score is one that is more than two standard deviations from the mean. Use
the values for μ and σ from parts (d) and (e).

$z = (x - \mu)/\sigma$
$z_8 = (8 - 5)/2.121$
$= 1.41$

Since 8 is 1.41<2 standard deviations from the mean, it would not be an unusual result.

(2) A score is unusual if the probability of getting that result or a more extreme result is less
than or equal to 0.05.

binomial: n = 50 and p = 0.10

normal approximation appropriate since
$np = 50(0.10) = 5 \geq 5$
$nq = 50(0.90) = 45 \geq 5$
Use the values for μ and σ
from parts (d) and (e).

$P(x \geq 8)$
$= P(x > 7.5)$
$= P(z > 1.18)$
$= 1 - 0.8810$
$= 0.1190$

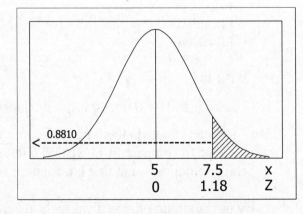

Since 0.1190 > 0.05, getting 8 left-handers in a group of 50 is not an unusual event.

Chapter 7

Estimates and Sample Sizes

7-2 Estimating a Population Proportion

1. The confidence level was not stated. The most common level of confidence is 95%, and sometimes that level is carelessly assumed without actually being stated.

3. By including a statement of the maximum likely error, a confidence interval provides information about the accuracy of an estimate.

5. For 99% confidence, $\alpha = 1-0.99 = 0.01$ and $\alpha/2 = 0.01/2 = 0.005$.
For the upper 0.005, A = 0.9950 and z = 2.575.
$z_{\alpha/2} = z_{0.005} = 2.575$

7. For $\alpha = 0.10$, $\alpha/2 = 0.10/2 = 0.05$.
For the upper 0.05, A = 0.9500 and z = 1.645.
$z_{\alpha/2} = z_{0.05} = 1.645$

9. Let L = the lower confidence limit; U = the upper confidence limit.
$\hat{p} = (L+U)/2 = (0.200+0.500)/2 = 0.700/2 = 0.350$
$E = (U-L)/2 = (0.500-0.200)/2 = 0.300/2 = 0.150$
The interval can be expressed as 0.350 ± 0.150.

11. Let L = the lower confidence limit; U = the upper confidence limit.
$\hat{p} = (L+U)/2 = (0.437+0.529)/2 = 0.966/2 = 0.483$
$E = (U-L)/2 = (0.529-0.437)/2 = 0.092/2 = 0.046$
The interval can be expressed as 0.483 ± 0.046.

13. Let L = the lower confidence limit; U = the upper confidence limit.
$\hat{p} = (L+U)/2 = (0.320+0.420)/2 = 0.740/2 = 0.370$
$E = (U-L)/2 = (0.420-0.320)/2 = 0.100/2 = 0.050$

15. Let L = the lower confidence limit; U = the upper confidence limit.
$\hat{p} = (L+U)/2 = (0.433+0.527)/2 = 0.960/2 = 0.480$
$E = (U-L)/2 = (0.527-0.433)/2 = 0.094/2 = 0.047$

IMPORTANT NOTE: When calculating $E = z_{\alpha/2}\sqrt{\hat{p}\hat{q}/n}$ do not round off in the middle of the problem. This, and the subsequent calculations of $U = \hat{p} + E$ and $L = \hat{p} - E$ may accomplished conveniently on most calculators having a memory as follows.
(1) Calculate $\hat{p} = x/n$ and STORE the value.
(2) Calculate E as $1 - \text{RECALL} = * \text{RECALL} = \div n = \sqrt{} * z_{\alpha/2} =$
(3) With the value for E showing on the display, the upper confidence limit U can be calculated by using + RECALL =.
(4) With the value for U showing on the display, the lower confidence limit L can be calculated by using − RECALL ± + RECALL.
THE MANUAL USES THIS PROCEDURE, AND ROUNDS THE FINAL ANSWER TO 3 SIGNIFICANT DIGITS, EVEN THOUGH IT REPORTS INTERMEDIATE STEPS WITH A FINITE

NUMBER OF DECIMAL PLACES. If the above procedure does not work on your calculator, or to find out if some other procedure would be more efficient on your calculator, ask your instructor for assistance. You must become familiar with your own calculator – and be sure to do your homework on the same calculator you will use for the exams.

17. $\alpha = 0.05$ and $z_{\alpha/2} = z_{0.025} = 1.96$; $\hat{p} = x/n = 400/1000 = 0.40$

$E = z_{\alpha/2}\sqrt{\hat{p}\hat{q}/n} = 1.96\sqrt{(0.40)(0.60)/1000} = 0.0304$

19. $\alpha = 0.02$ and $z_{\alpha/2} = z_{0.01} = 2.33$; $\hat{p} = x/n = [492]/1230 = 0.40$

$E = z_{\alpha/2}\sqrt{\hat{p}\hat{q}/n} = 2.33\sqrt{(0.40)(0.60)/1230} = 0.0325$

NOTE: The value x=[492] was not given. In truth, any $486 \le x \le 498$ rounds to the given $\hat{p} = x/1230 = 40\%$. For want of a more precise value, $\hat{p} = 0.40$ is used in the calculation of E.

21. $\alpha = 0.05$ and $z_{\alpha/2} = z_{0.025} = 1.96$; $\hat{p} = x/n = 40/200 = 0.2000$

$$\hat{p} \pm z_{\alpha/2}\sqrt{\hat{p}\hat{q}/n}$$
$$0.2000 \pm 1.96\sqrt{(0.2000)(0.8000)/200}$$

0.2000 ± 0.0554

$0.145 < p < 0.255$

23. $\alpha = 0.01$ and $z_{\alpha/2} = z_{0.005} = 2.575$; $\hat{p} = x/n = 109/1236 = 0.0882$

$$\hat{p} \pm z_{\alpha/2}\sqrt{\hat{p}\hat{q}/n}$$
$$0.0882 \pm 2.575\sqrt{(0.0882)(0.9118)/1236}$$

0.0882 ± 0.0207

$0.0674 < p < 0.109$

25. $\alpha = 0.05$, $z_{\alpha/2} = z_{0.025} = 1.96$ and $E = 0.045$; $\hat{p}$ unknown, use $\hat{p} = 0.5$

$n = [(z_{\alpha/2})^2\hat{p}\hat{q}]/E^2$

$\quad = [(1.96)^2(0.5)(0.5)]/(0.045)^2$

$\quad = 474.27$, rounded up to 475

27. $\alpha = 0.01$, $z_{\alpha/2} = z_{0.005} = 2.575$ and $E = 0.02$; $\hat{p}$ estimated to be 0.14

$n = [(z_{\alpha/2})^2\hat{p}\hat{q}]/E^2$

$\quad = [(2.575)^2(0.14)(0.86)]/(0.02)^2$

$\quad = 1995.82$, rounded up to 1996

29. Let x = the number of girls born using the method.

a. $\hat{p} = x/n = 525/574 = 0.9146$, rounded to 0.915

b. $\alpha = 0.05$, $z_{\alpha/2} = z_{0.025} = 1.96$

$$\hat{p} \pm z_{\alpha/2}\sqrt{\hat{p}\hat{q}/n}$$
$$0.9146 \pm 1.96\sqrt{(0.9146)(0.0854)/574}$$

0.9146 ± 0.0229

$0.892 < p < 0.937$

c. Yes. Since 0.5 is not within the confidence interval, and below the interval, we can be 95% certain that the method is effective.

31. Let x = the number of deaths in the week before Thanksgiving.

   a. $\hat{p}$ = x/n = 6062/12000 = 0.5052, rounded to 0.505

   b. α = 0.05, $z_{\alpha/2} = z_{0.025}$ = 1.96

$$\hat{p} \pm z_{\alpha/2}\sqrt{\hat{p}\hat{q}/n}$$

    0.5052 ± 1.96$\sqrt{(0.5052)(0.4948)/12000}$

    0.5052 ± 0.0089

    0.496 < p < 0.514

   c. No. Since 0.5 is within the confidence interval, there is no evidence that people can temporarily postpone their death in such circumstances.

33. Let x = the number of yellow peas

   a. α = 0.05, $z_{\alpha/2} = z_{0.025}$ = 1.96 and $\hat{p}$ = x/n = 152/(428+152) = 152/580 = 0.2621

$$\hat{p} \pm z_{\alpha/2}\sqrt{\hat{p}\hat{q}/n}$$

    0.2621 ± 1.96$\sqrt{(0.2621)(0.7379)/580}$

    0.2621 ± 0.0358

    0.226 < p < 0.298 or 22.6% < p < 29.8%

   b. No. Since 0.25 is within the confidence interval, it is a reasonable possibility for the true population proportion. The results do not contradict the theory.

35. Let x = the number that develop those types of cancer.

   a. α = 0.05, $z_{\alpha/2} = z_{0.025}$ = 1.96 and $\hat{p}$ = x/n = 135/420095 = 0.0003214

$$\hat{p} \pm z_{\alpha/2}\sqrt{\hat{p}\hat{q}/n}$$

    0.0003214 ± 1.96$\sqrt{(0.0003214)(0.9996786)/420095}$

    0.0003214 ± 0.0000542

    0.000267 < p < 0.000376 or 0.0267% < p < 0.0376%

   b. No. Since 0.0340% = 0.000340 is within the confidence interval, it is a reasonable possibility for the true population value. The results do not provide evidence that cell phone users have a different cancer rate than the general population.

37. Let x = the number who say they use the Internet.

   α = 0.05, $z_{\alpha/2} = z_{0.025}$ = 1.96 and $\hat{p}$ = x/n = [2198]/3011 = 0.73

   NOTE: The value x=[2198] was not given. In truth, any 2183 ≤ x ≤ 2213 rounds to the given $\hat{p}$ = x/3011 = 73%. For want of a more precise value, $\hat{p}$ = 0.73 is used in the calculations. Technically, this should limit the exercise to two significant digit accuracy

$$\hat{p} \pm z_{\alpha/2}\sqrt{\hat{p}\hat{q}/n}$$

   0.73 ± 1.96$\sqrt{(0.73)(0.27)/3011}$

   0.73 ± 0.0159

   0.714 < p < 0.746

   No. Since 0.75 is not within the confidence interval, it is not likely to be the correct value of the population proportion and should not be reported as such. In this particular exercise, however, the above NOTE indicates that the third significant digit in the confidence interval endpoints is not reliable – and if $\hat{p}$ is really 2213/3011 = 0.73497, for example, the confidence interval is 0.719 < p < 0.751 and 75% is acceptable.

39. Let x = the number who indicate the outbreak would deter them from taking a cruise.

$\alpha = 0.05$, $z_{\alpha/2} = z_{0.025} = 1.96$ and $\hat{p} = x/n = [21302]/34358 = 0.62$

NOTE: The value x=[21302] was not given. In truth, any $21131 \le x \le 21473$ rounds to the given $\hat{p} = x/34358 = 62\%$. For want of a more precise value, $\hat{p} = 0.62$ is used in the calculations. Technically, this should limit the exercise to two significant digit accuracy.

$\hat{p} \pm z_{\alpha/2}\sqrt{\hat{p}\hat{q}/n}$

$0.62 \pm 1.96\sqrt{(0.62)(0.38)/34258}$

0.62 ± 0.0051

$0.615 < p < 0.625$

No. Since the sample is a voluntary response sample, the respondents are not likely to be representative of the population.

41. $\alpha = 0.01$, $z_{\alpha/2} = z_{0.005} = 2.575$ and $E = 0.02$

a. $\hat{p}$ unknown, use $\hat{p} = 0.5$

$n = [(z_{\alpha/2})^2\hat{p}\hat{q}]/E^2 = [(2.575)^2(0.5)(0.5)]/(0.02)^2 = 4144.14$, rounded up to 4145

b. $\hat{p}$ estimated to be 0.73

$n = [(z_{\alpha/2})^2\hat{p}\hat{q}]/E^2 = [(2.575)^2(0.73)(0.27)]/(0.02)^2 = 3267.24$, rounded up to 3268

43. $\alpha = 0.05$, $z_{\alpha/2} = z_{0.025} = 1.96$ and $E = 0.03$; $\hat{p}$ unknown, use $\hat{p} = 0.5$

$n = [(z_{\alpha/2})^2\hat{p}\hat{q}]/E^2 = [(1.96)^2(0.5)(0.5)]/(0.03)^2 = 1067.11$, rounded up to 1068

45. Let x = the number of green M&M's.

$\alpha = 0.05$, $z_{\alpha/2} = z_{0.025} = 1.96$ and $\hat{p} = x/n = 19/100 = 0.19$

$\hat{p} \pm z_{\alpha/2}\sqrt{\hat{p}\hat{q}/n}$

$0.1900 \pm 1.96\sqrt{(0.19)(0.81)/100}$

0.1900 ± 0.0769

$0.113 < p < 0.267$ or $11.3\% < p < 26.7\%$

Yes. Since 0.160 is within the confidence interval, this result is consistent with the claim that the true population proportion is 16%.

47. Let x = the number of days with precipitation.

$\alpha = 0.05$, $z_{\alpha/2} = z_{0.025} = 1.96$

Wednesdays: $\hat{p} = x/n = 16/53 = 0.3019$.   Sundays: $\hat{p} = x/n = 15/52 = 0.2885$

$\hat{p} \pm z_{\alpha/2}\sqrt{\hat{p}\hat{q}/n}$ $\hat{p} \pm z_{\alpha/2}\sqrt{\hat{p}\hat{q}/n}$

$0.3019 \pm 1.96\sqrt{(0.3019)(0.6981)/53}$ $0.2885 \pm 1.96\sqrt{(0.2885)(0.7115)/52}$

0.3019 ± 0.1236 0.2885 ± 0.1231

$0.178 < p < 0.425$ $0.165 < p < 0.412$

The confidence intervals are similar. It does not appear to rain more on either day.

49. $\alpha = 0.05$, $z_{\alpha/2} = z_{0.025} = 1.96$ and $E = 0.03$; $\hat{p}$ unknown, use $\hat{p} = 0.5$

$$n = \frac{N\hat{p}\hat{q}[z_{\alpha/2}]^2}{\hat{p}\hat{q}[z_{\alpha/2}]^2 + (N-1)E^2}$$

$$= \frac{(12784)(0.5)(0.5)[1.96]^2}{(0.5)(0.5)[1.96]^2 + (12783)(0.03)^2} = \frac{12277.8}{12.4651} = 984.97, \text{ rounded up to } 985$$

No. The sample size is not too much lower than the n=1068 required for a population of millions of people.

51. $\alpha = 0.05$, $z_{\alpha/2} = z_{0.025} = 1.96$ and $\hat{p} = x/n = 3/8 = 0.3750$

$$\hat{p} \pm z_{\alpha/2}\sqrt{\hat{p}\hat{q}/n}$$
$$0.3750 \pm 1.96\sqrt{(0.3750)(0.6250)/8}$$

0.3750 ± 0.3355

$0.0395 < p < 0.710$

Yes. The results are "reasonably close" – being shifted down 4.5% from the correct interval $0.085 < p < 0.755$. But depending on the context, such an error could be serious.

53. a. If $\hat{p} = x/n = 0/n = 0$, then

   (1) np $\approx 0 < 5$, and the normal approximation to the binomial does not apply.

   (2) $E = z_{\alpha/2}\sqrt{\hat{p}\hat{q}/n} = 0$, and there is no meaningful interval.

b. Since $\hat{p} = x/n = 0/20 = 0$, use $3/n = 3/20 = 0.15$ as the 95% upper bound for p.

NOTE: The corresponding interval would be $0 \le p < 0.15$. Do not use $0 < p < 0.15$, because the failure to observe any successes in the sample does not rule out p=0 as the true population proportion.

7-3 Estimating a Population Mean: σ Known

1. A point estimate is a single value used to estimate a population parameter. If the parameter in question is the mean of a population, the best point estimate is the mean of a random sample from that population.

3. It is estimated that the mean height of U.S. women is 63.195 inches. This result comes from the Third National Health and Nutrition Examination Survey of the U.S. Department of Health and Human Services. It is based on an in-depth study of 40 women and assumes a population standard deviation of 2.5 inches. The estimate has a margin of error of 0.775 inches with a 95% level of confidence. In other words, 95% of all such studies can be expected to produce estimates that are within 0.775 inches of the true population mean height of all U.S. women.

5. For 90% confidence, $\alpha = 1-0.90 = 0.10$ and $\alpha/2 = 0.10/2 = 0.05$.
For the upper 0.05, $A = 0.9500$ and $z = 1.645$.
$z_{\alpha/2} = z_{0.05} = 1.645$

7. For $\alpha = 0.20$, $\alpha/2 = 0.20/2 = 0.10$.
For the upper 0.10, $A = 0.9000$ and $z = 1.28$.
$z_{\alpha/2} = z_{0.10} = 1.28$

9. Since σ is known and n>30, the methods of this section may be used.
$\alpha = 0.05$, $z_{\alpha/2} = z_{0.025} = 1.96$
$E = z_{\alpha/2}\sigma/\sqrt{n} = 1.96(68)/\sqrt{50} = 18.8$ FICO units
   $\bar{x} \pm E$
677.0 ± 18.8
$658.2 < \mu < 695.8$ (FICO units)
NOTE: The above interval assumes $\bar{x} = 677.0$. Technically, the failure to report $\bar{x}$ to tenths limits the endpoints of the confidence interval to whole number accuracy.

11. Since n<30 and the population is far from normal, the methods of this section may not be used.

13. $\alpha = 0.05$, $z_{\alpha/2} = z_{0.025} = 1.96$

    $n = [z_{\alpha/2} \cdot \sigma/E]^2$

      $= [(1.96)(68)/(3)]^2$

      $= 1973.73$, rounded up to 1974

15. $\alpha = 0.01$, $z_{\alpha/2} = z_{0.005} = 2.575$

    $n = [z_{\alpha/2} \cdot \sigma/E]^2$

      $= [(2.575)(0.212)/(0.010)]^2$

      $= 2980.07$, rounded up to 2981

17. $\overline{x} = 21.12$ mg

19. $E = (U - L)/2 = (22.387 - 19.853)/2 = 1.267$

    21.12 ± 1.267 (mg)

21. a. $\overline{x} = 146.22$ lbs

    b. $\alpha = 0.05$, $z_{\alpha/2} = z_{0.025} = 1.96$

        $\overline{x} \pm z_{\alpha/2} \cdot \sigma/\sqrt{n}$

      $146.22 \pm 1.96(30.86)/\sqrt{40}$

      146.22 ± 9.56

      $136.66 < \mu < 155.78$ (lbs)

23. a. $\overline{x} = 58.3$ seconds

    b. $\alpha = 0.05$, $z_{\alpha/2} = z_{0.025} = 1.96$

        $\overline{x} \pm z_{\alpha/2} \cdot \sigma/\sqrt{n}$

      $58.3 \pm 1.96(9.5)/\sqrt{40}$

      58.3 ± 2.9

      $55.5 < \mu < 61.2$ (seconds)

    c. Yes. Since the confidence interval contains 60 seconds, it is reasonable to assume that the sample mean was reasonably close to 60 seconds – and it was, in fact, 58.3 seconds.

25. a. $\alpha = 0.05$, $z_{\alpha/2} = z_{0.025} = 1.96$ b. $\alpha = 0.01$, $z_{\alpha/2} = z_{0.005} = 2.575$

      $\overline{x} \pm z_{\alpha/2} \cdot \sigma/\sqrt{n}$ $\overline{x} \pm z_{\alpha/2} \cdot \sigma/\sqrt{n}$

    $1522 \pm 1.96(333)/\sqrt{125}$ $1522 \pm 2.575(333)/\sqrt{125}$

    1522 ± 58 1522 ± 77

    $1464 < \mu < 1580$ $1445 < \mu < 1599$

    c. The 99% confidence interval in part (b) is wider than the 95% confidence interval in part (a). For an interval to have more confidence associated with it, it must be wider to allow for more possibilities.

27. summary statistics: $n = 14$ $\Sigma x = 1875$ $\overline{x} = 133.93$

    $\alpha = 0.05$, $z_{\alpha/2} = z_{0.025} = 1.96$

        $\overline{x} \pm z_{\alpha/2} \cdot \sigma/\sqrt{n}$

    $133.93 \pm 1.96(10)/\sqrt{14}$

    133.93 ± 5.24

    $128.7 < \mu < 139.2$ (mmHg)

Ideally, there is a sense in which all the measurements should be the same – and in that case there would be no need for a confidence interval. It is unclear what the given $\sigma = 10$ represents in this situation. Is it the true standard deviation in the values of all people in the population (in which case it would not be appropriate in this context where only a single person is

29. summary statistics: n = 35   Σx = 4305   x̄ = 123.00

$\alpha = 0.05$, $z_{\alpha/2} = z_{0.025} = 1.96$

$$\bar{x} \pm z_{\alpha/2} \cdot \sigma/\sqrt{n}$$

$123.00 \pm 1.96(100)/\sqrt{35}$

123.00 ± 33.13

$89.9 < \mu < 156.1$ (million dollars)

31. $\alpha = 0.05$, $z_{\alpha/2} = z_{0.025} = 1.96$

$n = [z_{\alpha/2} \cdot \sigma/E]^2$

$\quad = [(1.96)(15)/(5)]^2$

$\quad = 34.57$, rounded up to 35

33. $\alpha = 0.05$, $z_{\alpha/2} = z_{0.025} = 1.96$

$n = [z_{\alpha/2} \cdot \sigma/E]^2$

$\quad = [(1.96)(10.6)/(0.25)]^2$

$\quad = 6906.27$, rounded up to 6907

The sample size is too large to be practical.

35. $\alpha = 0.05$, $z_{\alpha/2} = z_{0.025} = 1.96$

Using the range rule of thumb: R = 40,000 – 0 = 40,000, and σ ≈ R/4 = 40,000/4 = 10,000.

$n = [z_{\alpha/2} \cdot \sigma/E]^2 = [(1.96)(10,000)/(100)]^2 = 38416$, rounded up to 217

37. Since n/N = 125/200 = 0.625 > 0.05, use the finite population correction factor.

$\alpha = 0.05$, $z_{\alpha/2} = z_{0.025} = 1.96$

$$\bar{x} \pm [z_{\alpha/2}\sigma/\sqrt{n}] \cdot \sqrt{(N-n)/(N-1)}$$

$1522 \pm [1.96(333)/\sqrt{125}] \cdot \sqrt{(200-125)/(200-1)}$

$1522 \pm [58.3774] \cdot [0.6139]$

1522 ± 36

$1486 < \mu < 1558$

The confidence interval becomes narrower because the sample is a larger portion of the population. As n approaches N, the length of the confidence interval shrinks to 0 – because when n=N the true mean μ can be determined with certainty.

7-4 Estimating a Population Mean: σ Not Known

1. According to the point estimate ("average"), the parameter of interest is a population mean. But according to the margin of error ("percentage points"), the parameter of interest is a population proportion. It is possible that the margin of error the paper intended to communicate was 1% of $483 (or $4.83, which in a 95% confidence interval would correspond to a sample standard deviation of $226.57) – but the proper units for the margin of error in a situation like this are "dollars" and not "percentage points."

3. No; the estimate will not be good for at least two reasons. First, the sample is a convenience sample using the state of California, and California residents may not be representative of then entire country. Secondly, any survey that involves self-reporting (especially of financial information) is suspect because people tend to report favorable rather than accurate data.

5. σ unknown, normal population, n=23: use t with df =22

$\alpha = 0.05$, $t_{df,\alpha/2} = t_{22,0.025} = 2.074$

IMPORTANT NOTE: This manual uses the following conventions.

(1) The designation "df" stands for "degrees of freedom."

(2) Since the t value depends on the degrees of freedom, a subscript may be used to clarify which t distribution is being used. For df =15 and $\alpha/2$ =0.025, for example, one may indicate $t_{15,\alpha/2} = 2.132$. As with the z distribution, it is also acceptable to use the actual numerical value within the subscript and indicate $t_{15,.025} = 2.132$.

(3) Always use the closest entry in Table A-3. When the desired df is exactly halfway between the two nearest tabled values, be conservative and choose the one with the lower df.

(4) As the degrees of freedom increase, the t distribution approaches the standard normal distribution – and the "large" row of the t table actually gives z values. Consequently the z score for certain "popular" α and $\alpha/2$ values may be found by reading Table A-3 "frontwards" instead of Table A-2 "backwards." This is not only easier but also more accurate – since Table A-3 includes one more decimal place. Note the following examples.

   For "large" df and $\alpha/2 = 0.05$, $t_{\alpha/2} = 1.645 = z_{\alpha/2}$ (as found in the z table).

   For "large" df and $\alpha/2 = 0.01$, $t_{\alpha/2} = 2.326 = z_{\alpha/2}$ (more accurate than the 2.33 in the z table). This manual uses this technique from this point on. [For df = "large" and $\alpha/2 = 0.005$, $t_{\alpha/2} = 2.576 \neq 2.575 = z_{\alpha/2}$ (as found in the z table). This is a discrepancy caused by using different mathematical approximation techniques to construct the tables, and not a true difference. While 2.576 is the more standard value, his manual will continue to use 2.575.]

7. σ unknown, population not normal, n=6: neither normal nor t applies

9. σ known, population not normal, n=200: use z
   $\alpha = 0.10$, $z_{\alpha/2} = z_{0.05} = 1.645$

11. σ unknown, population normal, n=12: use t with df = 11
   $\alpha = 0.01$, $t_{df,\alpha/2} = t_{11,0.005} = 3.106$

13. σ unknown, normal distribution: use t with df = 19
   $\alpha = 0.05$, $t_{df,\,\alpha/2} = t_{19,0.025} = 2.093$

   a. $E = t_{\alpha/2} \cdot s/\sqrt{n}$
      $= 2.093(569)/\sqrt{20}$
      $= 266$ dollars

   b. $\bar{x} \pm E$
         9004 ± 266
         $8738 < \mu < 9270$ (dollars)

15. From the SPSS display: $8.0518 < \mu < 8.0903$ (grams)
    There is 95% confidence that the interval from 8.0518 grams to 8.0903 grams contains the true mean weight of all U.S. dollar coins in circulations.

17. a. $\bar{x} = 3.2$ mg/dL
    b. σ unknown, n > 30: use t with df=46 [45]
       $\alpha = 0.05$, $t_{df,\,\alpha/2} = t_{46,0.025} = 2.014$
          $\bar{x} \pm t_{\alpha/2} \cdot s/\sqrt{n}$
       $3.2 \pm 2.014(18.6)/\sqrt{47}$
       3.2 ± 5.5
       $-2.3 < \mu < 8.7$ (mg/dl)
       Since the confidence interval includes 0, there is a reasonable possibility that the true value is zero – i.e., that the Garlicin treatment has no effect on LDL cholesterol levels.

19. a. $\bar{x} = 98.20 \ °F$

   b. σ unknown, n > 30: use t with df=105 [100]

   $\alpha = 0.01, t_{df,\alpha/2} = t_{105,0.005} = 2.626$

   $$\bar{x} \pm t_{\alpha/2} \cdot s/\sqrt{n}$$

   $98.20 \pm 2.626(0.62)/\sqrt{106}$

   98.20 ± 0.16

   $98.04 < \mu < 98.36 \ (°F)$

   c. No, the confidence interval does not contain the value 98.6 °F. This suggests that the common belief that 98.6 °F is the normal body temperature may not be correct.

21. a. σ unknown, n > 30: use t with df=336 [300]   b. σ unknown, n > 30: use t with df=369 [400]

   $\alpha = 0.05, t_{df,\alpha/2} = t_{336,0.025} = 1.968$ $\alpha = 0.05, t_{df,\alpha/2} = t_{369,0.025} = 1.966$

   $$\bar{x} \pm t_{\alpha/2} \cdot s/\sqrt{n}$$ $$\bar{x} \pm t_{\alpha/2} \cdot s/\sqrt{n}$$

   $6.0 \pm 1.968(2.3)/\sqrt{337}$ $6.1 \pm 1.966(2.4)/\sqrt{370}$

   6.0 ± 0.2 1.6 ± 0.2

   $5.8 < \mu < 6.2 \ (days)$ $5.9 < \mu < 6.3 \ (days)$

   c. The two confidence intervals are very similar and overlap considerably. There is no evidence that the echinacea treatment is effective.

23. a. σ unknown, n ≤ 30: if approximately   b. σ unknown, n ≤ 30: if approximately
      normal distribution, use t with df=19   normal distribution, use t with df=19

   $\alpha = 0.05, t_{df,\alpha/2} = t_{19,0.025} = 2.093$ $\alpha = 0.05, t_{df,\alpha/2} = t_{99,0.025} = 2.093$

   $$\bar{x} \pm t_{\alpha/2} \cdot s/\sqrt{n}$$ $$\bar{x} \pm t_{\alpha/2} \cdot s/\sqrt{n}$$

   $5.0 \pm 2.093(2.4)/\sqrt{20}$ $4.7 \pm 2.093(2.9)/\sqrt{20}$

   5.0 ± 1.1 4.7 ± 1.4

   $3.9 < \mu < 6.1 \ (VAS units)$ $3.4 < \mu < 6.1 \ (VAS units)$

   c. The two confidence intervals are very similar and overlap considerably. There is no evidence that the magnet treatment is effective.

25. preliminary values: n = 6, $\Sigma x = 9.23$, $\Sigma x^2 = 32.5197$

   $\bar{x} = (\Sigma x)/n = (9.23)/6 = 1.538$

   $s^2 = [n(\Sigma x^2) - (\Sigma x)^2]/[n(n-1)]$

   $\quad = [6(32.5197) - (9.23)^2]/[6(5)] = 3.664$

   $s = 1.914$

   σ unknown (and assuming the distribution is approximately normal), use t with df=5

   $\alpha = 0.05, t_{df,\alpha/2} = t_{5,0.025} = 2.571$

   $$\bar{x} \pm t_{\alpha/2} \cdot s/\sqrt{n}$$

   $1.538 \pm 2.571(1.914)/\sqrt{6}$

   1.538 ± 2.009

   $-0.471 < \mu < 3.547$ [which should be adjusted, since negative values are not possible]

   $0 < \mu < 3.547 \ (micrograms/cubic meter)$

   Yes. The fact that 5 of the 6 sample values are below $\bar{x}$ raises a question about whether the data meet the requirement that the underlying distribution is normal.

27. preliminary values: $n = 10$, $\Sigma x = 204.0$, $\Sigma x^2 = 5494.72$

$\bar{x} = (\Sigma x)/n = (204.0)/10 = 20.40$

$s^2 = [n(\Sigma x^2) - (\Sigma x)^2]/[n(n-1)]$

$= [10(5494.72) - (204.0)^2]/[10(9)] = 148.124$

$s = 12.171$

a. σ unknown (and assuming the distribution is approximately normal), use t with df=9

$\alpha = 0.05$, $t_{df,\alpha/2} = t_{9,0.05} = 2.262$

$\bar{x} \pm t_{\alpha/2} \cdot s/\sqrt{n}$

$20.40 \pm 2.262(12.171)/\sqrt{10}$

20.40 ± 8.71

$11.7 < \mu < 29.1$ (million dollars)

b. No. Since the data are the top 10 salaries, they are not a random sample.

c. There is a sense in which the data are the population (i.e., the top ten salaries) and are not a sample of any population. Possible populations from which the data could be considered a sample (but not a representative sample appropriate for any statistical inference) would be the salaries of all TV personalities, the salaries of the top 10 salaries of TV personality for different years.

d. No. Since no population can be identified from which these data are a random sample, the confidence interval has no context and makes no sense.

29. preliminary values: $n = 12$, $\Sigma x = 52118$, $\Sigma x^2 = 228,072,688$

$\bar{x} = (\Sigma x)/n = (52118)/12 = 4343.17$

$s^2 = [n(\Sigma x^2) - (\Sigma x)^2]/[n(n-1)]$

$= [12(228072688) - (52118)^2]/[12(11)] = 155957.06$

$s = 394.91$

σ unknown (and assuming the distribution is approximately normal), use t with df=11

$\alpha = 0.05$, $t_{df,\alpha/2} = t_{12,0.025} = 2.201$

$\bar{x} \pm t_{\alpha/2} \cdot s/\sqrt{n}$

$4343.17 \pm 2.201(394.91)/\sqrt{12}$

4343.17 ± 250.91

$4092.2 < \mu < 4594.1$ (seconds)

31. a. preliminary values: $n = 25$, $\Sigma x = 31.4$, $\Sigma x^2 = 40.74$

$\bar{x} = (\Sigma x)/n = (31.4)/25 = 1.256$

$s^2 = [n(\Sigma x^2) - (\Sigma x)^2]/[n(n-1)]$

$= [25(40.74) - (31.4)^2]/[25(24)] = 32.54/600 = 0.0542$

$s = 0.2329$

σ unknown (and assuming the distribution is approximately normal), use t with df=24

$\alpha = 0.05$, $t_{df,\alpha/2} = t_{24,0.025} = 2.064$

$\bar{x} \pm t_{\alpha/2} \cdot s/\sqrt{n}$

$1.256 \pm 2.064(0.2329)/\sqrt{25}$

1.256 ± 0.096

$1.16 < \mu < 1.35$ (mg)

NOTE: The Minitab output for this exercise is given below.

```
Variable   N     Mean     StDev    SE Mean      95% CI
nicotine   25   1.25600   0.23288   0.04658   (1.15987, 1.35213)
```

b. preliminary values: n = 25, Σx = 22.9, $Σx^2$ = 22.45

$\overline{x}$ = (Σx)/n = (22.9)/25 = 0.916

$s^2 = [n(Σx^2) – (Σx)^2]/[n(n-1)]$

$= [25(22.45) – (22.9)^2]/[25(24)] = 36.84/600 = 0.0614$

s = 0.2478

σ unknown (and assuming the distribution is approximately normal), use t with df=24

α = 0.05, $t_{df,α/2}$ = $t_{24,0.025}$ = 2.064

$\overline{x} ± t_{α/2}·s/\sqrt{n}$

0.916 ± 2.064(0.2478)/$\sqrt{25}$

0.916 ± 0.102

0.81 < μ < 1.02   (mg)

NOTE: The Minitab output for this exercise is given below.

```
Variable    N      Mean     StDev    SE Mean       95% CI
nicotine   25   0.916000  0.247790  0.049558  (0.813717, 1.018283)
```

c. There is no overlap in the confidence intervals. Yes; since the CI for the filtered cigarettes is completely below the CI for the unfiltered cigarettes, the filters appear to be effective in reducing the amounts of nicotine.

33. preliminary values: n = 43, Σx = 2738, $Σx^2$ = 307,250

$\overline{x}$ = (Σx)/n = (2738)/43 = 63.674

$s^2 = [n(Σx^2) – (Σx)^2]/[n(n-1)]$

$= [43(307250) – (2738)^2]/[43(42)] = 3164.511$

s = 56.254

σ unknown and n>30, use t with df=42 [40]

α = 0.01, $t_{df,α/2}$ = $t_{42,0.005}$ = 2.704

$\overline{x} ± t_{α/2}·s/\sqrt{n}$

63.674 ± 2.704(56.254)/$\sqrt{43}$

63.674 ± 23.197

40.5 < μ < 86.9   (years)

Yes, the confidence interval changes considerably from the previous 52.3 < μ < 57.4.

Yes, apparently confidence interval limits can be very sensitive to outliers.

When apparent outliers are discovered in data sets they should be carefully examined to determine if an error has been made. If an error has been made that cannot be corrected, the value should be discarded. If the value appears to be valid, it may be informative to construct confidence intervals with and without the outlier.

35. <u>assuming a large population</u>

α = 0.05 & df=99 [100], $t_{df,α/2}$ = $t_{99,0.025}$ = 1.984

$E = t_{α/2}·s/\sqrt{n}$

= 1.984(0.0518)/$\sqrt{100}$

= 0.0103 g

$\overline{x} ± E$

0.8565 ± 0.0103

0.8462 < μ < 0.8668 (grams)

<u>using the finite population N = 465</u>

α = 0.05 & df=99 [100], $t_{df,α/2}$ = $t_{99,0.025}$ = 1.984

$E = [t_{α/2}·s/\sqrt{n}] × \sqrt{(N-n)/(N-1)}$

= [1.984(0.0518)/$\sqrt{100}$] × $\sqrt{365/464}$

= 0.0091 g

$\overline{x} ± E$

0.8565 ± 0.0091

0.8474 < μ < 0.8656 (grams)

The second confidence interval is narrower, reflecting the fact that there are more restrictions and less variability (and more certainty) in the finite population situation when n>.05N.

7-5 Estimating a Population Variance

1. We can be 95% confident that the interval from 0.0455grams to 0.0602 grams includes the true value of the standard deviation in the weights for the population of all M&M's.

3. No; the population of last two digits from 00 to 99 follows a uniform distribution and not a normal distribution. One of the requirements for using the methods of this section is that the population values have a distribution that is approximately normal – even if the sample size is large.

5. $\alpha = 0.05$ and $df = 8$ $\qquad \chi_L^2 = \chi_{8,0.975}^2 = 2.180 \qquad \chi_R^2 = \chi_{8,0.025}^2 = 17.535$

7. $\alpha = 0.01$ and $df = 80$ $\qquad \chi_L^2 = \chi_{80,0.995}^2 = 51.172 \qquad \chi_R^2 = \chi_{80,0.005}^2 = 116.321$

9. $\alpha = 0.05$ and $df = 29$; $\chi_L^2 = \chi_{df,1-\alpha/2}^2$ and $\chi_R^2 = \chi_{df,\alpha/2}^2$

$$(n-1)s^2/\chi_R^2 < \sigma^2 < (n-1)s^2/\chi_L^2$$
$$(29)(333)^2/45.722 < \sigma^2 < (29)(333)^2/16.047$$
$$70333.3 < \sigma^2 < 200397.6$$
$$265 < \sigma < 448$$

11. $\alpha = 0.01$ and $df = 6$; $\chi_L^2 = \chi_{df,1-\alpha/2}^2$ and $\chi_R^2 = \chi_{df,\alpha/2}^2$

$$(n-1)s^2/\chi_R^2 < \sigma^2 < (n-1)s^2/\chi_L^2$$
$$(6)(2.019)^2/18.548 < \sigma^2 < (6)(2.019)^2/0.676$$
$$1.3186 < \sigma^2 < 36.1807$$
$$1.148 < \sigma < 6.015 \text{ (cells/microliter)}$$

13. From the upper right section of Table 7-2, $n = 19,205$.
    No. This sample size is too large to be practical for most applications.

15. From the lower left section of Table 7-2, $n = 101$.
    Yes. This sample size is practical for most applications.

17. $\alpha = 0.05$ and $df = 189$; $\chi_L^2 = \chi_{df,1-\alpha/2}^2$ and $\chi_R^2 = \chi_{df,\alpha/2}^2$

$$(n-1)s^2/\chi_R^2 < \sigma^2 < (n-1)s^2/\chi_L^2$$
$$(189)(645)^2/228.9638 < \sigma^2 < (189)(645)^2/152.8222$$
$$343411 < \sigma^2 < 514511$$
$$586 < \sigma < 717 \text{ (grams)}$$

No. Since the confidence interval includes 696, it is a reasonable possibility for σ.

19. a. $\alpha = 0.05$ and $df = 22$; $\chi_L^2 = \chi_{df,1-\alpha/2}^2$ and $\chi_R^2 = \chi_{df,\alpha/2}^2$

$$(n-1)s^2/\chi_R^2 < \sigma^2 < (n-1)s^2/\chi_L^2$$
$$(22)(22.9)^2/36.781 < \sigma^2 < (22)(22.9)^2/10.982$$
$$313.67 < \sigma^2 < 1050.54$$
$$17.7 < \sigma < 32.4 \text{ (minutes)}$$

b. $\alpha = 0.05$ and df $= 11$; $\chi_L^2 = \chi_{df,1-\alpha/2}^2$ and $\chi_R^2 = \chi_{df,\alpha/2}^2$

$$(n-1)s^2/\chi_R^2 < \sigma^2 < (n-1)s^2/\chi_L^2$$
$$(11)(20.8)^2/21.920 < \sigma^2 < (11)(20.8)^2/3.816$$
$$217.11 < \sigma^2 < 1247.13$$
$$14.7 < \sigma < 35.3 \text{ (minutes)}$$

c. The two intervals are similar. No, there does not appear to be a difference in the variation of lengths of PG/PGF-13 movies and R movies.

21. preliminary values: $n = 12$, $\Sigma x = 52118$, $\Sigma x^2 = 228,072,688$

$\overline{x} = (\Sigma x)/n = (52118)/12 = 4343.2$
$s^2 = [n(\Sigma x^2) - (\Sigma x)^2]/[n(n-1)]$
$\quad = [12(228072688) - (52118)^2]/[12(11)] = 155,957.06$
$s = 394.91$

$\alpha = 0.01$ and df $= 11$; $\chi_L^2 = \chi_{df,1-\alpha/2}^2$ and $\chi_R^2 = \chi_{df,\alpha/2}^2$

$$(n-1)s^2/\chi_R^2 < \sigma^2 < (n-1)s^2/\chi_L^2$$
$$(11)(394.91)^2/26.757 < \sigma^2 < (11)(394.91)^2/2.603$$
$$64115.10 < \sigma^2 < 659057.88$$
$$253.2 < \sigma < 811.8 \text{ (seconds)}$$

23. preliminary values: $n = 6$, $\Sigma x = 9.23$, $\Sigma x^2 = 32.5197$

$\overline{x} = (\Sigma x)/n = (9.23)/6 = 1.538$
$s^2 = [n(\Sigma x^2) - (\Sigma x)^2]/[n(n-1)]$
$\quad = [6(32.5197) - (9.213)^2]/[6(5)] = 3.664$
$s = 1.914$

$\alpha = 0.05$ and df $= 5$; $\chi_L^2 = \chi_{df,1-\alpha/2}^2$ and $\chi_R^2 = \chi_{df,\alpha/2}^2$

$$(n-1)s^2/\chi_R^2 < \sigma^2 < (n-1)s^2/\chi_L^2$$
$$(5)(3.664)/12.833 < \sigma^2 < (5)(3.664)/0.831$$
$$1.4276 < \sigma^2 < 22.0468$$
$$1.195 < \sigma < 4.695 \text{ (micrograms per cubic meter)}$$

Yes. One of the requirements to use the methods of this section is that the original distribution be approximately normal, and the fact that 5 of the 6 sample values are less than the mean suggests that the original distribution is not normal.

25. preliminary values: $n = 100$, $\Sigma x = 70311$, $\Sigma x^2 = 50,278,497$

$\overline{x} = (\Sigma x)/n = (70311)/100 = 703.11$
$s^2 = [n(\Sigma x^2) - (\Sigma x)^2]/[n(n-1)]$
$\quad = [100(50278497) - (70311)^2]/[100(99)] = 8506.36$
$s = 92.23$

$\alpha = 0.05$ and df $= 99$ [100]; $\chi_L^2 = \chi_{df,1-\alpha/2}^2$ and $\chi_R^2 = \chi_{df,\alpha/2}^2$

$$(n-1)s^2/\chi_R^2 < \sigma^2 < (n-1)s^2/\chi_L^2$$
$$(99)(8506.36)/129.561 < \sigma^2 < (99)(8506.36)/74.222$$
$$6499.87 < \sigma^2 < 11346.09$$
$$80.6 < \sigma < 106.5 \text{ (FICO units)}$$

NOTE: The statistical portion of Excel yielded the following results.

Confidence Level	Lower Conf. Limit	Stan. Dev.	Upper Conf. Limit
0.95	80.979	92.23	107.141

27. Applying the given formula yields the following χ_L^2 and χ_R^2 values.

$$\chi^2 = (1/2)[\pm z_{\alpha/2} + \sqrt{2(df) - 1}\,]^2$$
$$= (1/2)[\pm 1.96 + \sqrt{2(189) - 1}\,]^2$$
$$= (1/2)[\pm 1.96 + 19.416]^2$$
$$= (1/2)[17.456]^2 \quad \text{and} \quad (1/2)[21.376]^2$$
$$= 152.3645 \qquad \text{and} \quad 228.4771$$

These are close to the 152.8222 and 228.9638 given in exercise #17.

Statistical Literacy and Critical Thinking

1. A point estimate is a single value calculated from sample data that is used to estimate the true value of a population characteristic, called the parameter. In this context the sample proportion that test positive is the best point estimate for the population proportion that would test positive. A confidence interval is a range of values that is likely, with some specific degree of confidence, to include the true value of the population parameter. The major advantage of the confidence interval over the point estimate is its ability to communicate a sense of the accuracy of the estimate.

2. We can be 95% confident that the interval from 2.62% to 4.99% contains the true percentage of all job applicants who would test positive for drug use.

3. The confidence level in Exercise 2 is 95%. In general, the confidence level specifies the proportion of times a given procedure to construct an interval estimate can be expected to produce an interval that will include the true value of the parameter.

4. The respondents are not likely to be representative of the general population for two reasons. The sample is a convenience sample, composed only of those who visit the AOL Web site. The sample is a voluntary response sample, composed only of those who take the time to self-select themselves to be in the survey. Convenience samples are typically not representative racially, socio-economically, etc. Voluntary response samples typically include mainly those with strong opinions on, or a personal interest in, the topic of the survey.

Chapter Quick Quiz

1. We can be 95% confident that the interval from 20.0 to 20.0 contains the true value of the population mean.

2. The interval includes some values greater than 50%, suggesting that the Republican may win; but the interval also includes some values less than 50%, suggesting that the Republican may lose. Statement (2), that the election is too close to call, best describes the results of the survey.

3. The critical value of $t_{\alpha/2}$ for n=20 and $\alpha = 0.05$ is $t_{19,0.025} = 2.093$.

4. The critical value of $z_{\alpha/2}$ for n=20 and $\alpha = 0.10$ is $z_{0.05} = 1.645$.

5. $\alpha = 0.05$, $z_{\alpha/2} = z_{0.025} = 1.96$ and $E = 0.02$; $\hat{p}$ unknown, use $\hat{p} = 0.5$

$n = [(z_{\alpha/2})^2 \hat{p}\hat{q}]/E^2$

$\quad = [(1.96)^2(0.5)(0.5)]/(0.02)^2$

$\quad = 2401$

6. $\hat{p} = x/n = 240/600 = 0.40$

7. $\alpha = 0.05$, $z_{\alpha/2} = z_{0.025} = 1.96$ and $\hat{p} = x/n = 240/600 = 0.4000$

$\qquad \hat{p} \pm z_{\alpha/2}\sqrt{\hat{p}\hat{q}/n}$

$0.4000 \pm 1.96\sqrt{(0.4000)(0.6000)/600}$

0.4000 ± 0.0392

$0.361 < p < 0.439$

8. σ unknown, $n > 30$: use t with df=35 and $\alpha = 0.05$, $t_{df,\,\alpha/2} = t_{35,0.025} = 2.014$

$\qquad \overline{x} \pm t_{\alpha/2} \cdot s/\sqrt{n}$

$40.0 \pm 2.030(10.0)/\sqrt{36}$

40.0 ± 3.4

$36.6 < \mu < 43.4$ (years)

9. σ known, $n > 30$: use z and $\alpha = 0.05$, $z_{\alpha/2} = z_{0.025} = 1.96$

$\qquad \overline{x} \pm z_{\alpha/2} \cdot \sigma/\sqrt{n}$

$40.0 \pm 1.96(10.0)/\sqrt{36}$

40.0 ± 3.3

$36.7 < \mu < 43.3$ (years)

10. $\alpha = 0.05$, $z_{\alpha/2} = z_{0.025} = 1.96$

$n = [z_{\alpha/2} \cdot \sigma/E]^2$

$\quad = [(1.96)(12)/(0.5)]^2$

$\quad = 2212.76$, rounded up to 2213

Review Exercises

1. $\alpha = 0.05$ and $z_{\alpha/2} = z_{0.025} = 1.96$; $\hat{p} = x/n = 589/745 = 0.7906$

$\qquad \hat{p} \pm z_{\alpha/2}\sqrt{\hat{p}\hat{q}/n}$

$0.7906 \pm 1.96\sqrt{(0.7906)(0.2094)/745}$

0.7906 ± 0.0292

$0.761 < p < 0.820$ or $76.1\% < p < 82.0\%$

We can be 95% confident that the interval from 76.1% to 82.0% contains the true percentage of all adults who believe that it is morally wrong to not report all income on their tax returns.

2. $\alpha = 0.01$, $z_{\alpha/2} = z_{0.005} = 2.575$ and $E = 0.02$; $\hat{p}$ unknown, use $\hat{p} = 0.5$

$n = [(z_{\alpha/2})^2 \hat{p}\hat{q}]/E^2$

$\quad = [(2.575)^2(0.5)(0.5)]/(0.02)^2$

$\quad = 4144.14$, rounded up to 4145

3. $\alpha = 0.01$, $z_{\alpha/2} = z_{0.005} = 2.575$

$n = [z_{\alpha/2} \cdot \sigma/E]^2$

$= [(2.575)(28785)/(500)]^2$

$= 21975.91$, rounded up to 21,976

No, the required sample size is too large to be practical. It appears that some re-thinking of the requirements is necessary.

4. σ unknown, $n > 30$: use t with df=36 and $\alpha = 0.01$, $t_{df, \alpha/2} = t_{36,0.005} = 2.719$

$\overline{x} \pm t_{\alpha/2} \cdot s/\sqrt{n}$

$2.4991 \pm 2.719(0.0165)/\sqrt{37}$

2.4991 ± 0.0074

$2.4917 < \mu < 2.5065$ (grams)

Since the above interval includes 2.5 grams, it appears on the surface that the manufacturing process is meeting the design specifications – but the interval may not be relevant to determine whether the pennies were manufactured according to specification if the sample came (as so it seems) from worn pennies in circulation.

5. preliminary values: $n = 5$, $\Sigma x = 3344$, $\Sigma x^2 = 2,470,638$

$\overline{x} = (\Sigma x)/n = (3344)/5 = 668.8$

$s^2 = [n(\Sigma x^2) - (\Sigma x)^2]/[n(n-1)]$

$= [5(2470638) - (3344)^2]/[5(4)] = 58542.7$

$s = 241.96$

σ unknown and $n=5$: assuming a normal distribution, use t with df=4

$\alpha = 0.05$, $t_{df,\alpha/2} = t_{4,0.025} = 2.776$

$\overline{x} \pm t_{\alpha/2} \cdot s/\sqrt{n}$

$668.8 \pm 2.776(241.96)/\sqrt{5}$

668.8 ± 300.4

$368.4 < \mu < 969.2$ (hic)

6. preliminary values: $n = 5$, $\Sigma x = 3344$, $\Sigma x^2 = 2,470,638$

$\overline{x} = (\Sigma x)/n = (3344)/5 = 668.8$

$s^2 = [n(\Sigma x^2) - (\Sigma x)^2]/[n(n-1)]$

$= [5(2470638) - (3344)^2]/[5(4)] = 58542.7$

$s = 241.96$

$\alpha = 0.05$ and df = 4; $\chi^2_L = \chi^2_{df,1-\alpha/2}$ and $\chi^2_R = \chi^2_{df,\alpha/2}$

$(n-1)s^2/\chi^2_R < \sigma^2 < (n-1)s^2/\chi^2_L$

$(4)(58542.7)/11.143 < \sigma^2 < (4)(58542.7)/0.484$

$21015.06 < \sigma^2 < 483823.97$

$145.0 < \sigma < 695.6$ (hic)

7. Let x = the number who believe that cloning should not be allowed.

a. $\hat{p} = x/n = 901/1012 = 0.8903$, rounded to 0.890

b. $\alpha = 0.05$, $z_{\alpha/2} = z_{0.025} = 1.96$

$\hat{p} \pm z_{\alpha/2} \sqrt{\hat{p}\hat{q}/n}$

$0.8903 \pm 1.96\sqrt{(0.8903)(0.1097)/1012}$

0.8903 ± 0.0193

$0.871 < p < 0.910$

c. Yes. Since the entire interval is above 50% = 0.50, there is strong evidence that the majority is opposed to such cloning.

8. a. $\alpha = 0.05$, $z_{\alpha/2} = z_{0.025} = 1.96$ and E = 0.04; $\hat{p}$ unknown, use $\hat{p} = 0.5$

$$n = [(z_{\alpha/2})^2 \hat{p}\hat{q}]/E^2$$
$$= [(1.96)^2(0.5)(0.5)]/(0.04)^2$$
$$= 600.25, \text{ rounded up to } 601$$

b. $\alpha = 0.05$, $z_{\alpha/2} = z_{0.025} = 1.96$

$$n = [z_{\alpha/2} \cdot \sigma/E]^2$$
$$= [(1.96)(14227)/(750)]^2$$
$$= 1382.34, \text{ rounded up to } 1383$$

c. To meet both criteria simultaneously , use the larger sample size of n=1383.

9. preliminary values: n = 8, $\Sigma x = 30.72$, $\Sigma x^2 = 160.2186$

a. $\overline{x} = (\Sigma x)/n = (30.72)/8 = 3.840$ lbs

b. σ unknown and n=8: assuming a normal distribution, use t with df=7

$$s^2 = [n(\Sigma x^2) - (\Sigma x)^2]/[n(n-1)]$$
$$= [8(160.2186) - (30.72)^2]/[8(7)] = 6.03626$$
$$s = 2.4569$$

$\alpha = 0.05$, $t_{df,\alpha/2} = t_{7,0.025} = 2.365$

$$\overline{x} \pm t_{\alpha/2} \cdot s/\sqrt{n}$$

$3.8400 \pm 2.365(2.4569)/\sqrt{8}$

3.8400 ± 2.0543

$1.786 < \mu < 5.894$ (lbs)

c. σ known and a normal distribution: use z

$\alpha = 0.05$, $z_{\alpha/2} = z_{0.025} = 1.96$

$$\overline{x} \pm z_{\alpha/2} \cdot \sigma/\sqrt{n}$$

$3.8400 \pm 1.96(3.108)/\sqrt{8}$

3.8400 ± 2.1537

$1.686 < \mu < 5.994$ (lbs)

10. preliminary values: n = 8, $\Sigma x = 30.72$, $\Sigma x^2 = 160.2186$

$\overline{x} = (\Sigma x)/n = (30.72)/8 = 3.840$

$s^2 = [n(\Sigma x^2) - (\Sigma x)^2]/[n(n-1)] = [8(160.2186) - (30.72)^2]/[8(7)] = 6.03626$

$s = 2.4569$

a. $\alpha = 0.05$ and df = 7; $\chi_L^2 = \chi_{df,1-\alpha/2}^2$ and $\chi_R^2 = \chi_{df,\alpha/2}^2$

$$(n-1)s^2/\chi_R^2 < \sigma^2 < (n-1)s^2/\chi_L^2$$

$(7)(6.03626)/16.013 < \sigma^2 < (7)(6.03626)/1.690$

$2.6387 < \sigma^2 < 25.0022$

$1.624 < \sigma < 5.000$ (lbs)

b. $\alpha = 0.05$ and df = 7; $\chi_L^2 = \chi_{df,1-\alpha/2}^2$ and $\chi_R^2 = \chi_{df,\alpha/2}^2$

$$(n-1)s^2/\chi_R^2 < \sigma^2 < (n-1)s^2/\chi_L^2$$

$(7)(6.03626)/16.013 < \sigma^2 < (7)(6.03626)/1.690$

$2.639 < \sigma^2 < 25.002$ (lbs^2)

Cumulative Review Exercises

1. scores in order: 103   105   110   119   119   123   125   125   127   128
   preliminary values: $n = 10$, $\Sigma x = 1184$, $\Sigma x^2 = 140948$
   a. $\bar{x} = (\Sigma x)/n = (1184)/10 = 118.4$ lbs
   b. $\tilde{x} = (x_5 + x_6)/2 = (119 + 123)/2 = 121.0$ lbs
   c. $s^2 = [n(\Sigma x^2) - (\Sigma x)^2]/[n(n-1)] = [10(140948) - (1184)^2]/[10(9)] = 84.711$
      $s = 9.2$ lbs

2. Ratio, since differences are meaningful and there is a meaningful zero.

3. From Exercise 1, $\bar{x} = 118.4$ and $s = 9.204$.
   $\alpha = 0.05$, $t_{df,\alpha/2} = t_{9,0.025} = 2.262$
   $$\bar{x} \pm t_{\alpha/2} \cdot s/\sqrt{n}$$
   $118.4 \pm 2.262(9.204)/\sqrt{10}$
   118.4 ± 6.6
   $111.8 < \mu < 125.0$ (lbs)

4. $\alpha = 0.05$, $z_{\alpha/2} = z_{0.05} = 1.96$
   $n = [z_{\alpha/2} \cdot \sigma/E]^2 = [(1.96)(7.5)/(2)]^2 = 54.0225$, rounded up to 55

5. Let D = an applicant tests positive for drugs.
   $P(D) = 0.038$
   a. $P(\bar{D}) = 1 - P(D)$
      $= 1 - 0.038$
      $= 0.962$
   b. $P(D_1 \text{ and } D_2) = P(D_1) \cdot P(D_2|D_1)$
      $= (0.038)(0.038) = 0.00144$
   c. binomial: $n=500$ and $p=0.038$
      normal approximation appropriate since
      $np = 500(0.038) = 19 \geq 5$
      $nq = 500(0.962) = 481 \geq 5$
      $\mu = np = 500(0.038) = 19$
      $\sigma = \sqrt{npq} = \sqrt{500(0.038)(0.962)} = 4.275$
      $P(x \geq 20)$
      $= P(x > 19.5)$
      $= P(z > 0.12)$
      $= 1 - 0.5478$
      $= 0.4522$

6. a. normal distribution
      $\mu = 21.1$
      $\sigma = 4.8$
      $P(x > 20.0)$
      $= P(z > -0.23)$
      $= 1 - 0.4090$
      $= 0.5910$

NOTE: Since ACT scores are whole numbers, another valid interpretation of part (a) is $P(x>20) = P_C(x>20.5) = P(z>-0.13) = 1 - 0.4483 = 0.5517$. Presumably the "20.0" was specified in the exercise to discourage this interpretation and to allow for a direct comparison to part (b).

b. normal distribution,
    since the original distribution is so
$\mu_{\bar{x}} = \mu = 21.1$

$\sigma_{\bar{x}} = \sigma/\sqrt{n} = 4.8/\sqrt{25} = 0.96$

$P(\bar{x}>20.0)$
    $= P(z>-1.15)$
    $= 1 - 0.1251$
    $= 0.8749$

c. normal distribution: $\mu = 21.1$, $\sigma = 4.8$
    For P_{90}, $A = 0.9000$ [0.8997]
        and $z = 1.28$ [from z table]
        or $z = 1.282$ [from last row of t table]
    $x = \mu + z\sigma$
        $= 21.1 + (1.282)(4.8)$
        $= 21.1 + 6.2$
        $= 27.3$

7. A simple random sample of size n from some population occurs when every sample of size n has an equal chance of being selected from that population. A voluntary response sample occurs when the subjects themselves decide whether to be included.

8. As grade point averages are typically reported to two decimal places, $R = 4.00 - 0.00 = 4.00$. The range rule of thumb states that $\sigma \approx R/4 = 4.00/4 = 1.000$.

9. Let C = getting a T-F question correct by random guessing.
    $P(C_1 \text{ and } C_2 \text{ and } \ldots \text{and } C_{12}) = P(C_1) \cdot P(C_2) \cdot \ldots \cdot P(C_{12})$
                $= (0.5)(0.5)\ldots(0.5) = (0.5)^{12} = 0.000244$
    Since $0.000244 > 0$, it is possible to get all 12 questions correct by random guessing.
    Since $0.000244 < 0.05$, it is unlikely to get all 12 questions correct by random guessing.

10. This is a convenience sample, composed of those friends who happen to be available. Convenience samples are typically not representative of the population in a variety of ways – e.g., racially, socio-economically, by gender, etc.

Chapter 8

Hypothesis Testing

8-2 Basics of Hypothesis Testing

1. Given the large sample size and the fact that 20% is so much less than 50%, it is apparent that any confidence interval for the proportion of bosses that are good communicators would fall entirely below 50%. Assuming the magazine has properly interpreted the survey, the results appear to support the claim that "less than 50% of bosses are good communicators" but not necessarily that "less than 50% of the people believe that bosses are good communicators" – those are two different statements that should not be confused.

 Given that the responders constitute a voluntary response sample, and not a random sample, it is likely that they are not representative of the population and consist largely of people with strong feelings on and/or a personal interest in the topic. The results should not be used to support the stated claim.

3. No. Since the claim that the mean is equal to a specific value must be the null hypothesis, the only possible conclusions are to reject that claim or to fail to reject that claim. Hypothesis testing cannot be used to support a claim that a parameter is equal to a particular value.

5. If the claim were not true, and $p \leq 0.5$, then getting 90 heads in a sample of 100 tosses would be an unusual event. There is sufficient evidence to support the claim.

7. If the claim were not true, and $\mu \geq 75$, then getting a mean pulse rate of 74.4 in a sample of students would not be an unusual event. There is not sufficient evidence to support the claim.

9. original claim: $\mu > \$60,000$
 H_o: $\mu = \$60,000$
 H_1: $\mu > \$60,000$

11. original claim: $\sigma = 0.62\ °F$
 H_o: $\sigma = 0.62\ °F$
 H_1: $\sigma \neq 0.62\ °F$

13. original claim: $\sigma < 40$ seconds
 H_o: $\sigma = 40$ seconds
 H_1: $\sigma < 40$ seconds

15. original claim: $p = 0.80$
 H_o: $p = 0.80$
 H_1: $p \neq 0.80$

17. Two-tailed test; place $\alpha/2 = 0.005$ in each tail.
 Use $A = 1-\alpha/2 = 1-0.0050 = 0.9950$ and $z = 2.575$.
 critical values: $\pm z_{\alpha/2} = \pm z_{0.005} = \pm 2.575$

19. Right-tailed test; place $\alpha = 0.02$ in the upper tail.
 Use $A = 1-\alpha = 1-0.000 = 0.9800$ [closest entry $= 0.9798$] and $z = 2.05$.
 critical value: $+z_\alpha = +z_{0.02} = +2.05$

21. Two-tailed test; place $\alpha/2 = 0.025$ in each tail.
Use $A = 1-\alpha/2 = 1-0.0250 = 0.9750$ and $z = 1.96$.
critical values: $\pm z_{\alpha/2} = \pm z_{0.025} = \pm 1.96$

23. Left-tailed test; place $\alpha = 0.005$ in the lower tail.
Use $A = \alpha = 0.0050$ and $z = -2.575$.
critical value: $-z_\alpha = -z_{0.005} = -2.575$

25. $\hat{p} = x/n = 152/580 = 0.262$

$z\hat{p} = (\hat{p} - p)/\sqrt{pq/n}$

$\quad = (0.262 - 0.250)/\sqrt{(0.25)(0.75)/580}$

$\quad = 0.012/0.0180 = 0.67$

27. $\hat{p} = x/n = 314/1122 = 0.280$

$z\hat{p} = (\hat{p} - p)/\sqrt{pq/n}$

$\quad = (0.280 - 0.250)/\sqrt{(0.25)(0.75)/1122}$

$\quad = 0.030/0.0129 = 2.31$

29. P-value
$\quad = P(z < -1.25)$
$\quad = 0.1056$
Since $0.1056 > 0.05$, fail to reject H_o.

31. P-value
$\quad = 2 \cdot P(z > 1.75)$
$\quad = 2 \cdot [1 - 0.9599]$
$\quad = 2 \cdot [0.0401]$
$\quad = 0.0802$
Since $0.0802 > 0.05$, fail to reject H_o.

33. P-value
$\quad = 2 \cdot P(z < -2.75)$
$\quad = 2 \cdot (0030)$
$\quad = 0.0060$
Since $0.0060 < 0.05$, reject H_o.

35. P-value
     $= P(z > 2.30)$
     $= 1 - 0.9893$
     $= 0.0107$
     Since $0.0107 < 0.05$, reject H_o.

37. There is not sufficient evidence to conclude that the percentage of blue M&M's is greater than 5%.

39. There is not sufficient evidence to reject the claim that the percentage of Americans who know their credit score is equal to 20%.

41. original claim: $p = 0.41$
     H_o: $p = 0.41$
     type I error: rejecting H_o when H_o is actually true
         rejecting the claim that the percentage of non-smokers exposed to secondhand smoke is 41% when that percentage actually is 41%
     type II error: failing to reject H_o when H_1 is actually true
         failing to reject the claim that the percentage of non-smokers exposed to secondhand smoke is 41% what that percentage is actually different from 41%

43. original claim: $p > 0.70$
     H_o: $p = 0.70$
     type I error: rejecting H_o when H_o is actually true
         rejecting the claim that the percentage of college students who use alcohol is 70% when that percentage actually is 70%
     type II error: failing to reject H_o when H_1 is actually true
         failing to reject the claim that the percentage of college students who use alcohol is 70% when that percentage is actually greater than 70%

45. a. Not necessarily. Being 95% confident of a conclusion (i.e., rejecting the null hypothesis at the 0.05 level of significance) does not guarantee being 99% confident of that same conclusion (i.e., rejecting the null hypothesis at the 0.01 level of significance). Rejecting at the 0.01 level requires more evidence (i.e. a larger difference between the null hypothesis and the observed data) than rejecting at the 0.05 level. More simply: Rejection at the 0.05 level of significance means that the observed sample result was among the most extreme 5% of the possible results, but not necessarily among the most extreme 1% of the possible results as required for rejection at the 0.01 level of significance.
    b. Yes. Being 99% confident of a conclusion (i.e., rejecting the null hypothesis at the 0.01 level of significance) guarantees being 95% confident of that same conclusion (i.e., rejecting the null hypothesis at the 0.05 level of significance). Rejecting at the 0.05 level requires less evidence (i.e. a smaller difference between the null hypothesis and the observed data) than rejecting at the 0.01 level. More simply, rejection at the 0.01 level of significance means that the observed sample result was among the most extreme 1% of the possible results, and so it must be included among the most extreme 5% of the possible results as required for rejection at the 0.05 level of significance.

47. The test of hypothesis is given below and illustrated by the figure at the right.

H_o: $p = 0.50$

H_1: $p > 0.50$

$\alpha = 0.05$

C.V. $z = z_\alpha = 1.645$

The c corresponding to z = 1.645 is

found by solving $z_c = (c-p)/\sqrt{pq/n}$

for c as follows:

$c = p + z_c \cdot \sqrt{pq/n}$

$= 0.50 + (1.645) \cdot \sqrt{(0.50)(0.50)/64}$

$= 0.50 + (1.645) \cdot (0.0625)$

$= 0.6028$

a. The power calculations are given below
   and illustrated by the figure at the right.

power = P(rejecting H_o|H_o is false)

$= P(\hat{p} > .6028 | p = 0.65)$

$= P(z > -0.79)$

$= 1 - P(z < -0.79)$

$= 1 - 0.2148 = 0.7852$

The z corresponding to c= 0.6028 is
found as follows:

$z_c = (c-p)/\sqrt{pq/n}$

$= (0.6028-0.65)/\sqrt{(0.65)(0.35)/64}$

$= -0.0472/0.0596 = -0.79$

b. The shaded region in the bottom graph represents the probability of getting a sample
   proportion greater than 0.6028 (i.e., of rejecting the null hypothesis that the proportion is
   0.50) whenever the population proportion is actually 0.65. This is precisely the power of
   the test (i.e., the probability of correctly rejecting a false null hypothesis when a specified
   alternative is true).

8-3 Testing a Claim About a Proportion

NOTE: To reinforce the concept that all z scores are standardized rescalings obtained by
subtracting the mean and dividing by the standard deviation, the manual uses the "usual" z formula
written to apply to $\hat{p}$'s

$z\hat{p} = (\hat{p} - \mu\hat{p})/\sigma\hat{p}$.

When the normal approximation to the binomial applies, the $\hat{p}$'s are normally distributed with

$\mu\hat{p} = p$ and $\sigma\hat{p} = \sqrt{pq/n}$.

And so the formula for the z statistic may also be written as $z\hat{p} = (\hat{p} - p)/\sqrt{pq/n}$.

In addition, the manual continues to use the more accurate $z_{0.01} = 2.326$ taken from the "large"
row of the t table rather than the 2.33 obtained by reading the z table backwards.

1. There were $1261 + 491 + 384 = 2136$ total responses. The sample proportion of yes responses
   is $\hat{p} = x/n = 491/2136 = 0.230$. The symbol $\hat{p}$ is used to represent a sample proportion.

3. The value of the sample proportion is $\hat{p} = x/n = 123/280 = 0.439$. Since $0.979 > 0.05$, the observed results could easily occur by chance whenever the null hypothesis is true. There is not sufficient evidence to reject the hypothesis that the proportion of correct responses is $p = 0.50$ (i.e., the proportion expected by chance alone) in favor of the claim that $p > 0.50$.

5. a. $\hat{p} = x/n = 530/1000 = 0.530$

   $z\hat{p} = (\hat{p} - \mu\hat{p})/\sigma\hat{p}$

   $= (\hat{p} - p)/\sqrt{pq/n} = (0.530 - 0.50)/\sqrt{(0.50)(0.50)/1000} = 0.030/0.0158 = 1.90$

   b. $z = \pm z_{\alpha/2} = \pm z_{0.005} = \pm 2.575$

   c. P-value $= 2 \cdot P(z > 1.90) = 2 \cdot (1 - 0.9713) = 2 \cdot (0.0287) = 0.0574$

   d. Do not reject H_o; there is not sufficient evidence to reject the claim that the percentage of all college applications that are submitted online is 50%.

   e. No. A hypothesis test will either "reject" or "fail to reject" a claim that a population parameter is equal to a specified value.

7. original claim: $p = 0.75$

   $\hat{p} = x/n = 1640/2246 = 0.730$

   Ho: $p = 0.75$
   H$_1$: $p \neq 0.75$
   $\alpha = 0.05$
   C.V. $z = \pm z_{\alpha/2} = 1.96$
   calculations:

   $z\hat{p} = (\hat{p} - \mu\hat{p})/\sigma\hat{p} = -2.17$ [TI-83/84 Plus]
   P-value $= 0.0301$ [TI-83/84 Plus]

   conclusion:

   Reject H_o; there is sufficient evidence to reject the claim that $p = 0.75$ and conclude that $p \neq 0.75$ (in fact, that $p < 0.75$). There is sufficient evidence to reject the claim that the proportion of adults who use cell phones while driving is 75%.

NOTE: The "in fact, that $p < 0.75$" in the conclusion for Exercise 7 is given in parentheses. Some instructors prefer not to suggest one-sided conclusions from two sided tests. This manual follows the common sense approach consistent with the text's interpretation of two-sided confidence intervals that are entirely above or below a specified value.

9. original claim: $p = 0.75$

   $\hat{p} = x/n = 589/745 = 0.791$

   Ho: $p = 0.75$
   H$_1$: $p \neq 0.75$
   $\alpha = 0.01$
   C.V. $z = \pm z_{\alpha/2} = \pm z_{0.005} = \pm 2.575$
   calculations:

   $z\hat{p} = (\hat{p} - \mu\hat{p})/\sigma\hat{p}$

   $= (0.791 - 0.75)/\sqrt{(0.75)(0.25)/745}$

   $= 0.041/0.01586 = 2.56$

   P-value $= 2 \cdot P(z > 2.56) = 2 \cdot (1 - 0.9948) = 2 \cdot (0.0052) = 0.0104$

   conclusion:

   Do not reject H_o; there is not sufficient evidence to reject the claim that $p = 0.75$. There is not sufficient evidence to reject the claim that the proportion of adults who say it is morally wrong to not report all income is 75%.

11. original claim: $p > 1/3$

$\hat{p} = x/n = 327/839 = 0.390$

Ho: $p = 1/3$
H_1: $p > 1/3$
$\alpha = 0.01$
C.V. $z = z_\alpha = z_{0.01} = 2.326$
calculations:

$z\hat{p} = (\hat{p} - \mu\hat{p})/\sigma\hat{p}$

$= (0.390 - 0.333)/\sqrt{(1/3)(2/3)/839}$

$= 0.056/0.1627 = 3.47$

P-value $= P(z>3.47) = 1 - 0.9997 = 0.0003$

conclusion:

Reject H_o; there is sufficient evidence to conclude that $p > 1/3$. There is sufficient evidence to support the claim that the proportion of tennis challenges that are successful is greater than /3. It appears that the referees are erring on more than 1/3 of the challenged calls – which is not an enviable record, even if the proportion of challenged calls is quite small.

13. original claim: $p > 0.06$

$\hat{p} = x/n = 72/724 = 0.099$

Ho: $p = 0.06$
H_1: $p > 0.06$
$\alpha = 0.05$
C.V. $z = z_\alpha = z_{0.05} = 1.645$
calculations:

$z\hat{p} = (\hat{p} - \mu\hat{p})/\sigma\hat{p}$

$= (0.099 - 0.06)/\sqrt{(0.06)(0.94)/724}$

$= 0.039/0.008826 = 4.47$

P-value $= P(z>4.47) = 1 - 0.9999 = 0.0001$

conclusion:

Reject H_o; there is not sufficient evidence to conclude that $p > 0.06$. There is sufficient evidence to support the claim that the proportion of Tamiflu recipients who experience nausea is greater than the 6% rate experienced by those who took the placebo. Yes; nausea does appear to be a legitimate concern for recipients of Tamiflu.

15. original claim: $p \neq 0.000340$

$\hat{p} = x/n = 135/420,095 = 0.000321$

Ho: $p = 0.000340$
H_1: $p \neq 0.000340$
$\alpha = 0.005$
C.V. $z = \pm z_{\alpha/2} = \pm z_{0.0025} = \pm 2.81$
calculations:

$z\hat{p} = (\hat{p} - \mu\hat{p})/\sigma\hat{p}$

$= (0.000321 - 0.000340)/\sqrt{\dfrac{(0.000340)(0.999660)}{420,095}}$

$= -0.000019/0.0000284 = -0.66$

P-value $= 2 \cdot P(z<-0.66) = 2 \cdot (0.2546) = 0.5092$

conclusion:

Do not reject H$_o$; there is not sufficient evidence to conclude that p $\neq$ 0.000340. There is not sufficient evidence to support the claim that the proportion cell phone users who develop such cancers is different from the rate for people who do not use cell phones. No; based on this study, cell phone users have no extra reasons to be concerned about cancer of the brain or nervous system.

17. original claim: p < 0.20

$\hat{p}$ = x/n = 1299/(1299 + 5686) = 1299/6985 = 0.186

Ho: p = 0.20
H$_1$: p < 0.20
α = 0.01
C.V. z = -z$_\alpha$ = -z$_{0.01}$ = -2.326
calculations:

$z\hat{p}$ = ($\hat{p}$ – μ$\hat{p}$)/σ$\hat{p}$

= (0.186 – 0.20)/$\sqrt{(0.20)(0.80)/6985}$

= -0.014/0.004786 = -2.93

P-value = P(z<-2.93) = 0.0017

conclusion:

Reject H$_o$; there is sufficient evidence to conclude that p < 0.20. There is sufficient evidence to support the claim that less than 20% of Michigan gas pumps are inaccurate. No; from the perspective of the consumer, the rate does not appear to be low enough. While the point estimate of 0.186 indicates the rate is lower than 20%, it should probably be about 1/10 of that.

19. original claim: p < 0.80

$\hat{p}$ = x/n = 74/98 = 0.755

Ho: p = 0.80
H$_1$: p < 0.80
α = 0.05
C.V. z = -z$_\alpha$ = -z$_{0.05}$ = -1.645
calculations:

$z\hat{p}$ = ($\hat{p}$ – μ$\hat{p}$)/σ$\hat{p}$

= (0.755 – 0.80)/$\sqrt{(0.80)(0.20)/98}$

= -0.45/0.04041 = -1.11

P-value = P(z<-1.11) = 0.1335

conclusion:

Do not reject H$_o$; there is not sufficient evidence to conclude that p < 0.80. There is not sufficient evidence to support the claim polygraph tests are correct less than 80% of the time. Yes; based on these results, polygraph test results should probably be prohibited as evidence in trials. Even though the point estimate of 75.5% accuracy does not support the less than 80% claim, the accuracy rate is still far too small to make conclusions beyond a reasonable doubt.

21. NOTE: The value for x is not given. In truth, any 726 ≤ x ≤ 744 rounds to the given $\hat{p}$ = x/5000 = 15%. For want of more precise information, use $\hat{p}$ = 0.15.

original claim: p < 0.20

$\hat{p}$ = x/n = x/5000 = 0.15

H_o: p = 0.20
H_1: p < 0.20
$\alpha = 0.01$
C.V. $z = -z_\alpha = -z_{0.01} = -2.326$
calculations:

$z\hat{p} = (\hat{p} - \mu\hat{p})/\sigma\hat{p}$

$= (0.15 - 0.20)/\sqrt{(0.20)(0.80)/5000}$

$= -0.05/0.005657 = -8.84$

P-value = P(z<-8.84) = 0.0001

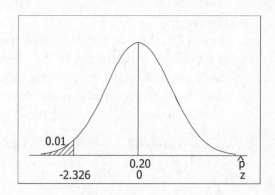

conclusion:

Reject H_o; there is sufficient evidence to conclude that p < 0.20. There is sufficient evidence to support the advertiser's claim that the proportion of households tuned to *60 Minutes* is less than 20%.

23. original claim: p = 0.50

use x = (.473)(150) = 70.95, rounded to 71

$\hat{p}$ = x/n = 71/150 = 0.473

H_o: p = 0.50
H_1: p ≠ 0.50
$\alpha = 0.05$ [assumed]
C.V. $z = \pm z_{\alpha/2} = \pm z_{0.025} = \pm 1.96$
calculations:

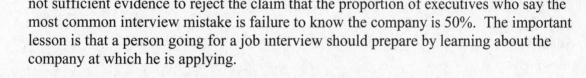

$z\hat{p} = (\hat{p} - \mu\hat{p})/\sigma\hat{p}$

$= (0.473 - 0.50)/\sqrt{(0.50)(0.50)/150}$

$= -0.27/0.04082 = -0.65$

P-value = 2·P(z<-0.65) = 2·(0.2578) = 0.5156

conclusion:

Do not reject H_o; there is not sufficient evidence to reject the claim that p = 0.50. There is not sufficient evidence to reject the claim that the proportion of executives who say the most common interview mistake is failure to know the company is 50%. The important lesson is that a person going for a job interview should prepare by learning about the company at which he is applying.

25. NOTE: The value for x is not given. In truth, any $2183 \leq x \leq 2213$ rounds to the given $\hat{p}$ = x/3011 = 73%. For want of more precise information, use $\hat{p} = 0.73$.

original claim: p = 0.75

$\hat{p}$ = x/n = x/3011 = 0.73

H_o: p = 0.75
H_1: p ≠ 0.75
$\alpha = 0.05$ [assumed]
C.V. $z = \pm z_{\alpha/2} = \pm z_{0.025} = \pm 1.96$
calculations:

$z\hat{p} = (\hat{p} - \mu\hat{p})/\sigma\hat{p}$

$= (0.73 - 0.75)/\sqrt{(0.75)(0.25)/3011}$

$= -0.02/0.007891 = -2.53$

P-value = 2·P(z<-2.53) = 2·(0.0057) = 0.0114

conclusion:

Reject H_o; there is sufficient evidence to reject the claim that $p = 0.75$ and conclude that $p \neq 0.75$ (in fact, that $p < 0.75$). There is sufficient evidence to reject the claim that the proportion of adults who use the Internet is ¾. While the difference between 0.73 and 0.75 may be of little practical significance, in the interest of accuracy the reporter should not write that ¾ of all adults use the Internet.

27. NOTE: The problem as stated is not possible. No whole number x yields x/59 = 43%, as 25/59 = 42.4% and 26/59 = 44.1%. Use the closest possible value, x = 25.

original claim: $p = 0.50$

$\hat{p} = x/n = 25/59 = 0.424$

Ho: $p = 0.50$
H_1: $p \neq 0.50$
$\alpha = 0.05$
C.V. $z = \pm z_{\alpha/2} = \pm z_{0.025} = \pm 1.96$
calculations:

$z\hat{p} = (\hat{p} - \mu\hat{p})/\sigma\hat{p}$

$= (0.424 - 0.50)/\sqrt{(0.50)(0.50)/59}$

$= -0.076/0.06509 = -1.17$

P-value $= 2 \cdot P(z < -1.17) = 2 \cdot (0.1210) = 0.2420$

conclusion:

Do not reject H_o; there is not sufficient evidence to reject the claim that $p = 0.50$. There is not sufficient evidence to reject the claim that women with 12 years of education or less have no ability to predict the gender of their babies. Conclude that the guesses of these women do not differ significantly from the results of random guesses.

29. NOTE: The value for x is not given. In truth, any $15{,}720 \leq x \leq 16{,}336$ rounds to the given $\hat{p} = x/61{,}647 = 26\%$. For want of more precise information, use $\hat{p} = 0.26$.

original claim: $p > 0.25$

$\hat{p} = x/n = x/61{,}647 = 0.26$

Ho: $p = 0.25$
H_1: $p > 0.25$
$\alpha = 0.05$
C.V. $z = z_{\alpha} = z_{0.05} = 1.645$
calculations:

$z\hat{p} = (\hat{p} - \mu\hat{p})/\sigma\hat{p}$

$= (0.26 - 0.25)/\sqrt{(0.25)(0.75)/61{,}647}$

$= 0.01/0.001744 = 5.73$

P-value $= P(z > 5.73) = 1 - 0.9999 = 0.0001$

conclusion:

Reject H_o; there is sufficient evidence to conclude that $p > 0.25$. There is sufficient evidence to support the claim that the proportion of employees who say that bosses scream at employees is greater than 25%. If the survey was done with a voluntary response sample, then the sample is not necessarily representative of the population and the above conclusion may not be valid.

31. NOTE: The value for x is not given. In truth, any $426 \leq x \leq 432$ rounds to the given
$\hat{p} = x/703 = 61\%$. For want of more precise information, use $\hat{p} = 0.61$.
original claim: p > 0.50

     $\hat{p} = x/n = x/703 = 0.61$

H$_0$: p = 0.50

H$_1$: p > 0.50

$\alpha = 0.05$

C.V. z = z$_\alpha$ = z$_{0.05}$ = 1.645

calculations:

     $z\hat{p} = (\hat{p} - \mu\hat{p})/\sigma\hat{p}$

        $= (0.61 - 0.50)/\sqrt{(0.50)(0.50)/703}$

        $= 0.11/0.01886 = 5.83$

     P-value = P(z>5.83) = 1 − 0.9999 = 0.0001

conclusion:

Reject H$_0$; there is sufficient evidence to conclude that p > 0.50. There is sufficient
evidence to support the claim that most (more than 50%) workers get their jobs through
networking. The results suggests that when it comes to getting a job, who you know may
be just as important as what you know.

33. There are 100 total M&M's, 13 of which are red.
original claim: p = 0.20

     $\hat{p} = x/n = 13/100 = 0.130$

H$_0$: p = 0.20

H$_1$: p ≠ 0.20

$\alpha = 0.05$ [assumed]

C.V. z = ±z$_{\alpha/2}$ = ±z$_{0.025}$ = ±1.96

calculations:

     $z\hat{p} = (\hat{p} - \mu\hat{p})/\sigma\hat{p}$

        $= (0.130 - 0.20)/\sqrt{(0.20)(0.80)/100}$

        $= -0.070/0.04000 = -1.75$

     P-value = 2·P(z<-1.75) = 2·(0.0401) = 0.0802

conclusion:

Do not reject H$_0$; there is not sufficient evidence to reject the claim that p = 0.20. There is
not sufficient evidence to reject the claim that the proportion of M&M's that are red is
20%.

35. There are 54 total bears, 35 of which are males.
original claim: p = 0.50

     $\hat{p} = x/n = 35/54 = 0.648$

H$_0$: p = 0.50

H$_1$: p ≠ 0.50

$\alpha = 0.05$

C.V. z = ±z$_{\alpha/2}$ = ±z$_{0.025}$ = ±1.96

calculations:

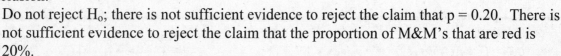

     $z\hat{p} = (\hat{p} - \mu\hat{p})/\sigma\hat{p}$

        $= (0.648 - 0.50)/\sqrt{(0.50)(0.50)/54}$

        $= 0.148/0.06804 = 2.18$

P-value = $2 \cdot P(z > 2.18) = 2 \cdot (1 - 0.9854) = 2 \cdot (0.0146) = 0.0292$

conclusion:

Reject H_o; there is sufficient evidence to reject the claim that $p = 0.50$ and conclude that $p \neq 0.50$ (in fact, that $p > 0.50$). There is sufficient evidence to reject the claim that the bears were selected from a population in which the percentage of males is equal to 50%.

37. There are 35 total movies, 12 of which are rated R.

original claim: $p = 0.55$

$\hat{p} = x/n = 12/35 = 0.343$

Ho: $p = 0.55$

H_1: $p \neq 0.55$

$\alpha = 0.01$

C.V. $z = \pm z_{\alpha/2} = \pm z_{0.005} = \pm 2.575$

```
MINITAB RESULT
Binomial with n = 35 and p = 0.55
  x  P( X <= x )
 12    0.0109367
```

calculations:

P-value = P(a result as extreme as or more extreme than the observed result| H_o is true)

$\quad = 2 \cdot P(12 \text{ or fewer occurrences}| p = 0.55)$

$\quad = 2 \cdot P(x \leq 12| p = 0.55)$

$\quad = 2 \cdot (0.0109367)$

$\quad = 0.0218$

conclusion:

Do not reject H_o; there is not sufficient evidence to reject the claim that $p = 0.55$. There is not sufficient evidence to reject the claim that the movies were selected from a population in which the percentage of R-rated movies is 55%.

39. original claim: $p = 0.10$

$\hat{p} = x/n = 0/50 = 0$

Yes, the methods of this section can be used. In general, the appropriateness of a test depends on the design of the experiment and not the particular results. For this problem, the normal distribution applies because

$np = 50(0.1) = 5 \geq 5$

$nq = 50(0.9) = 45 \geq 5$

H_o: $p = 0.10$

H_1: $p \neq 0.10$

$\alpha = 0.01$

C.V. $z = \pm z_{\alpha/2} = \pm z_{0.005} = \pm 2.575$

calculations:

$z\hat{p} = (\hat{p} - \mu\hat{p})/\sigma\hat{p}$

$\quad = (0 - 0.10)/\sqrt{(0.10)(0.90)/50}$

$\quad = -0.10/0.0424 = -2.36$

P-value = $2 \cdot P(z < -2.36) = 2 \cdot (0.0091) = 0.0182$

conclusion:

Do not reject H_o; there is not sufficient evidence to reject the claim that $p = 0.10$. There is not sufficient evidence to reject the claim that the proportion of M&M's that are blue is equal to 0.10.

8-4 Testing a Claim About a Mean: σ Known

1. In order to use the methods of this section to test the claim that $\mu = 1.5$,
   a. the sample must be a simple random sample.
   b. the population standard deviation σ must be known.
   c. the population distribution must be approximately normal (since $n \leq 30$).

3. A one-tailed test at the 0.01 level of significance rejects the null hypothesis if the sample statistic falls into the extreme 1% of the sampling distribution in the appropriate tail. The corresponding (two-sided) confidence interval test that places 1% each tail would be a 98% confidence interval.

5. original claim: $\mu = 5$ cm

   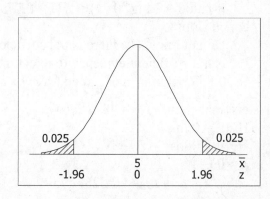

   H_o: $\mu = 5$ cm
   H_1: $\mu \neq 5$ cm
   $\alpha = 0.05$ [assumed]
   C.V. $z = \pm z_{\alpha/2} = \pm z_{0.025} = \pm 1.96$
   calculations:

   $$z_{\bar{x}} = (\bar{x} - \mu_{\bar{x}})/\sigma_{\bar{x}}$$
   $$= 1.34 \ [\text{TI-83/84}]$$
   P-value $= 0.1797$ [TI-83/84]

   conclusion:
   Do not reject H_o; there is not sufficient evidence to reject the claim that $\mu = 5$. There is not sufficient evidence to reject the claim that women have a mean wrist breadth equal to 5 cm.

7. original claim: $\mu > 210$ sec
   H_o: $\mu = 210$ sec
   H_1: $\mu > 210$ sec
   $\alpha = 0.05$
   C.V. $z = z_{\alpha} = z_{0.05} = 1.645$
   calculations:

   $$z_{\bar{x}} = (\bar{x} - \mu_{\bar{x}})/\sigma_{\bar{x}}$$
   $$= (252.5 - 210)/(54.5/\sqrt{40})$$
   $$= 42.5/8.6172 = 4.93$$
   P-value $= P(z>4.93) = 1 - 0.9999 = 0.0001$

   conclusion:
   Reject H_o; there is sufficient evidence to conclude that $\mu > 210$. There is sufficient evidence to support the claim that the sample is from a population of songs with a mean greater than 210 seconds. These results suggest that the advice given in the manual is not good advice.

9. original claim: $\mu = 0.8535$ grams

H_0: $\mu = 0.8535$ grams

H_1: $\mu \neq 0.8535$ grams

$\alpha = 0.05$

C.V. $z = \pm z_{\alpha/2} = \pm z_{0.025} = \pm 1.96$

calculations:

$z_{\bar{x}} = (\bar{x} - \mu_{\bar{x}})/\sigma_{\bar{x}}$

$\quad = (0.8635 - 0.8535)/(0.0565/\sqrt{19})$

$\quad = 0.0100/0.01296$

$\quad = 0.77$

P-value $= 2 \cdot P(z>0.77) = 2 \cdot (1 - 0.7794) = 2 \cdot (0.2206) = 0.4412$

conclusion:

Do not reject H_0; there is not sufficient evidence to reject the claim that $\mu = 0.8535$. There is not sufficient evidence to reject the claim that the mean weight of green M&M's is 0.8535 grams. Yes; green M&M's appear to have weights consistent with the label.

11. original claim: $\mu > 0$ lbs

H_0: $\mu = 0$ lbs

H_1: $\mu > 0$ lbs

$\alpha = 0.01$

C.V. $z = z_{\alpha} = z_{0.01} = 2.326$

calculations:

$z_{\bar{x}} = (\bar{x} - \mu_{\bar{x}})/\sigma_{\bar{x}}$

$\quad = (3.0 - 0)/(4.9/\sqrt{40})$

$\quad = 3.0/0.7748 = 3.87$

P-value $= P(z>3.87) = 1 - 0.9999 = 0.0001$

conclusion:

Reject H_0; there is sufficient evidence to conclude that $\mu > 0$. There is sufficient evidence to support the claim that the mean weight loss is greater than 0. The diet is effective in that the weight loss is statistically significant – but a mere 3.0 lbs weight loss after following the regimen for an entire year suggests the diet may have no practical significance.

13. original claim: $\mu \neq 91.4$ cm

H_0: $\mu = 91.4$ cm

H_1: $\mu \neq 91.4$ cm

$\alpha = 0.05$

C.V. $z = \pm z_{\alpha/2} = \pm z_{0.025} = \pm 1.96$

calculations:

$z_{\bar{x}} = (\bar{x} - \mu_{\bar{x}})/\sigma_{\bar{x}}$

$\quad = (92.8 - 91.4)/(3.6/\sqrt{36})$

$\quad = 1.4/0.6000$

$\quad = 2.33$

P-value $= 2 \cdot P(z>2.33) = 2 \cdot (1 - 0.9901) = 2 \cdot (0.0099) = 0.0198$

conclusion:

Reject H_0; there is sufficient evidence to conclude that $\mu \neq 91.4$ (in fact, that $\mu > 91.4$). There is not sufficient evidence to support the claim that males at her college have a sitting height different from 91.4. Yes; since she used a convenience sample (not a simple random sample), the requirements of this section are not met and the conclusion may not be valid.

15. original claim: $\mu < \$500{,}000$

H_o: $\mu = \$500{,}000$

H_1: $\mu < \$500{,}000$

$\alpha = 0.05$

C.V. $z = -z_\alpha = -z_{0.05} = -1.645$

calculations:

$z_{\bar{x}} = (\bar{x} - \mu_{\bar{x}})/\sigma_{\bar{x}}$

$= (415953 - 500000)/(463364/\sqrt{40})$

$= -84047/73264.2813 = -1.15$

P-value $= P(z < -1.15) = 0.1251$

conclusion:

Do not reject H_o; there is not sufficient evidence to conclude that $\mu < 500{,}000$. There is not sufficient evidence to support the claim that the mean salary of an NCAA football coach is less than $500,000.

17. original claim: $\mu \neq 235.8$ cm

H_o: $\mu = 235.8$ cm

H_1: $\mu \neq 235.8$ cm

$\alpha = 0.05$

C.V. $z = \pm z_{\alpha/2} = \pm z_{0.025} = \pm 1.96$

calculations:

$z_{\bar{x}} = (\bar{x} - \mu_{\bar{x}})/\sigma_{\bar{x}}$

$= (235.4 - 235.8)/(4.5/\sqrt{40})$

$= -0.4/0.7115 = -0.56$

P-value $= 2 \cdot P(z < -0.56) = 2 \cdot (0.2877) = 0.5754$

conclusion:

Do not reject H_o; there is not sufficient evidence to conclude that $\mu \neq 235.8$. There is not sufficient evidence to support the claim that the new baseballs have a bounce height different from the 235.8 cm of the old ones. No; the new baseballs are not significantly different.

19. summary statistics: $n = 12$ $\Sigma x = 9131$ $\Sigma x^2 = 6{,}985{,}297$ $\bar{x} = (\Sigma x)/n = 9131/12 = 760.9$

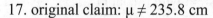

original claim: $\mu = 678$ FICO units

H_o: $\mu = 678$ FICO units

H_1: $\mu \neq 678$ FICO units

$\alpha = 0.05$

C.V. $z = \pm z_{\alpha/2} = \pm z_{0.025} = \pm 1.96$

calculations:

$z_{\bar{x}} = (\bar{x} - \mu_{\bar{x}})/\sigma_{\bar{x}}$

$= (760.9 - 678)/(58.3/\sqrt{12})$

$= 82.9/16.8298$

$= 4.93$

P-value $= 2 \cdot P(z > 4.93) = 2 \cdot (1 - 0.9999) = 2 \cdot (0.0001) = 0.0002$

conclusion:

Reject H_o; there is sufficient evidence to reject the claim that $\mu = 678$ and conclude that $\mu \neq 678$ (in fact, that $\mu > 678$). There is sufficient evidence to reject the claim that these FICO scores come from a population with a mean equal to 678.

21. The mean length of the n = 50 screws is $\bar{x}$ = 0.74682 inches.
    original claim: μ = 0.75 inches
    H_o: μ = 0.75 inches
    H_1: μ ≠ 0.75 inches
    α = 0.05
    C.V. z = ±$z_{α/2}$ = ±$z_{0.025}$ = ±1.96
    calculations:

$$z_{\bar{x}} = (\bar{x} - \mu_{\bar{x}})/\sigma_{\bar{x}}$$

$$= (0.74682 - 0.75)/(0.012/\sqrt{50})$$

$$= -0.00318/0.001697 = -1.87$$

P-value = 2·P(z<-1.87) = 2·(0.0307) = 0.0614

conclusion:

   Do not reject H_o; there is not sufficient evidence to reject the claim that μ = 0.75. There is not sufficient evidence to reject the claim that the screws have a mean length of ¾ in. as indicated on the label. Yes; the lengths appear to be consistent with the label.

23. a. In general, the power of a test is the probability of correctly rejecting the hull hypothesis when some specified alternative is actually correct. In this context, a power of 0.2296 indicates that when the true population mean is actually 170 lbs, the test has a 22.96% probability of correctly rejecting the null hypothesis that the mean is 166.3 lbs in order to conclude that the mean is greater than 166.3 lbs.

   b. In general, β = P(type II error)
                    = P(not rejecting H_o|H_o is false)
                    = 1 – P(rejecting H_o|H_o is false)
      In this context, β = 1 – P(rejecting H_o:μ = 166.3| μ = 170))
                    = 1 – 0.2296
                    = 0.7704

8-5 Testing a Claim About a Mean: σ Not Known

1. Yes; since n ≤ 30, the sample should be from a population that is approximately normally distributed. We consider the normality requirement to be satisfied for such data if there are no outliers and the histogram of the sample data is approximately bell-shaped. More formally, a normal quantile plot could be used to determine whether the sample data are approximately normally distributed.

3. A t test is a hypothesis test that uses the Student t distribution, typically to perform a test about μ when the true value of σ is not know. The letter t is used because that was the notation chosen by William Gosset (1786-1937), the developer of the distribution. He wrote under the pseudonym Student, and t is most prominent letter in "student" (considering that "s" was already being used to denote the standard deviation).

5. Use t. When σ is unknown and the x's are approximately normally distributed, use t.

7. Neither the z nor the t applies. When σ is unknown and the x's are not normally distributed, sample sizes n ≤ 30 cannot be used with the techniques in this chapter.

NOTE: Exercises 9–12 may be worked as follows.

   table: find the correct df row in Table A-3, and see what values surround the given t.

   TI-83+: use tcdf(lower bound, upper bound, df) to get the probability between 2 values

For the remainder of this manual, exact P-values for the t distribution <u>USING THE UNROUNDED VALUES</u> of t will be given in TI-83+ calculator format as described above and used in exercises 9–12 . While the input and/or output formats may be different for other technologies [e.g., Minitab, Excel, Statdisk, etc], the final answers should be the same.

9. P-value = $P(t_{24} > 0.430)$
    table for area in one tail: [0.430 < 1.318] P- value > 0.10
    TI-83+: tcdf(0.430,99,24) = 0.3355

11. P-value = $2 \cdot P(t_8 < -1.905)$
    table for area in two tails: [-2.306 < -1.905 < -1.860] 0.05 < P- value < 0.10
    TI-83+: 2·tcdf(-99,-1.905,8) = 0.0932

13. original claim: $\mu > 210$ sec
    H_o: $\mu = 210$ sec
    H_1: $\mu > 210$ sec
    $\alpha = 0.05$ and df = 39
    C.V. $t = t_\alpha = t_{0.05} = 1.685$
    calculations:

       $t_{\bar{x}} = (\bar{x} - \mu)/s_{\bar{x}}$
          = 4.93 [Minitab]
       P-value = 0.000 [Minitab]
    conclusion:

       Reject H_o; there is sufficient evidence to conclude that $\mu > 210$. There is sufficient evidence to support that claim that the sample is from a population with a mean greater than 210 seconds. The results suggest that the advice in the manual is not good advice.

15. original claim: $\mu < 21.1$ mg
    H_o: $\mu = 21.1$ mg
    H_1: $\mu < 21.1$ mg
    $\alpha = 0.05$ and df = 24
    C.V. $t = -t_\alpha = -t_{0.05} = -1.711$
    calculations:

       $t_{\bar{x}} = (\bar{x} - \mu)/s_{\bar{x}}$
          $= (13.2 - 21.1)/(3.7/\sqrt{25})$
          = -7.9/0.7400
          = -10.676
    P-value = $P(t_{24} < -10.676)$ = tcdf(-99,-10.676,24) = 6.752E-11 = 0.00000000007
    conclusion:

       Reject H_o; there is sufficient evidence to conclude that $\mu < 21.1$. There is sufficient evidence to support the claim that filtered 100 mm cigarettes have a mean tar amount less than 21.1 mg. The results suggest that filters are effective in reducing the amount of tar.

17. original claim: $\mu = 2.5$ grams

H_o: $\mu = 2.5$ grams
H_1: $\mu \neq 2.5$ grams
$\alpha = 0.05$ and $df = 36$
C.V. $t = \pm t_{\alpha/2} = \pm t_{0.025} = \pm 2.028$
calculations:

$t_{\bar{x}} = (\bar{x} - \mu)/s_{\bar{x}}$

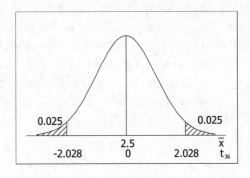

$\quad = (2.49910 - 2.5)/(0.01648/\sqrt{37})$

$\quad = -0.00090/0.002709$

$\quad = -0.332$

P-value $= 2 \cdot P(t_{36} < -1.756) = 2 \cdot \text{tcdf}(-99, -0.332, 36) = 0.7417$

conclusion:

Do not reject H_o; there is not sufficient evidence to reject the claim that $\mu = 2.5$. There is not sufficient evidence to reject the claim that the sample is from a population with mean weight 2.5 grams. Yes; the pennies appear to conform to the specifications.

19. original claim: $\mu > 4.5$ years

H_o: $\mu = 4.5$ years
H_1: $\mu > 4.5$ years
$\alpha = 0.05$ and $df = 80$
C.V. $t = t_{\alpha} = t_{0.05} = 1.664$
calculations:

$t_{\bar{x}} = (\bar{x} - \mu)/s_{\bar{x}}$

$\quad = (4.8 - 4.5)/(2.2/\sqrt{81})$

$\quad = 0.3/0.2444$

$\quad = 1.227$

P-value $= P(t_{80} > 1.227) = \text{tcdf}(1.227, 99, 80) = 0.1117$

conclusion:

Do not reject H_o; there is not sufficient evidence to conclude that $\mu > 4.5$. There is not sufficient evidence to support the claim that the mean time for all college students to earn their bachelor's degrees is greater than 4.5 years.

21. original claim: $\mu = 49.5$ cents

H_o: $\mu = 49.5$ cents
H_1: $\mu < 49.5$ cents
$\alpha = 0.01$ and $df = 99$
C.V. $t = -t_{\alpha} = -t_{0.01} = -2.364$
calculations:

$t_{\bar{x}} = (\bar{x} - \mu)/s_{\bar{x}}$

$\quad = (23.8 - 49.5)/(32.0/\sqrt{100})$

$\quad = -25.7/3.2000$

$\quad = -8.031$

P-value $= 2 \cdot P(t_{99} < -8.031) = 2 \cdot \text{tcdf}(-99, -8.031, 99) = 2.06E-12 = 0.000000000002$

conclusion:

Reject H_o; there is sufficient evidence to conclude that $\mu < 49.5$. There is sufficient evidence to conclude that the cents portion of all checks has a mean that is less than 49.5 cents. The results suggest that the cents portions of checks are not uniformly distributed from 0 to 99 cents.

23. original claim: $\mu = 8.00$ tons

H_o: $\mu = 8.00$ tons

H_1: $\mu \neq 8.00$ tons

$\alpha = 0.05$ and df $= 31$

C.V. $t = \pm t_{\alpha/2} = \pm t_{0.025} = \pm 2.040$

calculations:

$\quad t_{\bar{x}} = (\bar{x} - \mu)/s_{\bar{x}}$

$\quad\quad = (7.78 - 8.00)/(1.08/\sqrt{32})$

$\quad\quad = -0.22/0.1909$

$\quad\quad = -1.152$

P-value $= 2 \cdot P(t_{17} < -1.152) = 2 \cdot tcdf(-99, -1.152, 31) = 0.2580$

conclusion:

Do not reject H_o; there is not sufficient evidence to reject the claim that $\mu = 8.00$. There is not sufficient evidence to reject the claim that all cars have a mean greenhouse gas emission of 8.00 tons.

25. preliminary values: n $= 6$, $\Sigma x = 4222$, $\Sigma x^2 = 3,342,798$, $\bar{x} = 703.67$, $s^2 = 74383.47$, s $= 272.73$

original claim: $\mu < 1000$ hic

H_o: $\mu = 1000$ hic

H_1: $\mu < 1000$ hic

$\alpha = 0.01$ and df $= 5$

C.V. $t = -t_{\alpha} = -t_{0.01} = -3.365$

calculations:

$\quad t_{\bar{x}} = (\bar{x} - \mu)/s_{\bar{x}}$

$\quad\quad = (703.67 - 1000)/(272.73/\sqrt{6})$

$\quad\quad = -296.33/111.3429$

$\quad\quad = -2.661$

P-value $= P(t_5 < -2.661) = tcdf(-99, -2.661, 5) = 0.0224$

conclusion:

Do not reject H_o; there is not sufficient evidence to conclude that $\mu < 1000$. There is not sufficient evidence to support the claim that the population mean is less than 1000 hic. No; since one of the sample values is 1210, there is proof that not all of the child booster seats meet the specified requirement.

27. preliminary: n $= 5$, $\Sigma x = 32061$, $\Sigma x^2 = 220,431,831$, $\bar{x} = 6412.2$, $s^2 = 3,712,581.7$, s $= 1926.80$

original claim: $\mu = \$5000$

H_o: $\mu = \$5000$

H_1: $\mu \neq \$5000$

$\alpha = 0.05$ and df $= 4$

C.V. $t = \pm t_{\alpha/2} = \pm t_{0.025} = \pm 2.776$

calculations:

$\quad t_{\bar{x}} = (\bar{x} - \mu)/s_{\bar{x}}$

$\quad\quad = (6412.2 - 5000)/(1926.80/\sqrt{5})$

$\quad\quad = 1412.2/861.6927 = 1.639$

P-value $= 2 \cdot P(t_4 > 1.639) = 2 \cdot tcdf(1.639, 99, 4) = 0.1766$

conclusion:

Do not reject H_o; there is not sufficient evidence to reject the claim that $\mu = 5000$. There is not sufficient evidence to reject the claim that the mean damage cost is $5000.

29. summary statistics from statistical software: $n = 50$, $\bar{x} = 0.74682$, $s = 0.012322$
    original claim: $\mu = 0.75$ inches
    H_0: $\mu = 0.75$ inches
    H_1: $\mu \neq 0.75$ inches
    $\alpha = 0.05$ and $df = 49$
    C.V. $t = \pm t_{\alpha/2} = \pm t_{0.025} = \pm 2.009$
    calculations:

$$t_{\bar{x}} = (\bar{x} - \mu)/s_{\bar{x}}$$
$$= (0.74682 - 0.75)/(0.012322/\sqrt{50})$$
$$= -0.00318/0.001743$$
$$= -1.825$$

P-value $= 2 \cdot P(t_{49} < -1.825) = 2 \cdot \text{tcdf}(-99, -1.825, 49) = 0.0741$

conclusion:
    Do not reject H_0; there is not sufficient evidence to reject the claim that $\mu = 0.75$. There is not sufficient evidence to reject the claim that the screws have a mean length of ¾ in. as indicated on the label. Yes; the lengths appear to be consistent with the label.

31. summary statistics from statistical software: $n = 106$, $\bar{x} = 98.200$, $s = 0.6229$
    original claim: $\mu = 98.6$ °F
    H_0: $\mu = 98.6$ °F
    H_1: $\mu \neq 98.6$ °F
    $\alpha = 0.05$ [assumed] and $df = 105$
    C.V. $t = \pm t_{\alpha/2} = \pm t_{0.025} = \pm 1.984$
    calculations:

$$t_{\bar{x}} = (\bar{x} - \mu)/s_{\bar{x}}$$
$$= (98.2000 - 98.6)/(0.6229/\sqrt{106})$$
$$= -0.4000/0.06050$$
$$= -6.611$$

P-value $= 2 \cdot P(t_{105} < -6.611) = 2 \cdot \text{tcdf}(-99, -6.611, 105) = 1.627\text{E-}9 = 0.000000002$

conclusion:
    Reject H_0; there is sufficient evidence to reject the claim that $\mu = 98.6$ and conclude that $\mu \neq 98.6$ (in fact, that $\mu < 98.6$). There is sufficient evidence to reject the claim that the mean body temperature of the population is 98.6 °F. Yes; it appears there is sufficient evidence to conclude that the common belief is wrong.

33. The two methods may be compared as follows.

method of this section	alternative method
H_0: $\mu = 100$	H_0: $\mu = 100$
H_1: $\mu \neq 100$	H_1: $\mu \neq 100$
$\alpha = 0.05$ and $df = 31$	$\alpha = 0.05$
C.V. $t = \pm t_{\alpha/2} = \pm 2.040$	C.V. $z = \pm z_{\alpha/2} = \pm 1.960$
calculations:	calculations:

method of this section:
$$t_{\bar{x}} = (\bar{x} - \mu)/s_{\bar{x}}$$
$$= (105.3 - 100)/(15.0/\sqrt{32})$$
$$= 5.3/2.6517 = 1.999$$
P-value $= 2 \cdot P(t_{31} > 1.999)$
$$= 2 \cdot \text{tcdf}(1.999, 99, 31)$$
$$= 0.0545$$

alternative method:
$$z_{\bar{x}} = (\bar{x} - \mu)/\sigma_{\bar{x}}$$
$$= (105.3 - 100)/(15.0/\sqrt{32})$$
$$= 5.3/2.6517 = 1.999$$
P-Value) $= 2 \cdot P(z > 1.999)$
$$= 2 \cdot \text{normalcdf}(1.999, 99)$$
$$= 0.0456$$

conclusion:
    Do not reject H_o; there is not enough evidence to reject the claim that $\mu = 100$.

conclusion:
    Reject H_o; there is enough evidence to reject the claim that $\mu = 100$ and conclude that $\mu \neq 100$ (in fact, that $\mu > 100$).

The two methods lead to different conclusions – because of the unwarranted artificial precision created in the alternative method by assuming that the s value could be used for σ.

35. $A = z_\alpha \cdot [(8 \cdot df + 3)/(8 \cdot df + 1)]$ $t = \sqrt{df \cdot (e^{A \cdot A/df} - 1)}$

     $= 1.645 \cdot [(8 \cdot 74 + 3)/(8 \cdot 74 + 1)]$ $= \sqrt{74 \cdot (e^{(1.6505) \cdot (1.6505)/74} - 1)}$

     $= 1.645 \cdot [595/593]$ $= \sqrt{74 \cdot (e^{.0368} - 1)}$

     $= 1.6505$ $= \sqrt{74 \cdot (.03750)}$

                                   $= 1.6659$

This agrees exactly with the given $t_{74,.05} = 1.666$.

37. NOTE: Throughout this exercise the manual uses the s = 26.32 given in Example 1. Going back to the original Data Set 1 in Appendix B to find a more precise value will give slightly different answers (beginning at the third decimal place) in parts (a) and (b).
The test of hypothesis is given below
and illustrated by the figure at the right.
H_o: $\mu = 166.3$ lbs
H_1: $\mu > 166.3$ lbs
α = 0.05 and df = 39
C.V. $t = t_\alpha = t_{0.05} = 1.685$
The c corresponding to t = 1.685 is found by
solving $t_c = (c-\mu)/(s/\sqrt{n})$ or c as follows.

     $c = \mu + t_c \cdot (s/\sqrt{n})$

      $= 166.3 + 1.685(26.32/\sqrt{40})$

      $= 166.3 + 7.01$

      $= 173.31$

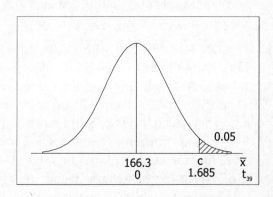

a. The power calculations are given below
and illustrated by the figure at the right.
The t corresponding to c= 173.31 is

     $t_c = (c-\mu)/(s/\sqrt{n})$

      $= (173.31 - 180)/(26.32/\sqrt{40})$

      $= -6.688/4.1616 = -1.607$

     power $= P(\text{rejecting } H_o | H_o \text{ is false})$

         $= P(\bar{x} > 173.31 | \mu = 180)$

         $= P(t_{39} > -1.607)$

         $= \text{tcdf}(-1.607, 99, 39) = 0.9419$

b. $\beta = P(\text{type II error}) = P(\text{ not rejecting } H_o | H_o \text{ is false})$

                 $= 1 - P(\text{rejecting } H_o | H_o \text{ is false})$

                 $= 1 - 0.9419$

                 $= 0.0481$

The test is very effective in recognizing that the mean is greater than 166.3 lbs when the mean is actually equal to 180 lbs, for it will do so 94.19% of the time and fail to do so only 4.81% of the time.

8-6 Testing a Claim About a Standard Deviation or Variance

1. The normality requirement for an hypothesis test about a standard deviation is stricter than the normality requirement for an hypothesis test about a mean. Deviations from normality that were tolerated when testing a claim about a mean may be serious enough to invalidate the results when testing a claim about a standard deviation.

3. No. Unlike tests and confidence intervals involving the mean, which do not require normality when n>30, test and confidence intervals involving standard deviations require approximate normality for all sample sizes.

5. a. test statistic: $\chi^2 = (n-1)s^2/\sigma^2$
$$= (24)(645)^2/(696)^2$$
$$= 20.612$$
   b. critical values for $\alpha = 0.05$ and df = 24:
$$\chi^2 = \chi^2_{1-\alpha/2} = \chi^2_{0.975} = 12.401$$
$$\chi^2 = \chi^2_{\alpha/2} = \chi^2_{0.025} = 39.364$$
   c. P-value limits: $15.659 < 20.612$
$$\text{P-value} > 0.20$$
   P-value exact: $2 \cdot \chi^2 \text{cdf}(0,20.612,24) = 0.6770$
   d. conclusion: Do not reject H_o; there is not sufficient to conclude that $\sigma \neq 696$

NOTES ON THE CHI-SQUARE DISTRIBUTION:
(1) Following the pattern used with the z and t distributions, this manual uses the closest entry from Table A-4 for the χ^2 as if it were the precise value necessary and does not use interpolation. This sacrifices little accuracy – and even interpolation does not yield precise values. When more accuracy is needed, refer either to more complete tables or to computer-produced values.
(2) The P-value portion of exercises 5–8 may be worked as follows.
   table: find the correct df row in Table A-4, and see what values surround the calculated χ^2
   TI-83/84+: use $\chi^2 \text{cdf}$(lower bound, upper bound, df) to get the probability between 2 values
For the remainder of this manual, exact P-values for the χ^2 distribution USING THE UNROUNDED VALUES of χ^2 will be given in TI-83/84+ calculator format as described above and used in exercises 5–8. While the input and/or output formats may be different for other technologies [e.g., Excel, Minitab, Statdisk, etc], the final answers should be the same. Since the χ^2 distribution tends to center around its degrees of freedom, calculated values less than df will be in the lower tail and calculated values greater than df will be in the upper tail – and those are the tails that should be considered when determining P-values.
(3) As with the t distribution, the df may be used as a subscript to identify which χ^2 distribution to use in the tables. In Exercise 5, for example, one could indicate
$$\chi^2_{24,0.975} = 12.401 \quad [\text{the } \chi^2 \text{ value with df} = 24 \text{ and } 0.975 \text{ in the upper tail}]$$
$$\chi^2_{24,0.025} = 39.364 \quad [\text{the } \chi^2 \text{ value with df} = 24 \text{ and } 0.025 \text{ in the upper tail}]$$

7. a. test statistic: $\chi^2 = (n-1)s^2/\sigma^2$
$$= (14)(4.8)^2/(3.5)^2$$
$$= 26.331$$

   b. critical value for $\alpha = 0.01$ and df = 14:
   $$\chi^2 = \chi_\alpha^2 = \chi_{0.01}^2 = 29.141$$

   c. P-value limits: $26.119 < 26.331 < 29.141$
   $$0.01 < \text{P-value} < 0.025$$
   P-value exact: $\chi^2 \text{cdf}(26.331,999,14) = 0.0235$

   d. conclusion: Do not reject H_o; there is not sufficient evidence to conclude that $\sigma > 3.5$.

NOTE: The formula $\chi^2 = (n-1)s^2/\sigma^2$ used in the calculations contains df = n-1. When the exact df needed in the problem does not appear in the table and the closest χ^2 value is used to determine the critical region, some instructors recommend using the same df in the calculations that were used to determine the C.V. This manual typically uses the closest entry to determine the C.V. and the n from the problem in the calculations – even though this introduces a slight discrepancy.

9. There are n=37 weights in the data set.
   original claim: $\sigma > 0.0230$ grams
   H_o: $\sigma = 0.0230$ grams
   H_1: $\sigma > 0.0230$ grams
   $\alpha = 0.05$ and df = 36
   C.V. $\chi^2 = \chi_\alpha^2 = \chi_{0.05}^2 = 55.758$

   calculations:
   $$\chi^2 = (n-1)s^2/\sigma^2$$
   $$= (36)(0.03910)^2/(0.0230)^2$$
   $$= 104.040$$
   P-value = $\chi^2 \text{cdf}(104.040,999,36) = 1.578\text{E-8} = 0.00000002$
   conclusion:

   Reject H_o; there is sufficient evidence to conclude that $\sigma > 0.0230$. There is sufficient evidence to support the claim that the weights of pre-1983 pennies have a standard deviation greater than 0.0230 grams. Based on these results, it appears that the weights pre-1983 pennies vary more than the weights of post-1983 pennies – which is partly, if not mostly, the result of being in circulation longer and having a wider variety in the amounts of wear.

GRAPHICS NOTE: While the χ^2 distribution is neither symmetric nor centered at zero, there are some important guidelines that help to keep χ^2 values in perspective. Loosely speaking, it "bunches up" around its df (which is actually its expected value) – and so the lower critical value will be less than df and the upper critical will be greater than df. Similarly, a calculated χ^2 value less than df falls in the lower tail and a calculated χ^2 value greater than df falls in the upper tail.

   To illustrate χ^2 tests of hypotheses, this manual uses a "generic" figure resembling a χ^2 distribution with df=4. Actually χ^2 distributions with df=1 and df=2 have no upper limit and approach the y axis asymptotically, while χ^2 distributions with df>30 are essentially symmetric and normal-looking. Because the χ^2 distribution is positively skewed, the expected value df is noted slightly to the right of the figure's peak.

11. original claim: $\sigma \neq 3.2$ mg

H_o: $\sigma = 3.2$ mg

H_1: $\sigma \neq 3.2$ mg

$\alpha = 0.05$ and df $= 24$

C.V. $\chi^2 = \chi^2_{1-\alpha/2} = \chi^2_{0.975} = 12.401$

$\chi^2 = \chi^2_{\alpha/2} = \chi^2_{0.025} = 39.364$

calculations:

$\chi^2 = (n-1)s^2/\sigma^2$

$= (24)(3.7)^2/(3.2)^2$

$= 32.086$

P-value $= 2 \cdot \chi^2 cdf(32.086,999,24) = 0.2498$

conclusion:

Do not reject H_o; there is not sufficient evidence to conclude that $\sigma \neq 3.2$. There is not sufficient evidence to support the claim that the tar content of such cigarettes has a standard deviation different from 3.2 mg.

13. There are n=9 heights.

original claim: $\sigma < 2.5$ inches

H_o: $\sigma = 2.5$ inches

H_1: $\sigma < 2.5$ inches

$\alpha = 0.05$ and df $= 8$

C.V. $\chi^2 = \chi^2_{1-\alpha} = \chi^2_{0.95} = 2.733$

calculations:

$\chi^2 = (n-1)s^2/\sigma^2$

$= (8)(1.5)^2/(2.5)^2$

$= 2.880$

P-value $= \chi^2 cdf(0,2.880,8) = 0.0583$

conclusion:

Do not reject H_o; there is not sufficient evidence to conclude that $\sigma < 2.5$. There is not sufficient evidence to support the claim that the heights of supermodels have a standard deviation that is less than 2.5 inches, the value for the general female population. The conclusion indicates that while supermodels may be more homogeneous than the general population, that cannot be concluded from this sample of size n=9.

15. original claim: $\sigma = 10$ beats/minute

H_o: $\sigma = 10$ beats/minute

H_1: $\sigma \neq 10$ beats/minute

$\alpha = 0.05$ and df $= 39$

C.V. $\chi^2 = \chi^2_{1-\alpha/2} = \chi^2_{0.975} = 24.433$

$\chi^2 = \chi^2_{\alpha/2} = \chi^2_{0.025} = 59.432$

calculations:

$\chi^2 = (n-1)s^2/\sigma^2$

$= (39)(12.5)^2/(10)^2$

$= 60.9375$

P-value $= 2 \cdot \chi^2 cdf(60.9375,999,39) = 0.0277$

conclusion:

Reject H_o; there is sufficient evidence to reject the claim that $\sigma = 10$ conclude that $\sigma \neq 10$ (in fact, that $\sigma > 10$). There is sufficient evidence to reject the claim that the pulse rates of women have a standard deviation equal to 10 beats/minute.

17. preliminary values: n = 10, Σx = 187.6, Σx² = 3532.04, x̄ = 18.76, s² = 1.4071, s = 1.1862

original claim: $\sigma = 1.34$

H_o: $\sigma = 1.34$

H_1: $\sigma \neq 1.34$

$\alpha = 0.01$ and df = 9

C.V. $\chi^2 = \chi^2_{1-\alpha/2} = \chi^2_{0.995} = 1.735$

$\chi^2 = \chi^2_{\alpha/2} = \chi^2_{0.005} = 23.589$

calculations:

$\chi^2 = (n-1)s^2/\sigma^2$

$= (9)(1.1862)^2/(1.34)^2$

$= 7.053$

P-value $= 2 \cdot \chi^2 cdf(0,7.053,9) = 0.7368$

conclusion:

Do not reject H_o; there is not sufficient evidence to reject the claim that $\sigma = 1.34$. There is not sufficient evidence to reject the claim that recent Miss America winners are from a population whose BMI values have a standard deviation of 1.34. No; recent winners do not appear to have MBI variation that is different from that of the 1920's and 1930's.

19. preliminary values: n = 12, Σx = -196, Σx² = 33452, x̄ = -16.3, s² = 2750.06, s = 52.441

original claim: $\sigma > 32.2$ ft

H_o: $\sigma = 32.2$ ft

H_1: $\sigma > 32.2$ ft

$\alpha = 0.05$ and df = 11

C.V. $\chi^2 = \chi^2_{\alpha} = \chi^2_{0.05} = 19.675$

calculations:

$\chi^2 = (n-1)s^2/\sigma^2$

$= (11)(52.441)^2/(32.2)^2$

$= 29.176$

P-value $= \chi^2 cdf(29.176,999,11) = 0.0021$

conclusion:

Reject H_o; there is sufficient evidence to conclude that $\sigma > 32.2$. There is sufficient evidence to conclude that the production method has errors with a standard deviation greater than 32.2 ft. A greater standard deviation means that the new altimeters will have more spread in the errors (i.e., generally larger errors) than in the past – which make the new method worse than the old method. Yes; the company should take action.

21. lower $\chi^2 = \frac{1}{2}(-z_{\alpha/2} + \sqrt{2 \cdot df - 1})^2$

$= \frac{1}{2}(-2.575 + \sqrt{2 \cdot (99) - 1})^2$

$= \frac{1}{2}(-2.575 + \sqrt{197})^2$

$= \frac{1}{2}(131.3469) = 65.673$

upper $\chi^2 = \frac{1}{2}(z_{\alpha/2} + \sqrt{2 \cdot df - 1})^2$

$= \frac{1}{2}(2.575 + \sqrt{2 \cdot (99) - 1})^2$

$= \frac{1}{2}(2.575 + \sqrt{197})^2$

$= \frac{1}{2}(275.9143) = 137.957$

Since the χ^2_{99} is shifted to the left (centered at df=99) compared to the χ^2_{100} (centered at df=100) used to give approximate values in Exercise 16, both values are slightly smaller as expected.

Statistical Literacy and Critical Thinking

1. For testing a claim that $\mu > 0$, that must be the alternative hypothesis – and the null hypothesis must be that $\mu = 0$. A P-value of 0.0091 indicates that the probability of obtaining these results when the null hypothesis is true is only 0.0091. Such a P-value calls for the rejection of the null hypothesis and support of the original claim that the mean rainfall amount is greater than 0 inches. The memory aid indicates that a small (e.g., less than 0.05) P-value calls for the rejection of the null hypothesis and that a large (i.e., greater than 0,05) P-value indicates there is not sufficient evidence to reject the null hypothesis.

2. Yes; since P-value = 0.0041 < 0.05, there is statistical significance. No; since 0.019 inches is such a trivial amount (and is it even possible to measure heights of males that accurately?), there is not practical significance.

3. A voluntary response sample is one in which the respondents themselves decide whether or not to participate. Since such samples tend to include mostly those with a special interest in the topic and/or strong feelings about the topic, they are not necessarily representative of the general population and should not be used to make inferences.

4. A procedure is robust against departures from normality if it works well (i.e., is "correct" $1-\alpha$ of the time) even when the sample data are from a population that does not follow a normal distribution. Yes; the t test of a population mean is robust against departures from normality. No; the $\chi 2$ test of a population standard deviation is not robust against departures from normality.

Chapter Quick Quiz

1. Since the claim that the proportion of males is greater than 0.5 does not contain the equality, it must be the alternative hypothesis. The null hypothesis is that the proportion of males is equal to 0.5. In symbolic form, H_o: p = 0.5 and H_1: p > 0.5.

2. The t distribution is appropriate for the indicated test. The others are not appropriate for the following reasons.
normal – used to perform a test about μ when σ is known..
chi-square – used to perform a test about σ.
binomial – used to perform an exact test about p.
uniform – not an appropriate sampling distribution for any sample statistic.

3. The chi-square distribution is appropriate for the indicated test. The others are not appropriate for the following reasons.
normal – used to perform a test about μ when σ is known..
t – used to perform a test about μ when σ is unknown.
binomial – used to perform an exact test about p.
uniform – not an appropriate sampling distribution for any sample statistic.

4. True; the null hypothesis is either rejected or not rejected, but it cannot be supported.

5. P-value = $2 \cdot P(z > 1.50) = 2 \cdot (1 - 0.9332) = 2 \cdot (0668) = 0.1336$

6. $\hat{p} = x/n = 30/100 = 0.30$

$z_{\hat{p}} = (\hat{p} - \mu_{\hat{p}})/\sigma_{\hat{p}} = (\hat{p} - p)/\sqrt{pq/n} = (0.30 - 0.40)/\sqrt{(0.40)(0.60)/100} = -0.10/0.04899 = -2.04$

7. The appropriate distribution is the t distribution with n-1 = 19 degrees of freedom. The critical values for the two-tailed test with α = 0.05 are $\pm t_{19,0.025} = \pm 2.093$.

8. P-value = $2 \cdot P(z>1.20) = 2 \cdot (1 - 0.8849) = 2 \cdot (1151) = 0.2302$

9. Do not reject H_o; there is not sufficient evidence to conclude that $p < 0.25$. There is not sufficient evidence to support the claim that $p < 0.25$.

10. False. Since hypothesis tests deal with a sample and not the entire population, there is always the chance of an error. When the null hypothesis is true, for example, we expect to incorrectly conclude that it is false α of the time.

Review Exercises

1. original claim: $p < 0.25$

$\hat{p} = x/n = 261/1088 = 0.240$

Ho: $p = 0.25$
H_1: $p < 0.25$
$\alpha = 0.05$
C.V. $z = -z_\alpha = -z_{0.05} = -1.645$
calculations:

$z\hat{p} = (\hat{p} - \mu\hat{p})/\sigma\hat{p}$

$= (0.240 - 0.25)/\sqrt{(0.25)(0.75)/1088}$

$= -0.01011/0.01313 = -0.77$

P-value $= P(z<-0.77) = 0.2206$

conclusion:

Do not reject H_o; there is not sufficient evidence to conclude that $p < 0.25$. There is not sufficient evidence to support the claim that less than ¼ of such adults smoke.

2. original claim: $p > 0.50$

$\hat{p} = x/n = 802/1486 = 0.540$

Ho: $p = 0.50$
H_1: $p > 0.50$
$\alpha = 0.01$
C.V. $z = z_\alpha = z_{0.01} = 2.326$
calculations:

$z\hat{p} = (\hat{p} - \mu\hat{p})/\sigma\hat{p}$

$= (0.540 - 0.50)/\sqrt{(0.50)(0.50)/1486}$

$= 0.03970/0.01297 = 3.06$

P-value $= P(z>3.06) = 1 - 0.9989 = 0.0011$

conclusion:

Reject H_o; there is sufficient evidence to conclude that $p > 0.50$. There is sufficient to support the claim that most college students seeking bachelor's degrees earn those degrees within five years.

NOTE: Although it was not specifically stated so, it must be assumed that the "1486 college students who are seeking bachelor's degrees" were followed for five years before the study was completed.

3. original claim: $\mu < 3700$ lbs
   H_o: $\mu = 3700$ lbs
   H_1: $\mu < 3700$ lbs
   $\alpha = 0.01$ and df $= 31$
   C.V. $t = -t_\alpha = -t_{0.01} = -2.453$
   calculations:

$$t_{\bar{x}} = (\bar{x} - \mu)/s_{\bar{x}}$$
$$= (3605.3 - 3700)/(501.7/\sqrt{32})$$
$$= -94.7/88.6889$$
$$= -1.068$$

   P-value $= P(t_9 < -1.068) = $ tcdf$(-99, -1.068, 31) = 0.1469$
conclusion:

   Do not reject H_o; there is not sufficient evidence to conclude that $\mu < 3700$. There is not sufficient evidence to support the claim that the mean weight of cars is less than 3700 lbs. While the mean weight of cars might be a factor in determining long run wear and tear, the most relevant factor for determining the required strength for highways is the weight of the heaviest vehicle that will be using the highway.

4. original claim: $\mu < 3700$ lbs
   H_o: $\mu = 3700$ lbs
   H_1: $\mu < 3700$ lbs
   $\alpha = 0.01$
   C.V. $z = -z_\alpha = -z_{0.01} = -2.326$
   calculations:

$$z_{\bar{x}} = (\bar{x} - \mu_{\bar{x}})/\sigma_{\bar{x}}$$
$$= (3605.3 - 3700)/(520/\sqrt{32})$$
$$= -94.7/91.9239$$
$$= -1.03$$

   P-value $= P(z < -1.03) = 0.1515$
conclusion:

   Do not reject H_o; there is not sufficient evidence to conclude that $\mu < 3700$. There is not sufficient evidence to support the claim that the mean weight of cars is less than 3700 lbs. While the mean weight of cars might be a factor in determining long run wear and tear, the most relevant factor for determining the required strength for highways is the weight of the heaviest vehicle that will be using the highway.

5. original claim: $p < 0.20$
   $\hat{p} = x/n = 5787/30617 = 0.1890$

   Ho: $p = 0.20$
   H_1: $p < 0.20$
   $\alpha = 0.01$
   C.V. $z = -z_\alpha = -z_{0.01} = -2.326$
   calculations:

$$z\hat{p} = (\hat{p} - \mu\hat{p})/\sigma\hat{p}$$
$$= (0.1890 - 0.20)/\sqrt{(0.20)(0.80)/30617}$$
$$= -0.01099/0.002286 = -4.81$$

   P-value $= P(z < -4.81) = 0.0001$

conclusion:

> Reject H_o; there is sufficient evidence to conclude that $p < 0.20$. There is sufficient evidence to support the claim that fewer than 20% of adults consumed herbs within the past 12 months.

6. original claim: $\mu < 281.8$ lbs

H_o: $\mu = 281.8$ lbs

H_1: $\mu < 281.8$ lbs

$\alpha = 0.01$ and df $= 174$

C.V. $t = -t_\alpha = -t_{0.01} = -2.345$

calculations:

$$t_{\bar{x}} = (\bar{x} - \mu)/s_{\bar{x}}$$
$$= (267.1 - 281.8)/(22.1/\sqrt{175})$$
$$= -14.7/1.6706$$
$$= -8.799$$

P-value $= P(t_9 < -8.799) = \text{tcdf}(-99,-8.799,174) \approx 0$

conclusion:

> Reject H_o; there is sufficient evidence to conclude that $\mu < 281.8$. There is sufficient evidence to support the claim that the thinner cans have a mean axial load that is less than 281.8 lbs. Yes; even though the thinner cans are not as strong as the thicker cans currently in use, they apparently can easily withstand the necessary pressure of 158 to 165 lbs.

7. original claim: $\mu = 74$

H_o: $\mu = 74$

H_1: $\mu \neq 74$

$\alpha = 0.05$

C.V. $z = \pm z_{\alpha/2} = \pm z_{0.025} = \pm 1.96$

calculations:

$$z_{\bar{x}} = (\bar{x} - \mu_{\bar{x}})/\sigma_{\bar{x}}$$
$$= (74.4 - 74)/(12.5/\sqrt{100})$$
$$= 0.4/1.25$$
$$= 0.32$$

P-value $= 2 \cdot P(z > 0.32) = 2 \cdot (1 - 0.6255) = 2 \cdot (0.3745) = 0.7490$

conclusion:

> Do not reject H_o; there is not sufficient evidence to reject the claim that $\mu = 74$. There is not sufficient evidence to reject the claim that the sample comes from a population with a mean equal to 74. Yes; based on these results, the calculator's random number generator appears to be working correctly.

8. original claim: $\mu = 74$

H_o: $\mu = 74$

H_1: $\mu \neq 74$

$\alpha = 0.05$ and df $= 99$

C.V. $t = \pm t_{\alpha/2} = \pm t_{0.025} = \pm 1.984$

calculations:

$$t_{\bar{x}} = (\bar{x} - \mu)/s_{\bar{x}}$$
$$= (74.4 - 74)/(11.7/\sqrt{100})$$
$$= 0.4/1.17$$
$$= 0.342$$

P-value $= 2 \cdot P(t_{99} > 0.342) = 2 \cdot \text{tcdf}(0.342,99,99) = 0.7332$

conclusion:

Do not reject H_o; there is not sufficient evidence to reject the claim that $\mu = 74$. There is not sufficient evidence to reject the claim that the sample comes from a population with a mean equal to 74. Yes; based on these results, the calculator's random number generator appears to be working correctly.

9. original claim: $\sigma = 12.5$

H_o: $\sigma = 12.5$

H_1: $\sigma \neq 12.5$

$\alpha = 0.05$ and df $= 995$

C.V. $\chi^2 = \chi^2_{1-\alpha/2} = \chi^2_{0.975} = 74.222$

$\chi^2 = \chi^2_{\alpha/2} = \chi^2_{0.025} = 129.561$

calculations:

$\chi^2 = (n-1)s^2/\sigma^2$

$= (99)(11.7)^2/(12.5)^2$

$= 86.734$

P-value $= 2 \cdot \chi^2 cdf(0, 86.734, 99) = 0.3883$

conclusion:

Do not reject H_o; there is not sufficient evidence to reject the claim that $\sigma = 12.5$. There is not sufficient evidence to reject the claim that the sample comes from a population with a standard deviation equal to 12.5.

10. original claim: $\sigma < 520$

H_o: $\sigma = 520$

H_1: $\sigma < 520$

$\alpha = 0.01$ and df $= 31$

C.V. $\chi^2 = \chi^2_{1-\alpha} = \chi^2_{0.99} = 14.954$

calculations:

$\chi^2 = (n-1)s^2/\sigma^2$

$= (31)(501.7)^2/(520)^2$

$= 28.856$

P-value $= \chi^2 cdf(0, 28.856, 31) = 0.4233$

conclusion:

Do not reject H_o; there is not sufficient evidence to conclude that $\sigma < 520$. There is not sufficient evidence to support the claim that the standard deviation of the weights of cars is less than 520 lbs.

Cumulative Review Exercises

1. scores in numerical order: 10.54 10.75 10.82 10.93 10.94 10.97 11.06 11.07 11.08

summary statistics: n $= 9$, $\Sigma x = 98.16$, $\Sigma x^2 = 1070.8508$

a. $\overline{x} = (\Sigma x)/n = (98.16)/9 = 10.907$ seconds

b. $\tilde{x} = 10.94$ seconds

c. $s^2 = [n(\Sigma x^2) - (\Sigma x)^2]/[n(n-1)] = [9(1070.8508) - (98.16)^2]/[9(8)] = 2.2716/72 = 0.03155$

$s = 0.178$ seconds

d. $s^2 = 0.03155$, rounded to 0.032 seconds^2

e. R $= 11.08 - 10.54 = 0.54$ seconds

2. a. Ratio, since differences are meaningful and there is a meaningful zero.
   b. Continuous, since time can be any value on a continuum.
   c. No; the times were not selected at random from some population, but they were determined by being the winning times for 9 consecutive Olympics.
   d. The sample statistics in Exercise 1 do not consider the chronological order of the data.
   e. A time series plot will reveal tendencies over time, while still giving a general idea (by looking from the vertical axis) of the central tendency and variation of the data.

3. σ unknown, use t with df=8
   $\alpha = 0.05$, $t_{df,\alpha/2} = t_{8,0.025} = 2.306$
   $$\bar{x} \pm t_{\alpha/2} \cdot s/\sqrt{n}$$
   $10.907 \pm 2.306(0.178)/\sqrt{9}$
   10.907 ± 0.136
   $10.770 < \mu < 11.043$ (seconds)
   No; this result cannot be used to estimate winning times in the future because there is a pattern of decreasing times and no fixed population mean.

4. original claim: $\mu < 11$ seconds
   H_o: $\mu = 11$ seconds
   H_1: $\mu < 11$ seconds
   $\alpha = 0.05$ and df = 8
   C.V. $t = -t_\alpha = -t_{0.05} = -1.860$
   calculations:

   $$t_{\bar{x}} = (\bar{x} - \mu)/s_{\bar{x}}$$
   $= (10.907 - 11)/(0.178/\sqrt{9})$
   $= -0.0933/0.05921$
   $= -1.576$
   P-value = $P(t_9 < -1.576)$ = tcdf(-99,-1.576,8) = 0.0768
   conclusion:
   Do not reject H_o; there is not sufficient evidence to conclude that $\mu < 11$. There is not sufficient evidence to support the claim that the mean winning time is less than 11 seconds. Since there is a decreasing pattern of winning times and no fixed population mean, this result does not apply to future times – but even though the mean for past winning times is not significantly less than 11 seconds, the decreasing pattern makes it likely that future winning times will be less than 11 seconds.

5. Some parts of this exercise are better answered from the frequency distribution representation at the right, where the x column gives the (approximate) midpoint value of each bar.
   a. Yes; the histogram is approximately bell-shaped.
   b. $\Sigma f = 100$
   c. Using the first two midpoints, $0.72 - 0.70 = 0.02$ grams
   d. Using the formula for summarized grouped data,
      $\bar{x} \approx (\Sigma f \cdot x)/(\Sigma f) = 85.70/100 = 0.857$ grams
   e. No. All we can tell, for example, is that there are 30 values in the class centered at 0.86 grams, but we cannot determine their individual weights.

x	f	f·x
0.70	1	0.70
0.72	0	0.00
0.74	1	0.74
0.76	2	1.52
0.78	5	3.90
0.80	8	6.40
0.82	8	6.56
0.84	13	10.92
0.86	30	25.80
0.88	15	13.20
0.90	3	2.70
0.92	4	3.68
0.94	6	5.64
0.96	1	0.96
0.98	2	1.96
1.00	0	0.00
1.02	1	1.02
	100	85.70

6. Using a vertical scale that does not start at zero exaggerates differences between the classes and gives a distorted impression of the data. The visual impression of the first class (x=1), for example, is that it contains about twice as many data points as the second class (x=2) – and in reality those values are 16 and 13.

x	f	f·x
1	16	16
2	13	26
3	22	66
4	21	84
5	13	65
6	15	90
	100	347

7. The requested frequency distribution is given by the first two columns of the table at the right. Using the formula for grouped data,

   $\overline{x} = (\Sigma f \cdot x)/(\Sigma f) = 347/100 = 3.47$

8. For one test of hypothesis at the $\alpha = 0.05$ level, the probability of making an error by rejecting a true null hypothesis is given by $P(E) = 0.05$. For two such independent tests, $P(E_1 \text{ and } E_2) = P(E_1) \cdot P(E_2|E_1) = (0.05)(0.05) = 0.0025$.

9. For each part, use the formula $z = (x - \mu)/\sigma$
   a. $P(x<700) = P(z<-1.18) = 0.1190$
   b. $P(x>750) = P(z>0.33) = 1 - 0.6293 = 0.3707 - 37.07\%$
   c. The $\overline{x}$'s are normally distributed with $\mu_{\overline{x}} = \mu = 739$ and $\sigma_{\overline{x}} = \sigma/\sqrt{n} = 33/\sqrt{50} = 4.6669$.

      $P(\overline{x}<730) = P(z<-1.93) = 0.0268$
   d. From Table A-2, the z score with 0.9000 [closest entry is 0.8997] below it is 1.28.

      $x = \mu + z\sigma = 739 + (1.28)(33) = 781.24$ mm

10. The sample heights appear to come from a normal distribution. The histogram shows a distribution that is approximately bell-shaped, and the normal quantile plot indicates that the data points approximate a straight line.

NOTE: Other representations of the data may give different impressions, and so it is always a good idea not to base a conclusion about normality on a single figure. The reasonable representation of the data given at the right, for example, suggests that the data may be bimodal – perhaps representing two different subpopulations. While frequency distributions and histograms are sensitive to how the classes are defined, the normal quantile plot is not subject to such arbitrary decisions.

height (mm)	frequency
725 – 749	2
750 – 774	6
775 – 799	3
800 – 824	7
825 – 846	2
	20

Chapter 9

Inferences from Two Samples

9-2 Inferences About Two Proportions

1. There are two requirements for using the methods of this section, and each of them is violated.
   (1) The samples should be two sample random samples that are independent. These samples are convenience samples, not simple random samples. These samples are likely not independent. Since she surveyed her friends, she may well have males and females that are dating each other (or least that associate with each other) – and people tend to associate with those that have similar behaviors.
   (2) The number of successes for each sample should be at least 5, and the number of failures for each sample should be at least 5. This is not true for the males, for which x=4.
   NOTE: This is the same requirement from previous chapters for using the normal distribution to approximate the binomial that required $np \geq 5$ and $nq \geq 5$.
   Using $\hat{p} = x/n$ to estimate p and $\hat{q} = 1 - x/n = (n-x)/n$ to estimate q,

$$n\hat{p} \geq 5 \qquad\qquad n\hat{q} \geq 5$$
$$n[x/n] \geq 5 \qquad\qquad n[(n-x)/n] \geq 5$$
$$x \geq 5 \qquad\qquad (n-x) \geq 5$$

   These inequalities state that the number of successes must be greater than 5, and the number of failures must be greater than 5.

3. In this context,

   $\hat{p}_1 = 15/1583 = 0.00948$

   $\hat{p}_2 = 8/157 = 0.05096$

   $\overline{p} = (15+8)/(1583+157) = 23/1740 = 0.01322$

   p_1 denotes the rue proportion of all Zocor users who experience headaches
   p_2 denotes the true proportion of all placebo users who experience headaches

5. $x = (0.158)(8834) = 1395.772 \approx 1396$
   NOTE: Any $1392 \leq x \leq 1400$ rounds to $x/8834 = 15.8\%$.

7. $\hat{p}_1 = x_1/n_1 = 13/36 = 0.361$ $\qquad\qquad$ $\hat{p}_1 - \hat{p}_2 = 0.361 - 0.519 = -0.157$

   $\hat{p}_2 = x_1/n_1 = 14/27 = 0.519$

   a. $\overline{p} = (x_1+x_2)/(n_1+n_2) = (13+14)/(36+27) = 27/63 = 0.429$

   b. $z_{\hat{p}_1 - \hat{p}_2} = (\hat{p}_1 - \hat{p}_2 - \mu_{\hat{p}_1 - \hat{p}_2})/\sigma_{\hat{p}_1 - \hat{p}_2}$

   $\qquad = (-0.517 - 0)/\sqrt{(0.429)(0.571)/36 + (0.429)(0.571)/27}$

   $\qquad = -0.517/0.1260$

   $\qquad = -1.25$

   c. for $\alpha = 0.05$, the critical values are $z = \pm z_{\alpha/2} = \pm z_{0.025} = \pm 1.96$

   d. P-value $= 2 \cdot P(z < -1.25) = 2 \cdot (0.1056) = 0.2112$

NOTE: While the intermediate steps for calculating $\sigma_{\hat{p}_1 - \hat{p}_2}$ and other values show decimals with a fixed number of decimal places, this manual always uses the exact values – i.e., the original fractions recalled from the calculator memory.

9. $\hat{p}_1 = x_1/n_1 = 13/36 = 0.361$ $\qquad\qquad$ $\hat{p}_1 - \hat{p}_2 = 0.361 - 0.519 = -0.157$

$\hat{p}_2 = x_1/n_1 = 14/27 = 0.519$

a. E = $z_{\alpha/2}\, \sigma_{\hat{p}_1 - \hat{p}_2}$

$\qquad$ = $1.96\sqrt{(0.361)(0.639)/36 + (0.519)(0.481)/27}$

$\qquad$ = 1.96(0.1251)

$\qquad$ = 0.2452

b. $(\hat{p}_1 - \hat{p}_2) \pm E$

$\qquad$ -0.1574 ± 0.2452

$\qquad$ $-0.4026 < p_1 - p_2 < 0.0878$

11. Let the Lipitor users group 1.

original claim: $p_1 - p_2 = 0$

H_o: $p_1 - p_2 = 0$

H_1: $p_1 - p_2 \neq 0$

$\alpha = 0.05$

C.V. $z = \pm z_{\alpha/2} = z_{0.025} = \pm 1.96$

calculations:

$\qquad$ $z_{\hat{p}_1 - \hat{p}_2} = (\hat{p}_1 - \hat{p}_2 - \mu_{\hat{p}_1 - \hat{p}_2})/\sigma_{\hat{p}_1 - \hat{p}_2}$

$\qquad\qquad$ = -0.73 [TI-83/84+]

$\qquad$ P-value = 0.4638 [TI-83/84+]

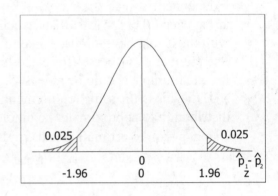

conclusion:

$\qquad$ Do not reject H_o; there is not sufficient evidence to reject the claim that $p_1 - p_2 = 0$. There is not sufficient evidence to reject the claim that the rate of infection was the same for those treated with Lipitor and those given a placebo.

13. Let the 1883 students be group 1.

$\qquad$ $\hat{p}_1 = x_1/n_1 = 171/560 = 0.305$ $\qquad\qquad$ $\hat{p}_1 - \hat{p}_2 = 0.305 - 0.365 = -0.060$

$\qquad$ $\hat{p}_2 = x_2/n_2 = 263/720 = 0.365$

$\qquad$ $\bar{p} = (171+263)/(560+720)$

$\qquad\qquad$ = 434/1280 = 0.339

original claim: $p_1 - p_2 < 0$

H_o: $p_1 - p_2 = 0$

H_1: $p_1 - p_2 < 0$

$\alpha = 0.05$

C.V. $z = -z_\alpha = -z_{0.05} = -1.645$

calculations:

$\qquad$ $z_{\hat{p}_1 - \hat{p}_2} = (\hat{p}_1 - \hat{p}_2 - \mu_{\hat{p}_1 - \hat{p}_2})/\sigma_{\hat{p}_1 - \hat{p}_2}$

$\qquad\qquad$ = $(-0.060 - 0)/\sqrt{(0.339)(0.661)/560 + (0.339)(0.661)/720}$

$\qquad\qquad$ = -0.060/0.02668

$\qquad\qquad$ = -2.25

$\qquad$ P-value = P(z<-2.25) = 0.0122

conclusion:

$\qquad$ Reject H_o; there is sufficient evidence to conclude that $p_1 - p_2 < 0$. There is sufficient evidence to support the claim that the proportion of college students using illegal drugs in 1993 was less than it is now.

15. Let those not wearing seat belts be group 1.

$\hat{p}_1 = x_1/n_1 = 31/2823 = 0.01098$ $\hat{p}_1 - \hat{p}_2 = 0.01098 - 0.00206 = 0.00892$

$\hat{p}_2 = x_2/n_2 = 16/7765 = 0.00206$ $\alpha = 0.10$

$(\hat{p}_1 - \hat{p}_2) \pm z_{\alpha/2}\, \sigma_{\hat{p}_1 - \hat{p}_2}$

$0.00892 \pm 1.645 \sqrt{(0.01098)(0.98902)/2823 + (0.00206)(0.99794)/7765}$

0.00892 ± 0.00334

$0.00558 < p_1 - p_2 < 0.01226$

Since the confidence interval does not include the value 0, it suggests that the two population proportions are not equal and that seat belts are effective because the proportion of non-users who are killed is greater than the proportion of users who are killed.

17. Let the women be group 1.

$\hat{p}_1 = x_1/n_1 = 347/386 = 0.899$ $\hat{p}_1 - \hat{p}_2 = 0.899 - 0.850 = 0.049$

$\hat{p}_2 = x_2/n_2 = 305/359 = 0.850$

$\bar{p} = (437+305)/(386+359)$

$\quad = 652/745 = 0.875$

original claim: $p_1 - p_2 \neq 0$

H_o: $p_1 - p_2 = 0$

H_1: $p_1 - p_2 \neq 0$

$\alpha = 0.05$

C.V. $z = \pm z_{\alpha/2} = \pm z_{0.025} = \pm 1.96$

calculations:

$z_{\hat{p}_1 - \hat{p}_2} = (\hat{p}_1 - \hat{p}_2 - \mu_{\hat{p}_1 - \hat{p}_2})/\sigma_{\hat{p}_1 - \hat{p}_2}$

$\quad = (0.049 - 0)/\sqrt{(0.875)(0.125)/386 + (0.875)(0.125)/359}$

$\quad = 0.049/0.02424$

$\quad = 2.04$

P-value $= 2 \cdot P(z > 2.04) = 2 \cdot (1 - 0.9793) = 2 \cdot (0207) = 0.0414$

conclusion:

Reject H_o; there is sufficient evidence to conclude that $p_1 - p_2 \neq 0$ (in fact, that $p_1 - p_2 > 0$). There is sufficient evidence to support the claim that the percentage of women who agree is different from the percentage of men who agree. Yes; there does appear to be a difference in the way that women and men feel about the issue.

19. Let the games with the roof closed be group 1.

$\hat{p}_1 = x_1/n_1 = 36/53 = 0.679$ $\hat{p}_1 - \hat{p}_2 = 0.679 - 0.577 = 0.102$

$\hat{p}_2 = x_2/n_2 = 15/26 = 0.577$

$\bar{p} = (36+15)/(53+26)$

$\quad = 51/79 = 0.646$

original claim: $p_1 - p_2 > 0$

H_o: $p_1 - p_2 = 0$

H_1: $p_1 - p_2 > 0$

$\alpha = 0.05$

C.V. $z = z_\alpha = z_{0.05} = 1.645$

calculations:

$$z_{\hat{p}_1-\hat{p}_2} = (\hat{p}_1-\hat{p}_2 - \mu_{\hat{p}_1-\hat{p}_2})/\sigma_{\hat{p}_1-\hat{p}_2}$$

$$= (0.102 - 0)/\sqrt{(0.646)(0.354)/53 + (0.646)(0.354)/26}$$

$$= 0.102/0.1145$$

$$\doteq 0.89$$

P-value = $P(z>0.89) = 1 - 0.8133 = 0.1867$

conclusion:

Do not reject H_o; there is not sufficient evidence to conclude that $p_1-p_2 > 0$. There is not sufficient evidence to support the claim that the proportion of wins is higher with the roof closed than with it open. No; the closed roof does not appear to be a significant advantage.

21. Let those treated with echinacea be group 1.

$\hat{p}_1 = x_1/n_1 = 40/45 = 0.889$ $\hat{p}_1 - \hat{p}_2 = 0.889 - 0.854 = 0.035$

$\hat{p}_2 = x_2/n_2 = 88/103 = 0.854$ $\alpha = 0.05$

$$(\hat{p}_1 - \hat{p}_2) \pm z_{\alpha/2}\sigma_{\hat{p}_1-\hat{p}_2}$$

$$0.0345 \pm 1.96\sqrt{(0.889)(0.111)/45 + (0.854)(0.146)/103}$$

$$0.0345 \pm 0.1143$$

$$-0.0798 < p_1-p_2 < 0.1489$$

Since the confidence interval includes the value 0, it suggests that there is no significant difference between the two population proportions. No; Echinacea does not appear to have any effect on the infection rate.

23. Let those on the Freedom of the Seas ship be group 1.

$\hat{p}_1 = x_1/n_1 = 338/3823 = 0.088$ $\hat{p}_1 - \hat{p}_2 = 0.088 - 0.167 = -0.079$

$\hat{p}_2 = x_2/n_2 = 276/1652 = 0.167$

$\overline{p} = (338+276)/(3823+1652)$

$\quad = 614/5475 = 0.112$

original claim: $p_1-p_2 < 0$

H_o: $p_1-p_2 = 0$

H_1: $p_1-p_2 < 0$

$\alpha = 0.01$

C.V. $z = -z_\alpha = -z_{0.01} = -2.326$

calculations:

$$z_{\hat{p}_1-\hat{p}_2} = (\hat{p}_1-\hat{p}_2 - \mu_{\hat{p}_1-\hat{p}_2})/\sigma_{\hat{p}_1-\hat{p}_2}$$

$$= (-0.079 - 0)/\sqrt{(0.112)(0.888)/3823 + (0.112)(0.888)/1652}$$

$$= -0.079/0.009291$$

$$= -8.47$$

P-value = $P(z<-8.47) = 0.0001$

conclusion:

Reject H_o; there is sufficient evidence to conclude that $p_1-p_2 < 0$. There is sufficient evidence to support the claim that the rate of sickness on the Freedom of the Seas ship is less than the rate on the Queen Elizabeth II. Yes; based on these results, it appears that infection rates can vary considerably.

25. Let the men be group 1.

$\hat{p}_1 = x_1/n_1 = 201/489 = 0.411$ $\hat{p}_1 - \hat{p}_2 = 0.411 - 0.360 = 0.051$

$\hat{p}_2 = x_2/n_2 = 126/350 = 0.360$ $\alpha = 0.01$

$(\hat{p}_1 - \hat{p}_2) \pm z_{\alpha/2}\, \sigma_{\hat{p}_1-\hat{p}_2}$

$0.0510 \pm 2.575 \sqrt{(0.411)(0.589)/489 + (0.360)(0.640)/350}$

0.0510 ± 0.0874

$-0.0364 < p_1-p_2 < 0.1385$

Since the confidence interval includes zero, it suggests there is no significant difference between the two population proportions. There does not appear to be a significant difference between rates at which challenges of men and women tennis players are upheld.

27. Let the females be group 1.

$\hat{p}_1 = x_1/n_1 = 1397/2739 = 0.510$ $\hat{p}_1 - \hat{p}_2 = 0.510 - 0.322 = 0.188$

$\hat{p}_2 = x_2/n_2 = 436/1352 = 0.322$

$\bar{p} = (1397+436)/(2739+1352)$

$\quad = 1833/4091 = 0.448$

original claim: $p_1-p_2 \neq 0$

H_o: $p_1-p_2 = 0$

H_1: $p_1-p_2 \neq 0$

$\alpha = 0.01$

C.V. $z = \pm z_{\alpha/2} = \pm z_{0.005} = \pm 2.575$

calculations:

$z_{\hat{p}_1-\hat{p}_2} = (\hat{p}_1-\hat{p}_2 - \mu_{\hat{p}_1-\hat{p}_2})/\sigma_{\hat{p}_1-\hat{p}_2}$

$\quad = (0.188 - 0)/\sqrt{(0.488)(0.512)/2739 + (0.488)(0.512)/1352}$

$\quad = 0.188/0.01653 = 11.35$

P-value = $2 \cdot P(z>11.35) = 2 \cdot (1 - 0.9999) = 2 \cdot (0.0001) = 0.0002$

conclusion:

Reject H_o; there is sufficient evidence to conclude that $p_1-p_2 \neq 0$ (in fact, that $p_1-p_2 > 0$). There is sufficient evidence to support the claim that female and male survivors have different rates of thyroid disease.

NOTE: The information given for exercises #29, #30, #33, #34, #35 and #36 is not sufficient to determine the exact values of x_1 and x_2. As a consequence, fractions should not be used and the usual level of accuracy given in the answers is not justified. Different methods for dealing with this lack of precision may result in answers slightly different from those given in the text or in this manual. The final conclusions of the tests of hypotheses or the inferences made from the confidence intervals are not affected by these minor differences.

29. Let the males be group 1.

$\hat{p}_1 = x_1/n_1 = x_1/731 = 0.69$ [as any $501 \leq x_1 \leq 508$ gives $\hat{p}_1 = 69\%$, use 0.69 in all calculations]

$\hat{p}_2 = x_2/n_2 = x_2/770 = 0.70$ [as any $536 \leq x_2 \leq 542$ gives $\hat{p}_2 = 70\%$, use 0.70 in all calculations]

$\hat{p}_1 - \hat{p}_2 = 0.69 - 0.70 = -0.01$ $\alpha = 0.10$

$(\hat{p}_1 - \hat{p}_2) \pm z_{\alpha/2}\, \sigma_{\hat{p}_1-\hat{p}_2}$

$-0.01 \pm 1.645 \sqrt{(0.69)(0.31)/731 + (0.70)(0.30)/770}$

-0.01 ± 0.0391

$-0.0491 < p_1-p_2 < 0.0291$

Since the confidence interval includes zero, there is no significant difference between the
proportion of "yes" answers from males and females

31. Let the 1976 returns be group 1.
NOTE: $x_1 = (.276)(250) = 69$ and $x_2 = (.073)(300) = 22$.
$\hat{p}_1 = x_1/n_1 = 69/250 = 0.276$ $\hat{p}_1 - \hat{p}_2 = 0.276 - 0.073 = 0.203$
$\hat{p}_2 = x_2/n_2 = 22/300 = 0.073$
$\bar{p} = (69+22)/(250+300)$
    $= 91/550 = 0.165$
original claim: $p_1-p_2 > 0$
H_o: $p_1-p_2 = 0$
H_1: $p_1-p_2 > 0$
$\alpha = 0.01$
C.V. $z = z_\alpha = z_{0.01} = 2.326$
calculations:

$$z_{\hat{p}_1-\hat{p}_2} = (\hat{p}_1-\hat{p}_2 - \mu_{\hat{p}_1-\hat{p}_2})/\sigma_{\hat{p}_1-\hat{p}_2}$$
$$= (0.203 - 0)/\sqrt{(0.165)(0.835)/250 + (0.165)(0.835)/300}$$
$$= 0.203/0.03182$$
$$= 6.37$$
P-value = $P(z>6.37) = 1 - 0.9999 = 0.0001$
conclusion:
Reject H_o; there is sufficient evidence to conclude that $p_1-p_2 > 0$. There is sufficient
evidence to support the claim that the percentage of returns designating funds for
campaigns was greater in 1976 than it is now.

33. Let the Viagra users be group 1.
$\hat{p}_1 = x_1/n_1 = x_1/734 = 0.16$ [as any $114 \le x_1 \le 121$ gives $\hat{p}_1 = 16\%$, use 0.16 in all calculations]
$\hat{p}_2 = x_2/n_2 = x_2/725 = 0.04$ [as any $26 \le x_2 \le 32$ gives $\hat{p}_2 = 4\%$, use 0.04 in all calculations]
$\hat{p}_1 - \hat{p}_2 = 0.16 - 0.04 = 0.12$
$\bar{p} = [(0.16)(734) + (0.04)(725)]/(734 + 725)$
    $= 146.44/1459 = 0.100$
original claim: $p_1-p_2 > 0$
H_o: $p_1-p_2 = 0$
H_1: $p_1-p_2 > 0$
$\alpha = 0.01$
C.V. $z = z_\alpha = z_{0.01} = 2.326$
calculations:
$$z_{\hat{p}_1-\hat{p}_2} = (\hat{p}_1-\hat{p}_2 - \mu_{\hat{p}_1-\hat{p}_2})/\sigma_{\hat{p}_1-\hat{p}_2}$$
$$= (0.12 - 0)/\sqrt{\frac{(0.100)(0.900)}{734} + \frac{(0.100)(0.900)}{725}}$$
$$= 0.12/0.01573 = 7.63$$
P-value = $P(z>7.63) = 1 - 0.9999 = 0.0001$
conclusion:
Reject H_o; there is sufficient evidence to conclude that $p_1-p_2 > 0$. There is sufficient
evidence to support the claim that the proportion of persons experiencing headaches is
greater for those treated with Viagra. Yes; headaches do appear to be a concern for those
who take Viagra.

35. Let the women be group 1 and use $n_1 = n_2 = 61647/2 = 30823.5$

$\hat{p}_1 = x_1/n_1 = 0.27$ [as many x_1 values give $\hat{p}_1 = 27\%$, use 0.27 in all calculations]

$\hat{p}_2 = x_2/n_2 = 0.25$ [as many x_2 values give $\hat{p}_2 = 25\%$, use 0.25 in all calculations]

$\hat{p}_1 - \hat{p}_2 = 0.27 - 0.25 = 0.02$ $\alpha = 0.05$

$(\hat{p}_1 - \hat{p}_2) \pm z_{\alpha/2}\, \sigma_{\hat{p}_1 - \hat{p}_2}$

$0.02 \pm 1.96 \sqrt{(0.27)(0.73)/30823.5 + (0.25)(0.75)/30823.5}$

0.02 ± 0.0069

$0.0131 < p_1 - p_2 < 0.0269$

Since the confidence interval does not include zero, there is a significant difference between the two proportions. Since the confidence interval includes only positive values, the proportion of women who say that female bosses are harshly critical is greater than the proportion of men who say so. Despite the large sample size, the fact that the respondents constitute a voluntary response sample means that the results should not be used to make inferences about the general population.

37. For all parts of this exercise

$\hat{p}_1 = x_1/n_1 = 112/200 = 0.560$ $\hat{p}_1 - \hat{p}_2 = 0.560 - 0.440 = 0.120$

$\hat{p}_2 = x_2/n_2 = 88/200 = 0.440$

$\bar{p} = (112+88)/(200+200)$

$= 200/400 = 0.500$

Use with $\alpha = 0.05$ for all parts.

a. $(\hat{p}_1 - \hat{p}_2) \pm z_{\alpha/2}\, \sigma_{\hat{p}_1 - \hat{p}_2}$

$0.1200 \pm 1.96 \sqrt{(0.560)(0.440)/200 + (0.440)(0.560)/200}$

0.1200 ± 0.0973

$0.0227 < p_1 - p_2 < 0.2173$

Since the interval does not include 0, conclude that p_1 and p_2 are different.

Since the interval lies entirely above 0, conclude that $p_1 - p_2 > 0$.

b. for group 1   for group 2

$\hat{p} \pm z_{\alpha/2} \sqrt{\hat{p}\hat{q}/n}$ $\hat{p} \pm z_{\alpha/2} \sqrt{\hat{p}\hat{q}/n}$

$0.560 \pm 1.96 \sqrt{(0.56)(0.44)/200}$ $0.440 \pm 1.96 \sqrt{(0.44)(0.56)/200}$

0.560 ± 0.069 0.440 ± 0.069

$0.491 < p < 0.629$ $0.371 < p < 0.509$

Since the intervals overlap, the implication is that p_1 and p_2 could have the same value.

c. original claim: $p_1 - p_2 = 0$

H_o: $p_1 - p_2 = 0$

H_1: $p_1 - p_2 \neq 0$

$\alpha = 0.05$

C.V. $z = \pm z_{\alpha/2} = \pm z_{0.025} = \pm 1.96$

calculations:

$z_{\hat{p}_1 - \hat{p}_2} = (\hat{p}_1 - \hat{p}_2 - \mu_{\hat{p}_1 - \hat{p}_2})/\sigma_{\hat{p}_1 - \hat{p}_2}$

$= (0.12 - 0)/\sqrt{\dfrac{(0.5625)(0.4375)}{40} + \dfrac{(0.5625)(0.4375)}{40}}$

$= 0.12/0.05$

$= 2.40$

P-value $= 2 \cdot P(z > 2.40) = 2 \cdot (1 - 0.9918) = 2 \cdot (0.0082) = 0.0164$

conclusion:

Reject H_o; there is sufficient evidence to reject the claim that p_1-p_2 = 0 and conclude that p_1-p_2 ≠ 0 (in fact, that p_1-p_2 > 0)

d. Based on the preceding results, conclude that p_1 and p_2 are unequal and that p_1 > p_2. The overlapping interval method of part (b) appears to be the least effective method for comparing two populations.

39. Let the females be group 1.

$\hat{p}_1 = x_1/n_1 = 1397/2739 = 0.510$ $\hat{p}_1 - \hat{p}_2 = 0.510 - 0.322 = 0.188$

$\hat{p}_2 = x_2/n_2 = 436/1352 = 0.322$

original claim: p_1-p_2 = 0.15

H_o: p_1-p_2 = 0.15

H_1: p_1-p_2 ≠ 0.15

α = 0.01

C.V. $z = \pm z_{\alpha/2} = \pm z_{0.005} = \pm 2.575$

calculations:

$$z_{\hat{p}_1-\hat{p}_2} = (\hat{p}_1-\hat{p}_2 - \mu_{\hat{p}_1-\hat{p}_2})/\sigma_{\hat{p}_1-\hat{p}_2}$$

$$= (0.188 - 0.15)/\sqrt{\frac{(0.510)(0.490)}{2739} + \frac{(0.322)(0.678)}{1352}}$$

$$= 0.038/0.01590$$

$$= 2.36$$

P-value = 2·P(z>2.36) = 2·(1 – 0.9909) = 2·(0.0091) = 0.0182

conclusion:

Do not reject H_o; there is not sufficient evidence to reject the claim that p_1-p_2 = 0.15. There is not sufficient evidence to reject the claim that the rate of thyroid disease among female survivors is 15 percentage points higher than that for male survivors.

9-3 Inferences About Two Means: Independent Samples

1. Reversing the designation of which sample is considered group1 and which sample is considered group 2 changes the sign of the point estimate and the signs of the endpoints of the interval estimate. The confidence interval using the new designation is $1.6 < \mu_1 - \mu_2 < 12.2$.

3. A one-tailed test of hypothesis at the 0.01 level of significance corresponds to a two-sided confidence interval at the 2(0.01) = 0.02 level of significance – i.e., to an interval with a confidence level of 98%.

5. Independent, since the men and women in the sample were selected randomly from their own populations with no connections between them.

7. Dependent, since cholesterol levels and determined by many factors that the Lipitor treatment cannot change. Treatments to lower cholesterol typically reduce everyone's levels by a certain amount, but persons who were high compared to the others before the treatment, for example, will likely still be high compared to the others after the treatment.

NOTE: To be consistent with the previous notation, reinforcing patterns and concepts presented in those sections, the manual uses the usual t formula written to apply to $\bar{x}_1$-$\bar{x}_2$'s

$$t_{\bar{x}_1-\bar{x}_2} = (\bar{x}_1-\bar{x}_2 - \mu_{\bar{x}_1-\bar{x}_2})/s_{\bar{x}_1-\bar{x}_2}, \text{ with } \mu_{\bar{x}_1-\bar{x}_2} = \mu_1 - \mu_2 \text{ and } s_{\bar{x}_1-\bar{x}_2} = \sqrt{s_1^2/n_1 + s_2^2/n_2}.$$

9. Let the children treated with low humidity be group 1.

$\bar{x}_1 - \bar{x}_2 = 0.98 - 1.09 = -0.11$

original claim: $\mu_1 - \mu_2 = 0$

H_o: $\mu_1 - \mu_2 = 0$

H_1: $\mu_1 - \mu_2 \neq 0$

$\alpha = 0.05$ and df = 45

C.V. $t = \pm t_{\alpha/2} = t_{0.025} = \pm 2.014$

calculations:

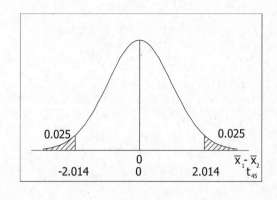

$$t_{\bar{x}_1 - \bar{x}_2} = (\bar{x}_1 - \bar{x}_2 - \mu_{\bar{x}_1 - \bar{x}_2}) / s_{\bar{x}_1 - \bar{x}_2}$$

$$= (-0.11 - 0) / \sqrt{(1.22)^2/46 + (1.11)^2/46}$$

$$= -0.11/0.2432 = -0.452$$

P-value = $2 \cdot$tcdf(-99,-0.452,45) = 0.6532

conclusion:

Do not reject H_o; there is not sufficient evidence to reject the claim that $\mu_1 - \mu_2 = 0$. There is not sufficient evidence to reject the claim that the two groups are from populations with the same mean. The results suggest that increasing the humidity does not have a significant effect on the treatment of croup.

11. Let the unfiltered cigarettes be group 1.

$\bar{x}_1 - \bar{x}_2 = 21.1 - 13.2 = 7.9$

$\alpha = 0.10$ and df = 24

$(\bar{x}_1 - \bar{x}_2) \pm t_{\alpha/2} s_{\bar{x}_1 - \bar{x}_2}$

$7.9 \pm 1.711 \sqrt{(3.2)^2/25 + (3.7)^2/25}$

7.9 ± 1.674

$6.2 < \mu_1 - \mu_2 < 9.6$ (mg)

Yes; since the confidence interval includes only positive values, the results suggest that the filtered cigarettes have less tar than the unfiltered ones.

13. Let the checks be group 1.

$\bar{x}_1 - \bar{x}_2 = 23.8 - 47.6 = -23.8$

original claim: $\mu_1 - \mu_2 < 0$ cents

H_o: $\mu_1 - \mu_2 = 0$ cents

H_1: $\mu_1 - \mu_2 < 0$ cents

$\alpha = 0.05$ and df = 99

C.V. $t = -t_{\alpha} = -t_{0.05} = -1.660$

calculations:

$$t_{\bar{x}_1 - \bar{x}_2} = (\bar{x}_1 - \bar{x}_2 - \mu_{\bar{x}_1 - \bar{x}_2}) / s_{\bar{x}_1 - \bar{x}_2}$$

$$= (-23.8 - 0) / \sqrt{(32.0)^2/100 + (33.5)^2/100}$$

$$= -23.8/4.6328$$

$$= -5.137$$

P-value = tcdf(-99,-5.137,99) = 7.000E-7 = 0.0000007

conclusion:

Reject H_o; there is sufficient evidence to conclude that $\mu_1 - \mu_2 < 0$. There is sufficient evidence to support the claim that the cents portions of the check amounts have a mean that is less than the cents portions of the credit card charges. One reason for the difference might be that check amounts include many more even dollar transactions (e.g., donations to church and charities) that have a cents portion of 0.

15. Let the supermodels be group 1, for which $n_1 = 9$.

$\overline{x}_1 - \overline{x}_2 = 70.0 - 63.2 = 6.8$

original claim: $\mu_1 - \mu_2 > 0$ inches

H_o: $\mu_1 - \mu_2 = 0$ inches

H_1: $\mu_1 - \mu_2 > 0$ inches

$\alpha = 0.01$ and df $= 8$

C.V. $t = t_\alpha = t_{0.01} = 2.896$

calculations:

$t_{\overline{x}_1-\overline{x}_2} = (\overline{x}_1 - \overline{x}_2 - \mu_{\overline{x}_1-\overline{x}_2})/s_{\overline{x}_1-\overline{x}_2}$

$\qquad = (6.8 - 0)/\sqrt{(1.5)^2/9 + (2.7)^2/40}$

$\qquad = 6.8/0.6575$

$\qquad = 10.343$

P-value $= \text{tcdf}(10.343,99,8) = 3.298\text{E-}6 = 0.000003$

conclusion:

Reject H_o; there is sufficient evidence to conclude that $\mu_1 - \mu_2 > 0$. There is sufficient evidence to support the claim that the mean height of supermodels is greater than the mean height of women who are not supermodels.

17. Let the four-cylinder cars be group 1.

$\overline{x}_1 - \overline{x}_2 = 137.5 - 136.3 = 1.2$

$\alpha = 0.10$ and df $= 11$

$(\overline{x}_1 - \overline{x}_2) \pm t_{\alpha/2} s_{\overline{x}_1-\overline{x}_2}$

$1.2 \pm 1.796 \sqrt{(5.8)^2/13 + (9.7)^2/12}$

1.2 ± 5.800

$-4.6 < \mu_1 - \mu_2 < 7.0$ (feet)

Since the confidence interval includes zero, there could be no difference between the two population means. No; there does not appear to be a significant difference between the mean braking distance of four-cylinder and six-cylinder cars.

19. Let the menthol cigarettes be group 1.

$\overline{x}_1 - \overline{x}_2 = 0.87 - 0.92 = -0.05$

original claim: $\mu_1 - \mu_2 \neq 0$ mg

H_o: $\mu_1 - \mu_2 = 0$ mg

H_1: $\mu_1 - \mu_2 \neq 0$ mg

$\alpha = 0.05$ and df $= 24$

C.V. $t = \pm t_{\alpha/2} = t_{0.025} = \pm 2.064$

calculations:

$t_{\overline{x}_1-\overline{x}_2} = (\overline{x}_1 - \overline{x}_2 - \mu_{\overline{x}_1-\overline{x}_2})/s_{\overline{x}_1-\overline{x}_2}$

$\qquad = (-0.05 - 0)/\sqrt{(0.24)^2/25 + (0.25)^2/25}$

$\qquad = -0.05/0.0693$

$\qquad = -0.721$

P-value $= 2 \cdot \text{tcdf}(-99,-0.721,24) = 0.4776$

conclusion:

Do not reject H_o; there is not sufficient evidence to conclude that $\mu_1 - \mu_2 \neq 0$. There is not sufficient evidence to support the claim that menthol cigarettes and non-menthol cigarettes have different amounts of nicotine. No; menthol does not appear to have an effect on the nicotine content.

21. Let the high-interest mortgages be group 1.

$\bar{x}_1 - \bar{x}_2 = 594.8 - 785.2 = -190.4$

original claim: $\mu_1 - \mu_2 < 0$ FICO units

H_o: $\mu_1 - \mu_2 = 0$ FICO units

H_1: $\mu_1 - \mu_2 < 0$ FICO units

$\alpha = 0.01$ and df = 39

C.V. $t = -t_\alpha = -t_{0.01} = -2.426$

calculations:

$$t_{\bar{x}_1 - \bar{x}_2} = (\bar{x}_1 - \bar{x}_2 - \mu_{\bar{x}_1 - \bar{x}_2})/s_{\bar{x}_1 - \bar{x}_2}$$

$$= (-190.4 - 0)/\sqrt{(12.2)^2/40 + (16.3)^2/40}$$

$$= -190.4/3.219$$

$$= -59.145$$

P-value = tcdf(-99,-59.145,39) ≈ 0

conclusion:

Reject H_o; there is sufficient evidence to conclude that $\mu_1 - \mu_2 < 0$. There is sufficient evidence to support the claim that the mean FICO score of borrowers with high-interest mortgages is lower than the mean FICO score of borrowers with low-interest mortgages. Yes; it appears that higher FICO scores are associated with lower interest rates – and, hence, with lower mortgage payments.

23. Let the unsuccessful applicants be group 1.

$\bar{x}_1 - \bar{x}_2 = 47.0 - 43.9 = 3.1$

original claim: $\mu_1 - \mu_2 > 0$ years

H_o: $\mu_1 - \mu_2 = 0$ years

H_1: $\mu_1 - \mu_2 > 0$ years

$\alpha = 0.05$ and df = 22

C.V. $t = t_\alpha = t_{0.05} = 1.717$

calculations:

$$t_{\bar{x}_1 - \bar{x}_2} = (\bar{x}_1 - \bar{x}_2 - \mu_{\bar{x}_1 - \bar{x}_2})/s_{\bar{x}_1 - \bar{x}_2}$$

$$= (3.1 - 0)/\sqrt{(7.2)^2/23 + (5.9)^2/30}$$

$$= 3.1/1.8478 = 1.678$$

P-value = tcdf(1.678,99,22) = 0.0538

conclusion:

Do not reject H_o; there is not sufficient evidence to conclude that $\mu_1 - \mu_2 > 0$. There is not sufficient evidence to support the claim that the unsuccessful applicants are from a population with a greater mean age than that of the successful candidates. No; based on these results there does not appear to be discrimination based on age.

25. Let the light marijuana users be group 1.

$\bar{x}_1 - \bar{x}_2 = 53.3 - 51.3 = 2.0$

original claim: $\mu_1 - \mu_2 > 0$ items

H_o: $\mu_1 - \mu_2 = 0$ items

H_1: $\mu_1 - \mu_2 > 0$ items

$\alpha = 0.01$ and df = 63

C.V. $t = t_\alpha = t_{0.01} = 2.390$

calculations:

$$t_{\overline{x}_1-\overline{x}_2} = (\overline{x}_1-\overline{x}_2 - \mu_{\overline{x}_1-\overline{x}_2})/s_{\overline{x}_1-\overline{x}_2}$$

$$= (2.0 - 0)/\sqrt{(3.6)^2/64 + (4.5)^2/65}$$

$$= 2.0/0.7170$$

$$= 2.790$$

P-value = tcdf(2.790,99,63) = 0.0035

conclusion:

Reject H_o; there is sufficient evidence to conclude that $\mu_1 - \mu_2 > 0$. There is sufficient evidence to support the claim that the heavy marijuana users have a lower mean number of recalled items than do light users. Yes; marijuana use should be of concern to college students – and an even more valuable study might be one comparing light users to those who do not use marijuana at all.

27. Let the magnet users be group 1.

$\overline{x}_1-\overline{x}_2 = 0.49 - 0.44 = 0.05$

original claim: $\mu_1 - \mu_2 > 0$ items

H_o: $\mu_1 - \mu_2 = 0$ items

H_1: $\mu_1 - \mu_2 > 0$ items

$\alpha = 0.05$ and df = 19

C.V. $t = t_\alpha = t_{0.05} = 1.729$

calculations:

$$t_{\overline{x}_1-\overline{x}_2} = (\overline{x}_1-\overline{x}_2 - \mu_{\overline{x}_1-\overline{x}_2})/s_{\overline{x}_1-\overline{x}_2}$$

$$= (0.05 - 0)/\sqrt{(0.96)^2/20 + (1.4)^2/20}$$

$$= 0.05/0.3796$$

$$= 0.132$$

P-value = tcdf(0.132,99,19) = 0.4483

conclusion:

Do not reject H_o; there is not sufficient evidence to conclude that $\mu_1 - \mu_2 > 0$. There is not sufficient evidence to support the claim that those treated with magnets have a greater mean reduction in pain than do those given a placebo-like treatment. No; it does not appear that magnets are effective in reducing back pain. If a much larger sample size achieved these same results, the calculated t could fall in the critical region and appear to provide evidence that the treatment is effective – but the observed difference would still be 0.05, and one would have to decide whether that statistically significant difference is of any practical significance.

NOTE: Exercises 29-32 involve raw data. Depending on how and when one rounds off, and/or the statistical software used to get any values, answers obtained may differ slightly from those given in the text or in this manual. The final conclusions of the tests of hypotheses or the inferences made from the confidence intervals are not affected by these minor differences.

29. Let the recent winners be group 1.

group 1: recent (n=10)   group 2: former (n=10)

$\Sigma x = 187.6$ $\Sigma x = 201.6$

$\Sigma x^2 = 3532.04$ $\Sigma x^2 = 4083.94$

$\overline{x} = 18.76$ $\overline{x} = 20.26$

$s^2 = 1.4071$ (s=1.186) $s^2 = 2.1871$ (s=1.479)

$\overline{x}_1-\overline{x}_2 = 18.76 - 20.16 = -1.40$

a. original claim: $\mu_1 - \mu_2 < 0$

   H_o: $\mu_1 - \mu_2 = 0$

   H_1: $\mu_1 - \mu_2 < 0$

   $\alpha = 0.05$ and df $= 9$

   C.V. $t = -t_\alpha = -t_{0.05} = -1.833$

   calculations:

$$t_{\overline{x}_1 - \overline{x}_2} = (\overline{x}_1 - \overline{x}_2 - \mu_{\overline{x}_1 - \overline{x}_2})/s_{\overline{x}_1 - \overline{x}_2}$$

$$= (-1.40 - 0)/\sqrt{1.4071/10 + 2.1871/10}$$

$$= -1.40/0.5995$$

$$= -2.335$$

   P-value $=$ tcdf$(-99, -2.335, 9) = 0.0222$

 conclusion:

   Reject H_o; there is sufficient evidence to conclude that $\mu_1 - \mu_2 < 0$. There is sufficient evidence to support the claim that recent winners have a lower mean BMI than winners from the 1920's and 1930's.

b. $\alpha = 0.10$ and df $= 9$

   $(\overline{x}_1 - \overline{x}_2) \pm t_{\alpha/2} s_{\overline{x}_1 - \overline{x}_2}$

   $-1.40 \pm 1.833 \sqrt{1.4071/10 + 2.1871/10}$

   -1.40 ± 1.0989

   $-2.50 < \mu_1 - \mu_2 < -0.30$

31. Let the Popes be group 1.

   <u>group 1: Popes (n=24)</u>   <u>group 2: Rulers (n=14)</u>

   $\Sigma x = 315$ $\Sigma x = 318$

   $\Sigma x^2 = 5981$ $\Sigma x^2 = 11722$

   $\overline{x} = 13.125$ $\overline{x} = 22.714$

   $s^2 = 80.2880$ (s=8.96) $s^2 = 346.0659$ (s=18.60)

  $\overline{x}_1 - \overline{x}_2 = 13.125 - 22.714 = -9.589$

a. original claim: $\mu_1 - \mu_2 < 0$ years

   H_o: $\mu_1 - \mu_2 = 0$ years

   H_1: $\mu_1 - \mu_2 < 0$ years

   $\alpha = 0.01$ and df $= 13$

   C.V. $t = -t_\alpha = -t_{0.01} = -2.650$

   calculations:

$$t_{\overline{x}_1 - \overline{x}_2} = (\overline{x}_1 - \overline{x}_2 - \mu_{\overline{x}_1 - \overline{x}_2})/s_{\overline{x}_1 - \overline{x}_2}$$

$$= (-9.589 - 0)/\sqrt{\frac{80.2880}{24} + \frac{346.0659}{14}}$$

$$= -9.589/5.2976$$

$$= -1.810$$

   P-value $=$ tcdf$(-99, -1.810, 9) = 0.0467$

 conclusion:

   Do not reject H_o; there is not sufficient evidence to conclude that $\mu_1 - \mu_2 < 0$. There is not sufficient evidence to support the claim that the mean longevity after election for popes is less than the mean longevity after coronation for British monarchs.

b. $\alpha = 0.02$ and df = 13

$$(\overline{x}_1 - \overline{x}_2) \pm t_{\alpha/2}\, s_{\overline{x}_1 - \overline{x}_2}$$

$-9.589 \pm 2.650\sqrt{80.2880/24 + 346.0659/14}$

-9.589 ± 14.038

$-23.6 < \mu_1 - \mu_2 < 4.4$ (years)

Since the confidence interval includes zero, the two population means could be equal.

NOTE: Exercises 33-36 involve large data sets. Depending on how and when one rounds off, and/or the statistical software used to get any values, answers obtained may differ slightly from those given in the text or in this manual. The final conclusions of the tests of hypotheses or the inferences made from the confidence intervals are not affected by these minor differences. This manual obtains the summary statistics from statistical software and then uses those (rounded) values to work the exercises by hand.

33. Let the PG & PG-13 movies be group 1. Statistical software yields the following summary.

group 1:   n = 23 $\overline{x}$ = 149.391   s = 108.019
group 2:   n = 12 $\overline{x}$ = 72.417   s = 57.820
$\overline{x}_1 - \overline{x}_2$ = 76.974

a. original claim: $\mu_1 - \mu_2 > 0$ million $

H_o: $\mu_1 - \mu_2 = 0$ million $

H_1: $\mu_1 - \mu_2 > 0$ million $

$\alpha = 0.01$ and df = 11

C.V. $t = t_\alpha = t_{0.01} = 2.718$

calculations:

$$t_{\overline{x}_1 - \overline{x}_2} = (\overline{x}_1 - \overline{x}_2 - \mu_{\overline{x}_1 - \overline{x}_2})/s_{\overline{x}_1 - \overline{x}_2}$$

$$= (76.974 - 0)/\sqrt{\frac{(108.019)^2}{23} + \frac{(57.820)^2}{12}}$$

$= 76.974/28.0340$

$= 2.746$

P-value = tcdf(2.746,99,11) = 0.0095

conclusion:

Reject H_o; there is sufficient evidence to conclude that $\mu_1 - \mu_2 > 0$. There is sufficient evidence to support the claim that that movies with PG or PG-13 ratings have a higher mean gross amount than movies with R ratings.

b. $\alpha = 0.02$ and df = 11

$$(\overline{x}_1 - \overline{x}_2) \pm t_{\alpha/2}\, s_{\overline{x}_1 - \overline{x}_2}$$

$76.974 \pm 2.718\sqrt{(108.019)^2/23 + (57.820)^2/12}$

76.974 ± 76.196

$0.8 < \mu_1 - \mu_2 < 153.2$ (million $)

Since the confidence interval includes only positive values, it suggests that the mean gross amount is higher for PG & PG-13 rated movies than for R rated movies – i.e., that PG & PG-13 movies are a better investment than R movies in that they have a larger gross income.

35. Let the home voltages be group 1. Statistical software yields the following summary.
    group 1: $n = 40$ $\bar{x} = 123.663$ $s = 0.240392$
    group 2: $n = 40$ $\bar{x} = 124.663$ $s = 0.288842$
    $\bar{x}_1 - \bar{x}_2 = -1.000$

original claim: $\mu_1 - \mu_2 = 0$ volts
H_o: $\mu_1 - \mu_2 = 0$ volts
H_1: $\mu_1 - \mu_2 \neq 0$ volts
$\alpha = 0.05$ and df $= 39$
C.V. $t = \pm t_{\alpha/2} = t_{0.025} = \pm 2.023$
calculations:

$$t_{\bar{x}_1 - \bar{x}_2} = (\bar{x}_1 - \bar{x}_2 - \mu_{\bar{x}_1 - \bar{x}_2})/s_{\bar{x}_1 - \bar{x}_2}$$
$$= (-1.000 - 0)/\sqrt{\frac{(0.240392)^2}{40} + \frac{(0.288842)^2}{40}}$$
$$= -1.000/0.05942$$
$$= -16.830$$

P-value $= 2 \cdot \text{tcdf}(-99, -16.830, 39) \approx 0$
conclusion:
> Reject H_o; there is sufficient evidence to reject the claim that $\mu_1 - \mu_2 = 0$. There is sufficient evidence to reject the claim that the samples of home voltages and generator voltages are from populations with the same mean. No; while the 1.000 volt difference is highly statistically significant (because of the consistency [low variability] of each source), it is likely of no practical significance or concern.

37. Let the children treated with low humidity be group 1.
    $\bar{x}_1 - \bar{x}_2 = 0.98 - 1.09 = -0.11$ df $= df_1 + df_2 = 45 + 45 = 90$

$$s_p^2 = [(df_1)s_1^2 + (df_2)s_2^2]/[df_1 + df_2]$$
$$= [(45)(1.22)^2 + (45)(1.11)^2]/[45 + 45]$$
$$= 122.4225/90 = 1.36025$$

original claim: $\mu_1 - \mu_2 = 0$
H_o: $\mu_1 - \mu_2 = 0$
H_1: $\mu_1 - \mu_2 \neq 0$
$\alpha = 0.05$ and df $= 90$
C.V. $t = \pm t_{\alpha/2} = t_{0.025} = \pm 1.987$
calculations:

$$t_{\bar{x}_1 - \bar{x}_2} = (\bar{x}_1 - \bar{x}_2 - \mu_{\bar{x}_1 - \bar{x}_2})/s_{\bar{x}_1 - \bar{x}_2}$$
$$= (-0.11 - 0)/\sqrt{1.36025/46 + 1.36025/46}$$
$$= -0.11/0.2432 = -0.452$$

P-value $= 2 \cdot \text{tcdf}(-99, -0.452, 90) = 0.6521$
conclusion:
> Do not reject H_o; there is not sufficient evidence to reject the claim that $\mu_1 - \mu_2 = 0$. There is not sufficient evidence to reject the claim that the two groups are from populations with the same mean. The results suggest that increasing the humidity does not have a significant effect on the treatment of croup.

*When $n_1 = n_2$, the calculated t statistic does not change at all. The only difference the assumption of equal standard deviation makes is in this instance is to change the df from 45 to 90 and the P-value from 0.6532 to 0.6521. The conclusion is unaffected.

39. Let the unfiltered cigarettes be group 1.

$\bar{x}_1 - \bar{x}_2 = 21.1 - 13.2 = 7.9$ $\qquad$ $df = df_1 + df_2 = 24 + 24 = 48$

$s_p^2 = [(df_1)s_1^2 + (df_2)s_2^2]/[df_1 + df_2]$

$\quad = [(24)(3.2)^2 + (24)(3.7)^2]/[24 + 24]$

$\quad = 574.32/48 = 11.965$

$\alpha = 0.10$ and $df = 48$

$\quad (\bar{x}_1 - \bar{x}_2) \pm t_{\alpha/2}\, s_{\bar{x}_1 - \bar{x}_2}$

$\quad 7.9 \pm 1.676 \sqrt{11.965/25 + 11.965/25}$

$\quad 7.9 \pm 1.640$

$\quad 6.3 < \mu_1 - \mu_2 < 9.5$ (mg)

Yes; since the confidence interval includes only positive values, the results suggest that the filtered cigarettes have less tar than the unfiltered ones.

*When $n_1 = n_2$ the value of $s_{\bar{x}_1 - \bar{x}_2}$ is unchanged. The only difference the assumption of equal standard deviation makes is in this instance is to change the df from 24 to 48 and the $t_{\alpha/2}$ from 1.711 to 1.676. This makes the interval slightly narrower, but the conclusion is unaffected.

41. Let the Popes be group 1.

group 1: Popes (n=24)	group 2: Rulers (n=14)
$\Sigma x = 315$	$\Sigma x = 2001$
$\Sigma x^2 = 5981$	$\Sigma x^2 = 2{,}901{,}433$
$\bar{x} = 13.125$	$\bar{x} = 142.929$
$s^2 = 80.2880$ (s=8.96)	$s^2 = 201187.1484$ (s=448.54)

$\bar{x}_1 - \bar{x}_2 = 13.125 - 142.929 = -129.804$

a. original claim: $\mu_1 - \mu_2 < 0$ years

H_o: $\mu_1 - \mu_2 = 0$ years

H_1: $\mu_1 - \mu_2 < 0$ years

$\alpha = 0.01$ and $df = 13$

C.V. $t = -t_\alpha = -t_{0.01} = -2.650$

calculations:

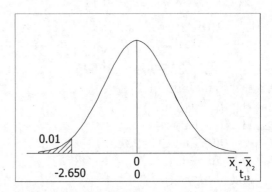

$\quad t_{\bar{x}_1 - \bar{x}_2} = (\bar{x}_1 - \bar{x}_2 - \mu_{\bar{x}_1 - \bar{x}_2})/s_{\bar{x}_1 - \bar{x}_2}$

$\quad\quad = (-129.804 - 0)/\sqrt{\dfrac{80.2880}{24} + \dfrac{201187.1484}{14}}$

$\quad\quad = -129.804/119.8910$

$\quad\quad = -1.083$

P-value = tcdf(-99,-1.083,9) = 0.1536

conclusion:

Do not reject H_o; there is not sufficient evidence to conclude that $\mu_1 - \mu_2 < 0$. There is not sufficient evidence to support the claim that the mean longevity after election for popes is less than the mean longevity after coronation for British monarchs.

b. $\alpha = 0.02$ and $df = 13$

$\quad (\bar{x}_1 - \bar{x}_2) \pm t_{\alpha/2}\, s_{\bar{x}_1 - \bar{x}_2}$

$\quad -129.804 \pm 2.650 \sqrt{80.2880/24 + 201187.1484/14}$

$\quad -129.804 \pm 317.711$

$\quad -447.5 < \mu_1 - \mu_2 < 187.9$ (years)

Since the confidence interval includes zero, the two population means could be equal.

*The conclusions from the test of hypothesis and the confidence remain the same, but the outlier has a dramatic effect on the calculations – greatly increasing the magnitude of the mean and standard deviation of group2, and the magnitude of the mean and standard deviation of the difference between the means. Although both $\overline{x}_1\text{-}\overline{x}_2$ and $s_{\overline{x}_1\text{-}\overline{x}_2}$ increase considerably, the calculated t statistic (which is a ratio) changes only from -1.810 to -1.083 – but the corresponding change in P-values from 0.0467 to 0.1536 is considerable. Since both $\overline{x}_1\text{-}\overline{x}_2$ and $s_{\overline{x}_1\text{-}\overline{x}_2}$ increase considerably, the change in the confidence interval is dramatic – with a shift in the center from to -10 years to -129 years and an increase in width from to 28 years to 635 years.

43. Let the treatment group be group 1.
$\overline{x}_1\text{-}\overline{x}_2 = 0.049 - 0.000 = 0.049$

original claim: $\mu_1 - \mu_2 = 0$
$H_o: \mu_1 - \mu_2 = 0$
$H_1: \mu_1 - \mu_2 \neq 0$
$\alpha = 0.05$ and df = 21
C.V. $t = \pm t_{\alpha/2} = t_{0.025} = \pm 2.080$
calculations:

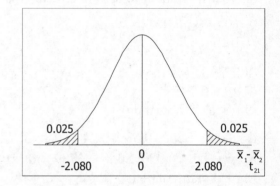

$t_{\overline{x}_1\text{-}\overline{x}_2} = (\overline{x}_1\text{-}\overline{x}_2 - \mu_{\overline{x}_1\text{-}\overline{x}_2})/s_{\overline{x}_1\text{-}\overline{x}_2}$

$= (0.049 - 0)/\sqrt{(0.015)^2/22 + (0)^2/22}$

$= 0.049/0.003198$

$= 15.322$

P-value $= 2 \cdot \text{tcdf}(15.322,99,21) = 7.14\text{E-}13 \approx 0$
conclusion:

Reject H_o; there is sufficient evidence to reject the claim that $\mu_1 - \mu_2 = 0$ and conclude that $\mu_1 - \mu_2 \neq 0$ (in fact, that $\mu_1 - \mu_2 > 0$). There is sufficient evidence to reject the claim that the two sample groups come from populations with the same mean.

The fact that there was no variation in the second sample did not affect the calculations or present any special problems. Since there is no variation in x_2, it is really equivalent to the constant value zero – and the test is mathematically equivalent to the one-sample test $H_o: \mu_1 = 0$, for which $t = (\overline{x}_1 - 0)/s_{\overline{x}_1}$

9-4 Inferences from Dependent Samples

1. The values of d = B–A are 6 -3 3 -13 6, for which n = 5 Σd = -1 Σd^2 = 259.
   a. $\overline{d} = (\Sigma d)/n = -1/5 = -0.2$ minutes
   b. $s_d^2 = [n(\Sigma d^2) - (\Sigma d)^2]/[n(n-1)] = [5(259) - (-1)^2]/[5(4)] = 1294/20 = 64.7$
      $s_d = 8.0$ minutes
   c. In general, μ_d represents the true mean of the differences from the population of matched pairs (which is mathematically equivalent to the true of the difference between the means of the two populations).

3. While the methods of this section can be used to produce numerical results, because there is paired data, those results will have no meaning. Because pulse rates and cholesterol levels measure different kinds of quantities and have different units, the differences between them have no meaningful interpretation and should not be calculated (either as matched pairs or as independent samples).

5. d = C-H: -8 -9 -8 -8
   summary: n = 4 $\Sigma d = -33$ $\Sigma d^2 = 273$
   a. $\bar{d} = (\Sigma d)/n = -33/4 = -8.3$ mpg
   b. $s_d^2 = [n(\Sigma d^2) - (\Sigma d)^2]/[n(n-1)]$
       $= [4(273) - (-33)^2]/[4(3)] = 3/12 = 0.25$
      $s_d = 0.5$ mpg
   c. $t_{\bar{d}} = (\bar{d} - \mu_{\bar{d}})/s_{\bar{d}}$
       $= (-8.25 - 0)/(0.5/\sqrt{4}) = -8.25/0.25 = -33.000$
   d. with df=3 and α=0.05, the critical values are t = $\pm t_{\alpha/2} = \pm t_{0.025} = \pm 3.182$

7. α = 0.05 and df = 4
   $\bar{d} \pm t_{\alpha/2} \cdot s_{\bar{d}}$

   $-8.25 \pm 3.182 \cdot (0.5/\sqrt{4})$
   -8.25 ± 0.7955
   $-9.0 < \mu_d < 7.5$ (mpg)

9. d = Apr – Sept: -0.53 -0.24 1.18 -0.72 0.36
   n = 5 $\Sigma d = 0.05$ $\Sigma d^2 = 2.3789$ $\bar{d} = 0.1$ $s_d = 0.7711$
   original claim: $\mu_d = 0$
   $H_0: \mu_d = 0$
   $H_1: \mu_d \neq 0$
   α = 0.05 and df = 4
   C.V. t = $\pm t_{\alpha/2} = \pm t_{0.025} = \pm 2.776$
   calculations
       $t_{\bar{d}} = (\bar{d} - \mu_{\bar{d}})/s_{\bar{d}}$
       $= (0.01 - 0)/(0.7711/\sqrt{5})$
       $= 0.01/0.3448$
       $= 0.029$
   P-value = 2·tcdf(0.029,99,4) = 0.9783
conclusion:
   Do not reject H_0; there is not sufficient evidence to reject the claim that $\mu_d = 0$. There is not sufficient evidence to reject the claim that the mean change in BMI for all students is equal to 0. No; BMI does not appear to change during the freshman year.

11. d = actress - actor: -34 -9 -25 -14 -8 -16 -31 -12 2 -9 -21 -14 -9 -41 6
    n = 15 $\Sigma d = -235$ $\Sigma d^2 = 6003$ $\bar{d} = -15.67$ $s_d = 12.877$
    original claim: $\mu_d < 0$
    $H_0: \mu_d = 0$
    $H_1: \mu_d < 0$
    α = 0.05 and df = 14
    C.V. t = $-t_{\alpha} = -t_{0.05} = -1.761$
    calculations
        $t_{\bar{d}} = (\bar{d} - \mu_{\bar{d}})/s_{\bar{d}}$
        $= (-15.67 - 0)/(12.877/\sqrt{15})$
        $= -15.67/5.3248$
        $= -4.712$
    P-value = tcdf(-99,-4.712,14) = 0.0000.

conclusion:

Reject H_o; there is sufficient evidence to conclude that $\mu_d < 0$. There is sufficient evidence to support the claim that best actresses are younger than best actors. Yes; the results suggest that (unless acting ability actually peaks earlier in women than in men) actresses and actors are evaluated by different criteria.

13. d = 8am − 12am: -1.0 -2.4 -1.2 -1.6 -0.8 1.6

$n = 6$ $\Sigma d = -5.4$ $\Sigma d^2 = 13.96$ $\overline{d} = -0.90$ $s_d = 1.349$

$\alpha = 0.05$ and df = 5

$\overline{d} \pm t_{\alpha/2} \cdot s_{\overline{d}}$

$-0.90 \pm 2.571 \cdot (1.349 / \sqrt{6})$

-0.90 ± 1.4160

$-2.32 < \mu_d < 0.52$ (°F)

Yes; since the confidence interval includes 0, body temperature is basically the same at both times.

15. $d = 6^{th} - 13^{th}$: -4 -6 -3 1 -1 -7

$n = 6$ $\Sigma d = -20$ $\Sigma d^2 = 112$ $\overline{d} = -3.333$ $s_d = 3.011$

original claim: $\mu_d = 0$ admissions

H_o: $\mu_d = 0$ admissions

H_1: $\mu_d \neq 0$ admissions

$\alpha = 0.05$ and df = 5

C.V. $t = \pm t_{\alpha/2} = \pm t_{0.025} = \pm 2.571$

calculations

$t_{\overline{d}} = (\overline{d} - \mu_{\overline{d}})/s_{\overline{d}}$

$= (-3.333 - 0)/(3.011/\sqrt{6})$

$= -3.333/1.229$

$= -2.712$

P-value = 2·tcdf(-99,-2.712,5) = 0.0422

0.025   0.025

0          $\overline{d}$

-2.571   0   2.571 t_s

conclusion:

Reject H_o; there is sufficient evidence to reject the claim that $\mu_d = 0$ and conclude that $\mu_d \neq 0$ (in fact, that $\mu_d < 0$). There is sufficient evidence to reject the claim that the numbers of hospital admissions from motor vehicle crashes are not affected when the 13th of a month falls on a Friday   and to conclude that Friday the 13th does appear to be unlucky!

17. d = front − rear: -544   -224   1450   -2159   2789   3573   700   -777   2748

$n = 9$ $\Sigma d = 7556$ $\Sigma d^2 = 36{,}299{,}976$ $\overline{d} = 839.56$ $s_d = 1935.081$

$\alpha = 0.05$ and df = 8

$\overline{d} \pm t_{\alpha/2} \cdot s_{\overline{d}}$

$839.56 \pm 2.306 \cdot (1935.081/\sqrt{9})$

839.56 ± 1487.4319

$-647.9 < \mu_d < 2327.0$ ($)

Since the confidence interval includes 0, there is no significant mean difference between front repair costs and rear repair costs.

19. d = old – new: 1 2 3 2 4 3 4 2 2 2 2 3 2 2 2 3 2 2 2 2

$n = 20$ $\Sigma d = 47$ $\Sigma d^2 = 121$ $\bar{d} = 2.35$ $s_d = 0.745$

original claim: $\mu_d > 0$ mpg

H_o: $\mu_d = 0$ mpg

H_1: $\mu_d > 0$ mpg

$\alpha = 0.01$ and df = 19

C.V. $t = t_\alpha = t_{0.01} = 2.539$

calculations

$$t_{\bar{d}} = (\bar{d} - \mu_{\bar{d}})/s_{\bar{d}}$$

$$= (2.35 - 0)/(0.745/\sqrt{20})$$

$$= 2.35/0.1666$$

$$= 14.104$$

P-value = tcdf(14.104,99,19) = 9.093E12 ≈ 0

conclusion:

Reject H_o; there is sufficient evidence to conclude that $\mu_d > 0$. There is sufficient evidence to support the claim that the old ratings are higher than the new ratings.

21. Minitab produces the following output for a t test on the data column consisting of c2-c4.

```
Test of mu = 0 vs not = 0
Variable   N     Mean     StDev    SE Mean        95% CI           T      P
home-UPS  40  0.075000  0.449929  0.071140  (-0.068894, 0.218894)  1.05  0.298
```

a. original claim: $\mu_d = 0$ volts

H_o: $\mu_d = 0$ volts

H_1: $\mu_d \neq 0$ volts

$\alpha = 0.05$ and df = 39

C.V. $t = \pm t_{\alpha/2} = \pm t_{0.025} = \pm 2.023$

calculations

$$t_{\bar{d}} = (\bar{d} - \mu_{\bar{d}})/s_{\bar{d}}$$

$$= (0.075 - 0)/(0.4499/\sqrt{40})$$

$$= 1.05 \text{ [Minitab]}$$

P-value = $2 \cdot P(t_{39} > 1.05) = 0.298$ [Minitab]

conclusion:

Do not reject H_o; there is not sufficient evidence to reject the claim that $\mu_d = 0$. There is not sufficient evidence to reject the claim that the paired sample values have differences that are from a population with a mean of 0 volts.

b. Even though the generator voltage amounts were recorded on the same day as the corresponding home voltage amounts, there is no physical justification for pairing the values. The gasoline-powered generator operates completely independently of the other sources, and so their data values represent independent samples.

23. Minitab produces the following output for a t interval on the data column consisting of c3-c4.

```
Variable   N    Mean     StDev   SE Mean       95% CI
pap-plas  62  7.51726  3.57036  0.45344  (6.61056, 8.42396)
```

$\alpha = 0.05$ and df = 61

$$\bar{d} \pm t_{\alpha/2} \cdot s_{\bar{d}}$$

$$7.517 \pm 2.000 \cdot (3.570/\sqrt{62})$$

$$6.611 < \mu_d < 8.424 \text{ (lbs) [Minitab]}$$

Since the confidence interval includes only positive values, there discarded paper appears to weigh more than the discarded plastic.

25. The following values were used in the analysis. Different readings from the graph may produce slightly different values and slightly differ numbers in the test of hypothesis.

right hand	97	116	116	165	191
left hand	171	191	196	207	224
d = right – left	-74	-75	-80	-42	-33

$n = 5$ $\Sigma d = -304$ $\Sigma d^2 = 20354$ $\overline{d} = -60.8$ $s_d = 21.626$

original claim: $\mu_d = 0$ milliseconds

H_o: $\mu_d = 0$ milliseconds

H_1: $\mu_d \neq 0$ milliseconds

$\alpha = 0.05$ and $df = 4$

C.V. $t = \pm t_{\alpha/2} = \pm t_{0.025} = \pm 2.776$

calculations

$$t_{\overline{d}} = (\overline{d} - \mu_{\overline{d}})/s_{\overline{d}}$$

$$= (-60.8 - 0)/(21.626/\sqrt{5})$$

$$= -60.8/9.6716$$

$$= -6.286$$

P-value = $2 \cdot tcdf(-99, -6.286, 4) = 0.0033$

conclusion:

Reject H_o; there is sufficient evidence to reject the claim that $\mu_d = 0$ and conclude that $\mu_d \neq 0$ (in fact, that $\mu_d < 0$). There is sufficient evidence to reject the claim that there is no difference between the reaction times of the right and left hands.

9-5 Comparing Variation in Two Samples

1. Since the F value is close to 1.0, the two sample standard deviations are very close in value.

3. When we say the F test is not robust against departures from normality, that means it performs poorly when one or both of the populations does not have a normal distribution. When we say it performs poorly, that means that the F test at the α level of significance will not reject a true null hypothesis approximately α of the time. Two alternatives that are more robust against departures from normality (but do not fare as well when the normality is met) are the count five test and the Levene-Brown-Forsyth test.

NOTE: The following conventions are used in this manual regarding the F test.

• The set of scores with the larger sample variance is designated group 1.

• If the desired df does not appear in Table A-5, the closest entry is used. For a desired entry exactly halfway between two tabled values, the conservative approach of using the smaller df is used – and 120 is used for all df larger than 120.

• Even though designating the scores with the larger sample variance as group 1 makes lower tail critical values unnecessary in two-tailed tests, the lower critical values are calculated exactly using technology and included in the accompanying graphs for completeness and consistency with the other tests. The technique for calculating lower critical values from Table A-5 is given in exercise #23 – it is exact for the df appearing in the table, but otherwise the closest entry must be used. The F distribution always "bunches up" around 1.0 regardless of the df values.

• Since the F value depends on two degrees of freedom, the df for group 1 (numerator) and group 2 (denominator) may be used with the F as a superscript and a subscript respectively to clarify which F distribution is being used.

• F is statistically defined to be $F = (s_1^2/\sigma_1^2)/(s_2^2/\sigma_2^2)$. Since all hypotheses in the text question the equality of σ_1^2 and σ_2^2, the calculation of F may be shortened to $F = s_1^2/s_2^2$

• Some problems are stated in terms of variance, and some are stated in terms of standard deviation – and some are stated simply in terms of "variation." Since $\sigma_1 = \sigma_2$ is equivalent to $\sigma_1^2 = \sigma_2^2$, this manual simply states all claims and hypotheses and conclusions using the variance.

5. Let the placebo group be group 1.

   original claim: $\sigma_1^2 > \sigma_2^2$

   H_o: $\sigma_1^2 = \sigma_2^2$

   H_1: $\sigma_1^2 > \sigma_2^2$

   $\alpha = 0.05$ and $df_1 = 15$, $df_2 = 15$

   C.V. $F = F_\alpha = F_{0.05} = 2.4034$

   calculations:

   $F = s_1^2/s_2^2$

   $= (728)^2/(669)^2$

   $= 1.1842$

   P-value = Fcdf(1.1842,99,15,15) = 0.3738

   conclusion:

   Do not reject H_o; there is not sufficient evidence to conclude that $\sigma_1^2 > \sigma_2^2$. There is not sufficient evidence to support the claim that the weights of babies born to mothers given placebos vary more than the weights of babies born to mothers given zinc supplements.

7. Let the thicker cans be group 1. Use technology to find the C.V. from the α.

   original claim: $\sigma_1^2 = \sigma_2^2$

   H_o: $\sigma_1^2 = \sigma_2^2$

   H_1: $\sigma_1^2 \neq \sigma_2^2$

   $\alpha = 0.01$ and $df_1 = 174$, $df_2 = 174$

   C.V. $F = F_{\alpha/2} = F_{0.005} = 1.4805$

   calculations:

   $F = s_1^2/s_2^2$

   $= (27.775)^2/(22.112)^2$

   $= 1.5777$ [TI-83/84+]

   P-value = $2 \cdot P(F_{174}^{174} > 1.5777) = 0.0028$ [TI-83/84+]

   conclusion:

   Reject H_o; there is sufficient evidence to reject the claim that $\sigma_1^2 = \sigma_2^2$ and   conclude that $\sigma_1^2 \neq \sigma_2^2$ (in fact, that $\sigma_1^2 > \sigma_2^2$). There is sufficient evidence to reject the claim that the two samples are from populations with the same standard deviation.

9. Let the Atkins group be group 1.

original claim: $\sigma_1^2 = \sigma_2^2$

H_o: $\sigma_1^2 = \sigma_2^2$

H_1: $\sigma_1^2 \neq \sigma_2^2$

$\alpha = 0.05$ and $df_1 = 39$, $df_2 = 39$

C.V. $F = F_{\alpha/2} = F_{0.025} = 1.8752$

calculations:

$F = s_1^2/s_2^2$

$\quad = (12)^2/(9)^2$

$\quad = 1.7778$

P-value $= 2 \cdot$ Fcdf(1.7778,99,39,39) $= 0.0762$

conclusion:

Do not reject H_o; there is not sufficient evidence to reject the claim that $\sigma_1^2 = \sigma_2^2$. There is not sufficient evidence to reject the claim that subjects from both treatment groups have ages with the same amount of variation.

If the treatment groups have different characteristics, then there is confounding – and we cannot tell if any observed differences between the groups are due to the treatments or to the different group characteristics.

11. Let the treatment group be group 1.

original claim: $\sigma_1^2 > \sigma_2^2$

H_o: $\sigma_1^2 = \sigma_2^2$

H_1: $\sigma_1^2 > \sigma_2^2$

$\alpha = 0.05$ and $df_1 = 21$, $df_2 = 21$

C.V. $F = F_\alpha = F_{0.05} = 2.0960$

calculations:

$F = s_1^2/s_2^2$

$\quad = (2.20)^2/(0.72)^2$

$\quad = 9.3364$

P-value $=$ Fcdf(9.3364,99,21,21) $= 1.679E\text{-}6 = 0.0000002$

conclusion:

Reject H_o; there is sufficient evidence to conclude that $\sigma_1^2 > \sigma_2^2$. There is sufficient evidence to support the claim that the treatment group has errors that vary more than the errors of the placebo group.

13. Let those receiving the sham treatment be group 1.

original claim: $\sigma_1^2 > \sigma_2^2$

H_o: $\sigma_1^2 = \sigma_2^2$

H_1: $\sigma_1^2 > \sigma_2^2$

$\alpha = 0.05$ and $df_1 = 19$, $df_2 = 19$

C.V. $F = F_\alpha = F_{0.05} = 2.1555$

calculations:

$F = s_1^2/s_2^2$

$\quad = (1.4)^2/(0.96)^2$

$\quad = 2.1267$

P-value = Fcdf(2.1267,99,19,19) = 0.0543

conclusion:

Do not reject H$_o$; there is not sufficient evidence to conclude that $\sigma_1^2 > \sigma_2^2$. There is not sufficient evidence to support the claim that those given a sham treatment have pain reductions that vary more than the pain reductions for those treated with magnets.

15. Let the Pennsylvania children be group 1.

group 1: PA (n=12)
Σx = 1771
Σx^2 = 262,615
$\overline{x}$ = 147.58
s^2 = 113.1742 (s=10.64)

group 2: NY (n=12)
Σx = 1637
Σx^2 = 223,613
$\overline{x}$ = 136.42
s^2 = 27.1742 (s=5.21)

original claim: $\sigma_1^2 > \sigma_2^2$

H$_o$: $\sigma_1^2 = \sigma_2^2$

H$_1$: $\sigma_1^2 > \sigma_2^2$

α = 0.05 and df$_1$ = 11, df$_2$ = 11

C.V. F = F$_\alpha$ = F$_{0.05}$ = 2.8536

calculations:

$F = s_1^2/s_2^2$

= 113.1742/ 27.1742

= 4.1648

P-value = Fcdf(4.1648,99,11,11) = 0.0130

conclusion:

Reject H$_o$; there is sufficient evidence to conclude that $\sigma_1^2 > \sigma_2^2$. There is sufficient evidence to support the claim that the amounts of Strontium-90 from Pennsylvania residents vary more than the amounts from New York residents.

17. Let the unsuccessful applicants be group 1.

group 1: unsuccesful (n=23)
Σx = 1080
Σx^2 = 51,860
$\overline{x}$ = 46.96
s^2 = 52.1344 (s=7.220)

group 2: successful (n=30)
Σx = 1318
Σx^2 = 58,908
$\overline{x}$ = 49.93
s^2 = 34.6161 (s=5.884)

original claim: $\sigma_1^2 = \sigma_2^2$

H$_o$: $\sigma_1^2 = \sigma_2^2$

H$_1$: $\sigma_1^2 \neq \sigma_2^2$

α = 0.05 and df$_1$ = 22, df$_2$ = 29

C.V. F = F$_{\alpha/2}$ = F$_{0.025}$ = 2.2131

calculations:

$F = s_1^2/s_2^2$

= 52.1344/34.6161 = 1.5061

P-value = 2· Fcdf(1.5061,99,22,29) = 0.2990

conclusion:

Do not reject H$_o$; there is not sufficient evidence to reject the claim that $\sigma_1^2 = \sigma_2^2$. There is not sufficient evidence to reject the claim that both samples are from populations having the same standard deviation among their ages.

NOTE: Exercises 19 and 20 involve large data sets. While the following Minitab output gives the P-value, which is all that is necessary to make a decision in a test of hypothesis, the general format does not always follow the text. In Exercise 19, for example, Minitab does not designate the group with the larger standard as group 1, and the test statistic falls in the lower tail of the F distribution. This manual uses the standard deviations from the Minitab output and completes the tests in the usual format.

19. The Minitab test of two variances produces the following output.

```
95% Bonferroni confidence intervals for standard deviations
sex   N    Lower     StDev    Upper
 F   35  4.60572   5.8629   8.0038
 M   32  8.45430  10.8709  15.0946
F-Test (normal distribution)
   Test statistic = 0.29, p-value = 0.001
Levene's Test (any continuous distribution)
   Test statistic = 6.43, p-value = 0.014
```

Let the males be group 1.

original claim: $\sigma_1^2 > \sigma_2^2$

$H_0: \sigma_1^2 = \sigma_2^2$

$H_1: \sigma_1^2 > \sigma_2^2$

$\alpha = 0.05$ and $df_1 = 31$, $df_2 = 34$

C.V. $F = F_\alpha = F_{0.05} = 1.8409$

calculations:

$F = s_1^2/s_2^2$

$= (10.8709)^2/(5.8629)^2$

$= 3.4380$

P-value = Fcdf(3.4380,99,31,34) = 0.0003

conclusion:

Reject H_0; there is sufficient evidence to conclude that $\sigma_1^2 > \sigma_2^2$. There is sufficient evidence to support the claim that the April weights of male college freshmen vary more than the April weights of female college freshmen.

21. a. The following Minitab commands identify the appropriate means and place the $n_1 = 40$ ordered MAD's for the pre-1964 quarters in c13 and the $n_2 = 40$ ordered MAD's for the post-1964 quarters in c14.

```
MTB > desc c6 c7
MTB > let c10 = abs(c6-6.1927)
MTB > let c11 = abs(c7-5.6393)
MTB > sort c10 c13
MTB > sort c11 c14
```

The 5 largest MAD's for the pre-1964 quarters are 0.1351 0.1488 0.1670 0.1742 0.1925. The largest MAD for the post-1964 quarters is 0.1397.

b. Based on the results in part (a), $c_1 = 4$ and $c_2 = 0$.

c. Since $n_1 = n_2$, the critical value is 5.

d. Since $c_1 < 5$ and $c_2 < 5$, fail to reject the null hypothesis of $\sigma_1^2 = \sigma_2^2$. There is not enough evidence to reject the claim that the weights of pre-1964 quarters and the weights of post-1964 quarters have the same variation.

23. For a two-tailed test $\alpha = 0.05$ and $df_1 = 9$ and $df_2 = 20$, the following F values apply.

$F_R = F_{20, \alpha/2}^9 = F_{20, 0.025}^9 = 2.8365$

$F_L = F_{20, 1-\alpha/2}^9 = F_{20, 0.975}^9 = 1/F_{9, 0.025}^{20} = 1/3.6669 = 0.2727$

Statistical Literacy and Critical Thinking

1. When a method is robust against departures from normality, it still performs well when the populations involved do have a normal distribution. When a method performs well, a test at the α level of significance will reject a true null hypothesis approximately α of the time and a confidence interval at the α level of significance will include the true value of the parameter approximately 1-α of the time. The F test for comparing two population standard deviations (or variances) is not robust against departures from normality.

2. These results do not tell us anything about how the general population feels about the word *ginormous*. Since the sample is a voluntary response sample, it is not necessarily representative of the general population. No methods of statistics should be used on this sample data to make inferences about the general population.

3. The samples are independent. Since the Coke and Pepsi cans were selected independently of each, they are not related in any way and any pairing of the cans would be arbitrary and artificial.

4. The described procedure is not valid for at least two reasons. Since the states have different population sizes, the mean of the 50 state means is not necessarily the correct value for the mean of the entire country – a weighted mean should be used. The construction of confidence intervals to estimate population parameters assumes one is working with random samples – but the mean ages obtained for each state are not random selections, and they are presumably the true values (as determined from census or other records) and not sample values subject to sampling variability.

Chapter Quick Quiz

1. Let the California teachers be group 1. Since the original claim does not contain the equality, it must become the alternative hypothesis.
   original claim: $p_1 > p_2$ [i.e., $p_1 - p_2 > 0$]
   H_o: $p_1 - p_2 = 0$
   H_1: $p_1 - p_2 > 0$

2. $\bar{p} = (x_1 + x_2)/(n_1 + n_2)$
   $= (20 + 55)/(50 + 100)$
   $= 75/150 = 0.50$

3. $\hat{p}_1 = x_1/n_1 = 20/50 = 0.40$
   $\hat{p}_2 = x_2/n_2 = 55/100 = 0.55$
   $\hat{p}_1 - \hat{p}_2 = 0.40 - 0.55 = -0.15$
   $z_{\hat{p}_1 \cdot \hat{p}_2} = (\hat{p}_1 - \hat{p}_2 - \mu_{\hat{p}_1 \cdot \hat{p}_2})/\sigma_{\hat{p}_1 \cdot \hat{p}_2}$
   $= (-0.15 - 0)/\sqrt{(0.5)(0.5)/50 + (0.5)(0.5)/100}$
   $= -0.15/0.08660$
   $= -1.73$

4. P-value $= 2 \cdot P(Z < -2.05) = 2 \cdot (0.0202) = 0.0404$

5. A P-value of 0.0001 is small enough to reject the null hypothesis for any reasonable level of significance. The final conclusion is as follows.
   There is sufficient evidence to support the claim that $\mu_1 > \mu_2$.

6. This is a test involving matched pairs.
   original claim: $\mu_d = 0$
       H_o: $\mu_d = 0$ inches
       H_1: $\mu_d \neq 0$ inches

7. This is a test involving independent samples. Let the California voters be group 1.
   original claim $\mu_1 < \mu_2$ [i.e., $\mu_1 - \mu_2 < 0$]
       H_o: $\mu_1 - \mu_2 = 0$ years
       H_1: $\mu_1 - \mu_2 < 0$ years

8. The F distribution is used to test a claim about two standard deviations or variances.

9. A P-value of 0.0009 is small enough to reject the null hypothesis for any reasonable level of significance. The final conclusion is as follows.
   Reject H_o; there is sufficient evidence to conclude that $\mu_1 - \mu_2 \neq 0$. There is sufficient evidence to support the claim that the two populations gave different means.

10. True. In any test of hypothesis, the null hypothesis must contain the condition of equality.

Review Exercises

1. Let those treated with surgery be group 1.
   $\hat{p}_1 = x_1/n_1 = 67/73 = 0.918$ $\hat{p}_1 - \hat{p}_2 = 0.918 - 0.723 = 0.195$
   $\hat{p}_2 = x_2/n_2 = 60/83 = 0.723$
   $\overline{p} = (67+60)/(73+83)$
       $= 127/156 = 0.814$
   original claim: $p_1 - p_2 > 0$
   H_o: $p_1 - p_2 = 0$
   H_1: $p_1 - p_2 > 0$
   $\alpha = 0.01$
   C.V. $z = z_\alpha = z_{0.01} = 2.326$
   calculations:

   $z_{\hat{p}_1 - \hat{p}_2} = (\hat{p}_1 - \hat{p}_2 - \mu_{\hat{p}_1 - \hat{p}_2})/\sigma_{\hat{p}_1 - \hat{p}_2}$
       $= (0.195 - 0)/\sqrt{(0.814)(0.186)/73 + (0.814)(0.186)/83}$
       $= 0.195/0.06242$
       $= 3.12$
   P-value $= P(z > 3.12) = 1 - 0.9991 = 0.0009$
   conclusion:
       Reject H_o; there is sufficient evidence to conclude that $p_1 - p_2 > 0$. There is sufficient evidence to support the claim that treatment of carpal tunnel syndrome with release surgery results in better outcomes than treatment with wrist splinting. In this context, surgery should generally be recommended instead of splinting.

2. Let those born to cocaine users be group 1.

$\bar{x}_1 - \bar{x}_2 = 7.3 - 8.2 = -0.9$

original claim: $\mu_1 - \mu_2 < 0$

H_0: $\mu_1 - \mu_2 = 0$

H_1: $\mu_1 - \mu_2 < 0$

$\alpha = 0.05$ and df = 185

C.V. $t = -t_\alpha = -t_{0.01} = -1.653$

calculations:

$$t_{\bar{x}_1 - \bar{x}_2} = (\bar{x}_1 - \bar{x}_2 - \mu_{\bar{x}_1 - \bar{x}_2})/s_{\bar{x}_1 - \bar{x}_2}$$

$$= (-0.9 - 0)/\sqrt{(3.0)^2/190 + (3.0)^2/186}$$

$$= -0.9/0.3094$$

$$= -2.908$$

P-value = tcdf(-99,-2.908,185) = 0.0020

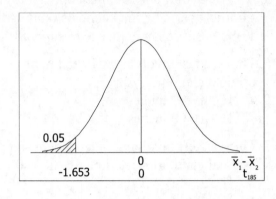

conclusion:

Reject H_0; there is sufficient evidence to conclude that $\mu_1 - \mu_2 < 0$. There is sufficient evidence to support the claim that prenatal cocaine exposure is associated with lower scores on the test of object assembly.

3. d = regular – kiln: -5.75 -1.25 -1.00 -5.00 0.00 0.25 2.25 -0.50 0.75 -1.50 -0.25

n = 11 $\Sigma d = -12.00$ $\Sigma d^2 = 68.8750$ $\bar{d} = -1.091$ $s_d = 2.3619$

a. original claim: $\mu_d = 0$ cwt/acre

H_0: $\mu_d = 0$ cwt/acre

H_1: $\mu_d \neq 0$ cwt/acre

$\alpha = 0.05$ and df = 10

C.V. $t = \pm t_{\alpha/2} = \pm t_{0.025} = \pm 2.228$

calculations

$$t_{\bar{d}} = (\bar{d} - \mu_{\bar{d}})/s_{\bar{d}}$$

$$= (-1.091 - 0)/(2.3619/\sqrt{11})$$

$$= -1.091/0.7121$$

$$= -1.532$$

P-value = 2·tcdf(-99,-1.532,10) = 0.1565

conclusion:

Do not reject H_0; there is not sufficient evidence to reject the claim that $\mu_d = 0$. There is not sufficient evidence to reject the claim that there is no difference between the yields from the two types of seed.

b. $\alpha = 0.05$ and df = 10

$$\bar{d} \pm t_{\alpha/2} \cdot s_{\bar{d}}$$

$-1.09 \pm 2.228 \cdot (2.3619/\sqrt{11})$

-1.09 ± 1.587

$-2.67 < \mu_d < 0.50$

c. We cannot be 95% certain that either type of seed is better.

4. Let the blinded abstracts be group 1.

$\hat{p}_1 = x_1/n_1 = x_1/13200 = 0.267$ [as any $3518 \le x_1 \le 3530$ gives $\hat{p}_1 = 26.7\%$, use 0.267 in all work]

$\hat{p}_2 = x_2/n_2 = x_2/13433 = 0.290$ [as any $3889 \le x_2 \le 3902$ gives $\hat{p}_2 = 29.0\%$, use 0.290 in all work]

$\hat{p}_1 - \hat{p}_2 = 0.267 - 0.290 = -0.023$

$$\bar{p} = \frac{(0.267)(13200) + (0.290)(13433)}{13200 + 13433}$$

$$= 7419.97/26633 = 0.279$$

original claim: $p_1 - p_2 = 0$

H_o: $p_1 - p_2 = 0$

H_1: $p_1 - p_2 \ne 0$

$\alpha = 0.01$

C.V. $z = \pm z_{\alpha/2} = \pm z_{0.005} = \pm 2.575$

calculations:

$z_{\hat{p}_1 - \hat{p}_2} = (\hat{p}_1 - \hat{p}_2 - \mu_{\hat{p}_1 - \hat{p}_2})/\sigma_{\hat{p}_1 - \hat{p}_2}$

$\qquad = (-0.023 - 0)/\sqrt{(0.279)(0.721)/13200 + (0.279)(0.721)/13433}$

$\qquad = -0.023/0.005494 = -4.19$

P-value $= 2 \cdot P(z < -4.10) = 2 \cdot (0.0001) = 0.0002$

conclusion:

Reject H_o; there is sufficient evidence to reject the claim that $p_1 - p_2 = 0$ and conclude that $p_1 - p_2 \ne 0$ (in fact, that $p_1 - p_2 < 0$). There is sufficient evidence to reject the claim that the acceptance rate is the same with or without blinding. When there is no blinding the reviewers know the names and institutions of the authors and may be (either subconsciously or knowingly) influenced by that knowledge.

5. Let the Rowling scores be group 1.

group 1: Rowling (n=12)	group 2: Tolstoy (n=12)
$\Sigma x = 969.0$	$\Sigma x = 793.8$
$\Sigma x^2 = 78487.82$	$\Sigma x^2 = 53189.10$
$\bar{x} = 80.75$	$\bar{x} = 66.15$
$s^2 = 21.915$ (s=4.681)	$s^2 = 61.748$ (s=7.858)

$\bar{x}_1 - \bar{x}_2 = 80.75 - 66.15 = 14.60$

original claim: $\mu_1 - \mu_2 > 0$

H_o: $\mu_1 - \mu_2 = 0$

H_1: $\mu_1 - \mu_2 > 0$

$\alpha = 0.05$ and df = 11

C.V. $t = t_\alpha = t_{0.05} = 1.796$

calculations:

$t_{\bar{x}_1 - \bar{x}_2} = (\bar{x}_1 - \bar{x}_2 - \mu_{\bar{x}_1 - \bar{x}_2})/s_{\bar{x}_1 - \bar{x}_2}$

$\qquad = (14.60 - 0)/\sqrt{21.915/12 + 61.748/12}$

$\qquad = 14.60/2.6404$

$\qquad = 5.529$

P-value $= \text{tcdf}(5.529, 99, 11) = 0.0001$

conclusion:

Reject H_o; there is sufficient evidence to conclude that $\mu_1 - \mu_2 > 0$. There is sufficient evidence to support the claim that *Harry Potter* is easier to read than *War and Peace*. Yes; this is the expected result.

6. d = before – after: 9   4   21   3   20   31   17   26   26   10   23   33
   n = 12 $\Sigma d = 223$ $\Sigma d^2 = 5267$ $\overline{d} = 18.58$ $s_d = 10.104$
   a. $\alpha = 0.01$ and df = 11
   $$\overline{d} \pm t_{\alpha/2} \cdot s_{\overline{d}}$$
   $18.58 \pm 3.106 \cdot (10.104/\sqrt{12})$
   18.58 ± 9.05917
   $9.5 < \mu_d < 27.6$ (mm Hg)

   b. original claim: $\mu_d > 0$ mm Hg
   H_o: $\mu_d = 0$ mm Hg
   H_1: $\mu_d > 0$ mm Hg
   $\alpha = 0.01$ [assumed] and df = 11
   C.V. $t = t_\alpha = t_{0.01} = 2.718$
   calculations

   $$t_{\overline{d}} = (\overline{d} - \mu_{\overline{d}})/s_{\overline{d}}$$
   $= (18.58 - 0)/(10.104/\sqrt{12})$
   $= 18.58/2.9167$
   $= 6.371$
   P-value = tcdf(6.371,99,11) = 2.643E-5 = 0.00003

   conclusion:
   Reject H_o; there is sufficient evidence to conclude that $\mu_d > 0$. There is sufficient evidence to support the claim that captopril is effective in lowering systolic blood pressure.

7. Let the men be group 1.
   $\hat{p}_1 = x_1/n_1 = 71/280 = 0.254$ $\hat{p}_1 - \hat{p}_2 = 0.254 - 0.200 = 0.054$
   $\hat{p}_2 = x_2/n_2 = 68/340 = 0.200$
   $\overline{p} = (71+68)/(280+340)$
   $\quad = 139/620 = 0.224$
   original claim: $p_1-p_2 > 0$
   H_o: $p_1-p_2 = 0$
   H_1: $p_1-p_2 > 0$
   $\alpha = 0.05$
   C.V. $z = z_\alpha = z_{0.05} = 1.645$
   calculations:
   $$z_{\hat{p}_1-\hat{p}_2} = (\hat{p}_1-\hat{p}_2 - \mu_{\hat{p}_1-\hat{p}_2})/\sigma_{\hat{p}_1-\hat{p}_2}$$
   $= (0.054 - 0)/\sqrt{(0.224)(0.776)/280 + (0.224)(0.776)/340}$
   $= 0.054/0.03366$
   $= 1.59$
   P-value = P(z>1.59) = 1 – 0.9441 = 0.0559

   conclusion:
   Do not reject H_o; there is not sufficient evidence to conclude that $p_1-p_2 > 0$. There is not sufficient evidence to support the claim that the proportion of men who smoke is greater than the proportion of women who smoke.

8. Let those with a high school diploma be group 1.

$\bar{x}_1 - \bar{x}_2 = 37622 - 77689 = -40067$

original claim: $\mu_1 - \mu_2 < \$0$

H_o: $\mu_1 - \mu_2 = \$0$

H_1: $\mu_1 - \mu_2 < \$0$

$\alpha = 0.01$ and df = 38

C.V. $t = -t_\alpha = -t_{0.01} = -2.429$

calculations:

$$t_{\bar{x}_1 - \bar{x}_2} = (\bar{x}_1 - \bar{x}_2 - \mu_{\bar{x}_1 - \bar{x}_2})/s_{\bar{x}_1 - \bar{x}_2}$$

$$= (-40067 - 0)/\sqrt{(14115)^2/80 + (24227)^2/39}$$

$$= -40067/4188.1203$$

$$= -9.567$$

P-value = tcdf(-99,-9.567,38) = 5.76E-12 ≈ 0

conclusion:

Reject H_o; there is sufficient evidence to conclude that $\mu_1 - \mu_2 < 0$. There is sufficient evidence to support the claim that workers with a only high school diploma have a lower mean annual income than workers with a bachelor's degree. Yes; since solving this problem contributes to a bachelor's degree, it contributes to a higher income.

9. Let those with a bachelor's degree be group 1.

original claim: $\sigma_1^2 = \sigma_2^2$

H_o: $\sigma_1^2 = \sigma_2^2$

H_1: $\sigma_1^2 \neq \sigma_2^2$

$\alpha = 0.05$ and $df_1 = 38$, $df_2 = 79$

C.V. $F = F_{\alpha/2} = F_{0.025} = 1.7440$

calculations:

$$F = s_1^2/s_2^2$$

$$= (24227)^2/(14115)^2$$

$$= 2.9460$$

P-value = 2· Fcdf(2.9460,99,38,79) = 0.0001

conclusion:

Reject H_o; there is sufficient evidence to reject the claim that $\sigma_1^2 = \sigma_2^2$ and conclude that $\sigma_1^2 \neq \sigma_2^2$ (in fact, that $\sigma_1^2 > \sigma_2^2$). There is sufficient evidence to reject the claim that the two samples are from populations with the same standard deviation. It appears that workers with a bachelor's degrees have incomes with more variation than the incomes of workers with only a high school diploma.

10. Let the treatment group be group 1.

original claim: $\sigma_1^2 = \sigma_2^2$

H_o: $\sigma_1^2 = \sigma_2^2$

H_1: $\sigma_1^2 \neq \sigma_2^2$

$\alpha = 0.05$ and $df_1 = 83$, $df_2 = 68$

C.V. $F = F_{\alpha/2} = F_{0.025} = 1.6668$

calculations:

$F = s_1^2 / s_2^2$

$\quad = (5.9)^2 / (5.2)^2 = 1.2874$

P-value $= 2 \cdot$ Fcdf$(1.2874,99,83,68) = 0.2825$

conclusion:

Do not reject H_o; there is not sufficient evidence to reject the claim that $\sigma_1^2 = \sigma_2^2$.   There is not sufficient evidence to reject the claim that the two samples are from populations with the same amount of variation.

Cumulative Review Exercises

1. a. The samples are dependent, because the data are from married couples.

   b. male scores in numerical order: 8 8 9 14 14 15 16 19 21 25

   summary statistics: $n = 10$, $\Sigma x = 149$, $\Sigma x^2 = 2509$

   $\bar{x} = (\Sigma x)/n = (149)/10 = 14.9$ thousand words per day

   $\tilde{x} = (14+15)/2 = 14.5$ thousand words per day

   $M = 8,14$ thousand words per day [bi-modal]

   $R = 25 - 8 = 17$ thousand words per day

   $s^2 = [n(\Sigma x^2) - (\Sigma x)^2]/[n(n-1)] = [10(2509) - (149)^2]/[10(9)] = 2889/90 = 32.1$

   $s = \sqrt{32.1} = 5.666$, rounded to 5.7 thousand words per day

   c. Because differences are meaningful and there is a meaningful zero, the data are at the ratio level of measurement.

2. $d =$ before $-$ after: 0   13   -22   -7   -6   -8   -20   -1   -10   -7

   $n = 10$ $\Sigma d = -68$ $\Sigma d^2 = 1352$ $\bar{d} = -6.8$ $s_d = 9.942$

   original claim: $\mu_d < 0$

   H_o: $\mu_d = 0$

   H_1: $\mu_d < 0$

   $\alpha = 0.05$ and $df = 9$

   C.V. $t = t_\alpha = t_{0.05} = -1.833$

   calculations

   $t_{\bar{d}} = (\bar{d} - \mu_{\bar{d}})/s_{\bar{d}}$

   $\quad = (-6.8 - 0)/(9.942/\sqrt{10})$

   $\quad = -6.8/3.1440$

   $\quad = -2.163$

   P-value $=$ tcdf$(-99,-2.163,9) = 0.0294$

conclusion:

   Reject H_o; there is sufficient evidence to conclude that $\mu_d < 0$. There is sufficient evidence to support the claim that , among couples, females are more talkative than males.

3. Let the males be group 1.

  group 1: males (n=10)   group 2: females (n=10)

  $\Sigma x = 149$ $\Sigma x = 217$

  $\Sigma x^2 = 2509$ $\Sigma x^2 = 5472$

  $\overline{x} = 14.9$ $\overline{x} = 21.7$

  $s^2 = 32.100$ $(s=5.67)$ $s^2 = 84.789$ $(s=9.208)$

$\overline{x}_1 - \overline{x}_2 = 14.9 - 21.7 = -6.8$

original claim: $\mu_1 - \mu_2 = 0$

H_o: $\mu_1 - \mu_2 = 0$

H_1: $\mu_1 - \mu_2 \neq 0$

$\alpha = 0.05$ and df $= 9$

C.V. $t = \pm t_{\alpha/2} = t_{0.025} = \pm 2.262$

calculations:

  $t_{\overline{x}_1 - \overline{x}_2} = (\overline{x}_1 - \overline{x}_2 - \mu_{\overline{x}_1 - \overline{x}_2})/s_{\overline{x}_1 - \overline{x}_2}$

   $= (-6.8 - 0)/\sqrt{32.100/10 + 84.789/10}$

   $= -6.8/3.4189$

   $= -1.989$

  P-value $= 2 \cdot$tcdf$(-99, -1.989, 9) = 0.0779$

conclusion:

  Do not reject H_o; there is not sufficient evidence to reject the claim that $\mu_1 - \mu_2 = 0$. There is not sufficient evidence to reject the claim that the two samples are from populations with the same mean.

4. σ unknown (and assuming the distribution is approximately normal), use t with df=9

  $\alpha = 0.05$, $t_{df, \alpha/2} = t_{9, 0.025} = 2.262$

   $\overline{x} \pm t_{\alpha/2} \cdot s/\sqrt{n}$

  $14.9 \pm 2.262(5.666)/\sqrt{10}$

  14.9 ± 4.0527

  $10.8 < \mu < 19.0$ (thousand words per day)

5. The requested frequency distribution is given at the right. The word count is in thousands of words per day.

word count	f
6 – 9	3
10 – 13	0
14 – 17	4
18 – 21	2
22 – 25	1
	10

6. For each part, use the formula $z = (x - \mu)/\sigma$.

Let x be the word count in thousands of words per day.

 a. $P(x > 17) = P(z > 0.33)$

    $= 1 - 0.6293$

    $= 0.3707$

 b. The $\overline{x}$'s are normally distributed with $\mu_{\overline{x}} = \mu = 15$ and $\sigma_{\overline{x}} = \sigma/\sqrt{n} = 6/\sqrt{9} = 2$.

  $P(\overline{x} > 17) = P(z > 1.00)$

     $= 1 - 0.8413$

     $= 0.1587$

c. From Table A-2, the z score with 0.9000 [closest entry is 0.8997]below it is 1.28.
$$x = \mu + z\sigma$$
$$= 15 + (1.28)(6)$$
$$= 22.68, \text{ or } 22,680 \text{ words per day}$$

7. $\alpha = 0.10$, $z_{\alpha/2} = z_{0.05} = 1.645$ and $E = 0.025$; $\hat{p}$ unknown, use $\hat{p} = 0.5$

$$n = [(z_{\alpha/2})^2 \hat{p}\hat{q}]/E^2$$
$$= [(1.645)^2(0.5)(0.5)]/(0.025)^2$$
$$= 1082.41, \text{ rounded up to } 1083$$

8. Let x = the number who have witnessed gender discrimination.
$$\hat{p} = x/n = 126/(126+205) = 126/331 = 0.3807$$
$$\alpha = 0.05, z_{\alpha/2} = z_{0.025} = 1.96$$
$$\hat{p} \pm z_{\alpha/2}\sqrt{\hat{p}\hat{q}/n}$$
$$0.3807 \pm 1.96\sqrt{(0.3807)(0.6193)/331}$$
$$0.3807 \pm 0.0523$$
$$0.328 < p < 0.433 \text{ or } 32.8\% < p < 43.3\%$$

9. This is a binomial problem with n=50 and p=0.50.
   a. $\mu = np = (50)(0.50) = 25.0$
   b. $\sigma^2 = npq = (50)(0.50)(0.50) = 12.5$
      $\sigma = 3.536$, rounded to 3.5
   c. normal approximation appropriate since
      $np = 50(0.50) = 25.0 \geq 5$
      $nq = 50(0.50) = 25.0 \geq 5$
      Use the formula $z = (x - \mu)/\sigma$.
      $$P(x \geq 20) = P_c(x > 19.5)$$
      $$= P(z > -1.56)$$
      $$= 1 - 0.0594$$
      $$= 0.9406$$

10. This is a binomial problem with n=20 and p=0.016
$$P(x) = \frac{n!}{(n-x)!x!}p^x q^{n-x}$$
$$P(x=0) = [20!/(20!0!)](0.016)^0(0.984)^{20}$$
$$= [1](1)(0.984)^{20}$$
$$= 0.724$$
No; since $0.724 \geq 0.05$ such an event is not unusual.

Chapter 10

Correlation and Regression

10-2 Correlation

1. a. r = the correlation in the sample. In this context, r is the linear correlation coefficient computed using the chosen paired (points in Super Bowl, number of new cars sold) values for the randomly selected years in the sample.
   b. ρ = the correlation in the population. In this context, ρ is the linear correlation coefficient computed using all the paired (points in Super Bowl, number of new cars sold) values for every year there has been a Super Bowl.
   c. Since there is no relationship between the number of points scored in a Super Bowl and the number of new cars sold that year, the estimated value of r is 0.

3. Correlation is the existence of a relationship between two variables – so that knowing the value of one of the variables allows a researcher to make a reasonable inference about the value of the other. Correlation measures only association and not causality. If there is an association between two variables, it may or may not be cause-and-effect – and if it is cause-and-effect, there is nothing in the mathematics of correlation analysis to identify which variable is the cause and which is the effect.

5. a. From Table A-6 for n = 62 [closest entry is n=60], C.V. = ±0.254. Therefore r = 0.758 indicates a significant (positive) linear correlation. Yes; there is sufficient evidence to support the claim that there is a linear correlation between the weight of discarded garbage and the household size.
   b. The proportion of the variation in household size that can be explained by the linear relationship between household size and weight of discarded garbage is $r^2 = (0.758)^2 = 0.575$, or 57.5%.

7. a. From Table A-6 for n = 40, C.V. = ±0.312. Therefore r = -0.202 does not indicate a significant linear correlation. No; there is not sufficient evidence to support the claim that there is a linear correlation between the heights and pulse rates of women.
   b. The proportion of the variation in the heights of women that can be explained by the linear relationship between their heights and pulse rates is $r^2 = (-0.202)^2 = 0.041$, or 4.1%.

9. a. Excel produces the following scatterplot.

x	y	xy	x^2	y^2
10	9.14	91.40	100	83.5396
8	8.14	65.12	64	66.2596
13	8.74	113.62	169	76.3876
9	8.77	78.93	81	76.9129
11	9.26	101.86	121	85.7476
14	8.10	113.40	196	65.61
6	6.13	36.78	36	37.5769
4	3.10	12.40	16	9.61
12	9.13	109.56	144	83.3569
7	7.26	50.82	49	52.7076
5	4.74	23.70	25	22.4676
99	82.51	797.59	1001	660.1763

b. See the chart above at the right, where n = 11.

$n(\Sigma xy) - (\Sigma x)(\Sigma y) = 11(797.59) - (99)(82.51) = 605.00$

$n(\Sigma x^2) - (\Sigma x)^2 = 11(1001) - (99)^2 = 1210$

$n(\Sigma y^2) - (\Sigma y)^2 = 11(660.1763) - (82.51)^2 = 454.0392$

$r = [n(\Sigma xy) - (\Sigma x)(\Sigma y)]/[\sqrt{n(\Sigma x^2) - (\Sigma x)^2} \sqrt{n(\Sigma y^2) - (\Sigma y)^2}]$

$= 605.00/[\sqrt{1210}\sqrt{454.0392}] = 0.816$

From Table A-6 for n = 11, C.V. = ±0.602. Therefore r = 0.816 indicates a significant (positive) linear correlation. Yes; there is sufficient evidence to support the claim that there is a linear correlation between the two variables.

c. The scatterplot indicates that the relationship between the variables is quadratic, not linear.

NOTE: In addition to the value of n, calculation of r requires five sums: Σx, Σy, Σx^2, Σy^2 and Σxy. As the sums can usually be found conveniently using a calculator and without constructing a chart as in exercise 9, the remaining exercises give only the values of the sums and do not show a chart. In addition, calculation of r involves three subcalculations.

(1) $n(\Sigma xy) - (\Sigma x)(\Sigma y)$ determines the sign of r. If large values of x are associated with large values of y, it will be positive. If large values of x are associated with small values of y, it will be negative. If not, a mistake has been made.

(2) $n(\Sigma x^2) - (\Sigma x)^2$ cannot be negative. If it is, a mistake had been made.

(3) $n(\Sigma y^2) - (\Sigma y)^2$ cannot be negative. If it is, a mistake had been made.

Finally, r must be between -1 and 1 inclusive. If not, a mistake has been made. If this or any of the previous mistakes occurs, stop immediately and find the error – continuing is a waste of effort.

11. The following table and summary statistics apply to all parts of this exercise.

    x: 1 1 1 2 2 2 3 3 3 10

    y: 1 2 3 1 2 3 1 2 3 10

    using all the points: n =10 $\Sigma x = 28$ $\Sigma y = 28$ $\Sigma xy = 136$ $\Sigma x^2 = 142$ $\Sigma y^2 = 142$

    without the outlier: n = 9 $\Sigma x = 18$ $\Sigma y = 18$ $\Sigma xy = 36$ $\Sigma x^2 = 42$ $\Sigma y^2 = 42$

a. There appears to be a strong positive linear correlation, with r close to 1.

b. $n(\Sigma xy) - (\Sigma x)(\Sigma y) = 10(136) - (28)(28) = 576$

$n(\Sigma x^2) - (\Sigma x)^2 = 10(142) - (28)^2 = 636$

$n(\Sigma y^2) - (\Sigma y)^2 = 10(142) - (28)^2 = 636$

$r = [n(\Sigma xy) - (\Sigma x)(\Sigma y)]/[\sqrt{n(\Sigma x^2) - (\Sigma x)^2} \sqrt{n(\Sigma y^2) - (\Sigma y)^2}]$

$= 576/[\sqrt{636}\sqrt{636}] = 0.906$

From Table A-6 for n = 10, assuming $\alpha = 0.05$, C.V. = ±0.632. Therefore r = 0.906 indicates a significant (positive) linear correlation. This agrees with the interpretation of the scatterplot.

c. There appears to be no linear correlation, with r close to 0.

$n(\Sigma xy) - (\Sigma x)(\Sigma y) = 9(36) - (18)(18) = 0$

$n(\Sigma x^2) - (\Sigma x)^2 = 9(42) - (18)^2 = 54$

$n(\Sigma y^2) - (\Sigma y)^2 = 9(42) - (18)^2 = 54$

$r = [n(\Sigma xy) - (\Sigma x)(\Sigma y)]/[\sqrt{n(\Sigma x^2) - (\Sigma x)^2} \sqrt{n(\Sigma y^2) - (\Sigma y)^2}]$

$= 0/[\sqrt{54}\sqrt{54}] = 0$

From Table A-6 for n = 9 assuming $\alpha = 0.05$, C.V. = ±0.666. Therefore r = 0 does not indicate a significant linear correlation. This agrees with the interpretation of the scatterplot.

d. The effect of a single pair of values can be dramatic, changing the conclusion entirely.

NOTE: In each of exercises 13-28 the first variable listed is designated x, and the second variable listed is designated y. In correlation problems the designation of x and y is arbitrary – so long as a person remains consistent after making the designation. In each test of hypothesis, the C.V. and test statistic are given in terms of t using the P-value Method. The usual t formula written for r is

$$t_r = (r - \mu_r)/s_r, \text{ where } \mu_r = \rho = 0 \text{ and } s_r = \sqrt{(1-r^2)/(n-2)} \text{ and df} = n-2.$$

Performing the test using the t statistic allows the calculation of exact P-values. <u>For the r method, the C.V. in terms of r is given in brackets and indicated on the accompanying graph – and the test statistic is simply r.</u> The two methods are mathematically equivalent and always agree.

The scatterplots for the following exercises were generated by Minitab. Scatterplots produced by other statistical software, while the x and y scales may be slightly different, will produce the same visual impression as to how closely the data cluster around a straight line.

13. a. n = 6

$\Sigma x = 742.7$

$\Sigma y = 6.50$

$\Sigma xy = 1067.910$

$\Sigma x^2 = 118115.51$

$\Sigma y^2 = 9.7700$

b. $n(\Sigma xy) - (\Sigma x)(\Sigma y) = 6(1067.910) - (742.7)(6.50) = 1579.910$

$n(\Sigma x^2) - (\Sigma x)^2 = 6(118115.51) - (742.7)^2 = 157,089.77$

$n(\Sigma y^2) - (\Sigma y)^2 = 6(9.7700) - (6.50)^2 = 16.3700$

$r = [n(\Sigma xy) - (\Sigma x)(\Sigma y)]/[\sqrt{n(\Sigma x^2) - (\Sigma x)^2} \sqrt{n(\Sigma y^2) - (\Sigma y)^2}]$

$= 1579.910/[\sqrt{157,089.77} \sqrt{16.3700}] = 0.985$

c. $H_o: \rho = 0$

$H_1: \rho \neq 0$

$\alpha = 0.05$ and df = 4

C.V. $t = \pm t_{\alpha/2} = \pm t_{0.025} = \pm 2.776$ [or $r = \pm 0.811$]

calculations:

$t_r = (r - \mu_r)/s_r$

$= (0.985 - 0)/\sqrt{[1 - (0.985)^2]/4}$

$= 0.985 /0.08556$

$= 11.504$

P-value = $2 \cdot \text{tcdf}(11.504, 99, 4) = 0.0003$

conclusion:

Reject H_o; there is sufficient evidence to conclude that $\rho \neq 0$ (in fact, that $\rho > 0$). Yes; there is sufficient evidence to support the claim of a linear correlation between the CPI and the cost of a slice of pizza.

15. a. n = 5

$\Sigma x = 455$

$\Sigma y = 816$

$\Sigma xy = 74937$

$\Sigma x^2 = 41923$

$\Sigma y^2 = 134362$

b. $n(\Sigma xy) - (\Sigma x)(\Sigma y) = 5(74937) - (455)(816) = 3405$

$n(\Sigma x^2) - (\Sigma x)^2 = 5(41923) - (255)^2 = 2590$

$n(\Sigma y^2) - (\Sigma y)^2 = 5(134362) - (816)^2 = 5954$

$r = [n(\Sigma xy) - (\Sigma x)(\Sigma y)]/[\sqrt{n(\Sigma x^2) - (\Sigma x)^2}\ \sqrt{n(\Sigma y^2) - (\Sigma y)^2}]$

$= 3405/[\sqrt{2590}\sqrt{5954}] = 0.867$

c. H_o: $\rho = 0$

H_1: $\rho \neq 0$

$\alpha = 0.05$ and df = 3

C.V. $t = \pm t_{\alpha/2} = \pm t_{0.025} = \pm 3.182$ [or $r = \pm 0.878$]

calculations:

$t_r = (r - \mu_r)/s_r$

$= (0.867 - 0)/\sqrt{[1 - (0.867)^2]/3}$

$= 0.867/0.2876$

$= 3.015$

P-value = 2·tcdf(3.015,99,3) = 0.0570

conclusion:

Do not reject H_o; there is not sufficient evidence to conclude that $\rho \neq 0$. No; there is not sufficient evidence to support the claim of a linear correlation between right and left arm systolic blood pressure measurements.

17. a. n = 6

$\Sigma x = 51.0$

$\Sigma y = 1108$

$\Sigma xy = 9639.0$

$\Sigma x^2 = 439.00$

$\Sigma y^2 = 214482$

b. $n(\Sigma xy) - (\Sigma x)(\Sigma y) = 6(9639.0) - (51.0)(1108) = 1326.0$

$n(\Sigma x^2) - (\Sigma x)^2 = 6(439.00) - (51.0)^2 = 33.00$

$n(\Sigma y^2) - (\Sigma y)^2 = 6(214482) - (1108)^2 = 59,228$

$r = [n(\Sigma xy) - (\Sigma x)(\Sigma y)]/[\sqrt{n(\Sigma x^2) - (\Sigma x)^2}\ \sqrt{n(\Sigma y^2) - (\Sigma y)^2}]$

$= 1326.0/[\sqrt{33.00}\sqrt{59,228}] = 0.948$

c. H_o: $\rho = 0$
   H_1: $\rho \neq 0$
   $\alpha = 0.05$ and df = 4
   C.V. $t = \pm t_{\alpha/2} = \pm t_{0.025} = \pm 2.776$ [or r = ± 0.811]
   calculations:

$t_r = (r - \mu_r)/s_r$

$= (0.948 - 0)/\sqrt{[1 - (0.948)^2]/4}$

$= 0.948/0.1592$

$= 5.956$

P-value = $2 \cdot$tcdf(5.956,99,4) = 0.0040

conclusion:

Reject H_o; there is sufficient evidence to conclude that $\rho \neq 0$ (in fact, that $\rho > 0$). Yes; there is sufficient evidence to support the claim of a linear correlation between the overhead widths of seals from photographs and the weights of the seals.

19. a. n = 7
   $\Sigma x = 1908$
   $\Sigma y = 4832$
   $\Sigma xy = 1340192$
   $\Sigma x^2 = 523336$
   $\Sigma y^2 = 3661094$

b. $n(\Sigma xy) - (\Sigma x)(\Sigma y) = 7(1340192) - (1908)(4832) = 161{,}888$
   $n(\Sigma x^2) - (\Sigma x)^2 = 7(523336) - (1908)^2 = 22{,}888$
   $n(\Sigma y^2) - (\Sigma y)^2 = 7(3661094) - (4832)^2 = 2{,}279{,}434$

$r = [n(\Sigma xy) - (\Sigma x)(\Sigma y)]/[\sqrt{n(\Sigma x^2) - (\Sigma x)^2} \sqrt{n(\Sigma y^2) - (\Sigma y)^2}]$

$= 161{,}888/[\sqrt{22{,}888}\sqrt{2{,}279{,}434}] = 0.709$

c. H_o: $\rho = 0$
   H_1: $\rho \neq 0$
   $\alpha = 0.05$ and df = 5
   C.V. $t = \pm t_{\alpha/2} = \pm t_{0.025} = \pm 2.571$ [or r = ± 0.754]
   calculations:

$t_r = (r - \mu_r)/s_r$

$= (0.709 - 0)/\sqrt{[1 - (0.709)^2]/5}$

$= 0.709/0.3155$

$= 2.247$

P-value = $2 \cdot$tcdf(2.247,99,5) = 0.0746

conclusion:

Do not reject H_o; there is not sufficient evidence to conclude that $\rho \neq 0$. No; there is not sufficient evidence to support the claim of a linear correlation between the costs of tickets purchased 30 days in advance and those purchased one day in advance.

21. a. n = 7
   $\Sigma x = 16890$
   $\Sigma y = 11303$
   $\Sigma xy = 24833485$
   $\Sigma x^2 = 53892334$
   $\Sigma y^2 = 23922183$

b. $n(\Sigma xy) - (\Sigma x)(\Sigma y) = 7(24833)(485) - (16890)(11303) = -17,073,275$
   $n(\Sigma x^2) - (\Sigma x)^2 = 7(53892334) - (16890)^2 = 91,974,238$
   $n(\Sigma y^2) - (\Sigma y)^2 = 7(23922183) - (11303)^2 = 39,697,472$

   $r = [n(\Sigma xy) - (\Sigma x)(\Sigma y)]/[\sqrt{n(\Sigma x^2) - (\Sigma x)^2} \sqrt{n(\Sigma y^2) - (\Sigma y)^2}]$

   $= -17,073,275/[\sqrt{91,974,238}\sqrt{39,697,472}] = -0.283$

c. $H_o: \rho = 0$
   $H_1: \rho \neq 0$
   $\alpha = 0.05$ and df = 5
   C.V. $t = \pm t_{\alpha/2} = \pm t_{0.025} = \pm 2.571$ [or r = ±0.754]
   calculations:

   $t_r = (r - \mu_r)/s_r$

   $= (-0.283 - 0)/\sqrt{[1 - (-0.283)^2]/5}$

   $= -0.283/0.4290$

   $= -0.659$

   P-value = 2·tcdf(-99,-0.659,5) = 0.5392

conclusion:
   Do not reject H_o; there is not sufficient evidence to conclude that $\rho \neq 0$. No; there is not sufficient evidence to support the claim of a linear correlation between the repair costs from full-front crashes and full-rear crashes.

23. a. n = 10
   $\Sigma x = 3377$
   $\Sigma y = 141.7$
   $\Sigma xy = 47888.6$
   $\Sigma x^2 = 1143757$
   $\Sigma y^2 = 2008.39$

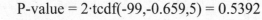

b. $n(\Sigma xy) - (\Sigma x)(\Sigma y) = 10(47888.6) - (3377)(141.7) = 365.1$
   $n(\Sigma x^2) - (\Sigma x)^2 = 10(1143757) - (3377)^2 = 33,441$
   $n(\Sigma y^2) - (\Sigma y)^2 = 10(2008.39) - (141.7)^2 = 5.01$

   $r = [n(\Sigma xy) - (\Sigma x)(\Sigma y)]/[\sqrt{n(\Sigma x^2) - (\Sigma x)^2} \sqrt{n(\Sigma y^2) - (\Sigma y)^2}]$

   $= 365.1/[\sqrt{33,441}\sqrt{5.01}] = 0.892$

c. H_o: $\rho = 0$
H_1: $\rho \neq 0$
$\alpha = 0.05$ and df = 8
C.V. $t = \pm t_{\alpha/2} = \pm t_{0.025} = \pm 2.306$ [or r = ± 0.632]
calculations:
   $t_r = (r - \mu_r)/s_r$
      $= (0.892 - 0)/\sqrt{[1 - (0.892)^2]/8}$
      $= 0.892/0.1598$
      $= 5.581$
   P-value = $2 \cdot \text{tcdf}(5.581,99,8) = 0.0005$
conclusion:

   Reject H_o; there is sufficient evidence to conclude that $\rho \neq 0$ (in fact, that $\rho > 0$). Yes; there is sufficient evidence to support the claim of a linear correlation between global temperature and the concentration of CO_2.

25. a. n = 7
   $\Sigma x = 154$
   $\Sigma y = 3.531$
   $\Sigma xy = 118.173$
   $\Sigma x^2 = 86016$
   $\Sigma y^2 = 1.807253$

b. $n(\Sigma xy) - (\Sigma x)(\Sigma y) = 7(118.173) - (154)(3.531) = 283.437$
   $n(\Sigma x^2) - (\Sigma x)^2 = 7(86016) - (152)^2 = 578{,}396$
   $n(\Sigma y^2) - (\Sigma y)^2 = 7(1.807253) - (3.531)^2 = 0.182810$
   $r = [n(\Sigma xy) - (\Sigma x)(\Sigma y)]/[\sqrt{n(\Sigma x^2) - (\Sigma x)^2} \sqrt{n(\Sigma y^2) - (\Sigma y)^2}]$
      $= 283.437/[\sqrt{578{,}396} \sqrt{0.182810}] = 0.872$
c. H_o: $\rho = 0$
   H_1: $\rho \neq 0$
   $\alpha = 0.05$ and df = 5
   C.V. $t = \pm t_{\alpha/2} = \pm t_{0.025} = \pm 2.571$ [or r = ± 0.754]
   calculations:
      $t_r = (r - \mu_r)/s_r$
         $= (0.872 - 0)/\sqrt{[1 - (0.872)^2]/5}$
         $= 0.872/0.2192$
         $= 3.977$
      P-value = $2 \cdot \text{tcdf}(3.977,99,5) = 0.0106$

conclusion:
   Reject H_o; there is sufficient evidence to conclude that $\rho \neq 0$ (in fact, that $\rho > 0$). Yes; there is sufficient evidence to support the claim of a linear correlation between a team's proportion of wins and its difference between numbers of runs scored and runs allowed.

27. a. n = 10
     $\Sigma x = 10821$
     $\Sigma y = 1028$
     $\Sigma xy = 1114491$
     $\Sigma x^2 = 11782515$
     $\Sigma y^2 = 107544$

b. $n(\Sigma xy) - (\Sigma x)(\Sigma y) = 8(1114491) - (10821)(1028) = 20{,}922$
   $n(\Sigma x^2) - (\Sigma x)^2 = 10(11782515) - (10821)^2 = 731{,}109$
   $n(\Sigma y^2) - (\Sigma y)^2 = 10(107544) - (1028)^2 = 18{,}656$

   $r = [n(\Sigma xy) - (\Sigma x)(\Sigma y)]/[\sqrt{n(\Sigma x^2) - (\Sigma x)^2} \sqrt{n(\Sigma y^2) - (\Sigma y)^2}]$

   $= 20{,}922/[\sqrt{731{,}109}\sqrt{18{,}656}] = 0.179$

c. $H_o: \rho = 0$
   $H_1: \rho \neq 0$
   $\alpha = 0.05$ and df = 8
   C.V. $t = \pm t_{\alpha/2} = \pm t_{0.025} = \pm 2.306$ [or $r = \pm 0.632$]
   calculations:
   $t_r = (r - \mu_r)/s_r$

   $= (0.179 - 0)/\sqrt{[1 - (0.179)^2]/8}$

   $= 0.179/0.3478$

   $= 0.515$

   P-value $= 2 \cdot \text{tcdf}(0.505,99,8) = 0.6205$
   conclusion:
   Do not reject H_o; there is not sufficient evidence to conclude that $\rho \neq 0$. No; there is not sufficient evidence to support the claim of a linear correlation between brain size and IQ score. No; it does not appear that people with larger brains are more intelligent.

NOTE: Exercises 29-32 involve large data sets from Appendix B. Use statistical software to find the sample correlation, and then proceed as usual using that value. Those using the P-value method to test an hypothesis about a correlation will be limited by the degree of accuracy with which the sample correlation is reported by the statistical software. This manual proceeds using the 3 decimal accuracy for r reported by Minitab as if it were the exact sample value.

29. For the n=35 paired sample values,
    the Minitab regression of c3 on c4 yields
    $r = 0.744$.

H_o: $\rho = 0$
H_1: $\rho \neq 0$
$\alpha = 0.05$ and df $= 33$
C.V. $t = \pm t_{\alpha/2} = \pm t_{0.025} = \pm 2.035$ [or $r = \pm 0.335$]
calculations:

$t_r = (r - \mu_r)/s_r$

$= (0.744 - 0)/\sqrt{[1 - (0.744)^2]/33}$

$= 0.744/0.1163$

$= 6.396$

P-value $= 2 \cdot \text{tcdf}(6.396, 99, 33)$

$= 3.018\text{E-}7 = 0.0000003$

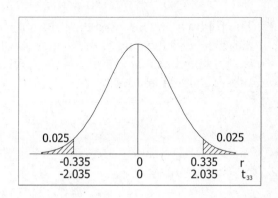

conclusion:

Reject H_o; there is sufficient evidence to conclude that $\rho \neq 0$ (in fact, that $\rho > 0$). Yes; there is sufficient evidence to support the claim of a linear correlation between a movie's budget amount and the amount that movie grosses.

31. For the n=56 paired sample values, the Minitab regression of c1 on c2 yields $r = 0.319$.

WORD COUNTS FOR COUPLES

female of the couple

male of the couple

H_o: $\rho = 0$
H_1: $\rho \neq 0$
$\alpha = 0.05$ and df $= 54$
C.V. $t = \pm t_{\alpha/2} = \pm t_{0.025} = \pm 2.009$ [or $r = \pm 0.254$]
calculations:

$t_r = (r - \mu_r)/s_r$

$= (0.319 - 0)/\sqrt{[1 - (0.319)^2]/54}$

$= 0.319/0.1290$

$= 2.473$

P-value $= 2 \cdot \text{tcdf}(2.473, 99, 54) = 0.0166$

conclusion:

Reject H_o; there is sufficient evidence to conclude that $\rho \neq 0$ (in fact, that $\rho > 0$). Yes; there is sufficient evidence to support the claim of a linear correlation between the numbers of words spoken by men and women who are a couple.

33. A significant linear correlation indicates that the factors are associated, not that there is a cause-and-effect relationship. Even if there is a cause-and-effect relationship, correlation analysis cannot identify which factor is the cause and which factor is the effect.

35. A significant linear correlation between group averages indicates nothing about the relationship between the individual scores – which may be uncorrelated, correlated in the opposite direction, or have different correlations in each of the groups.

37. The following table gives the values for y, x, x², log x, √x and 1/x. The rows at the bottom of the table give the sum of the values (Σv), the sum of the squares of the values (Σv²), the sum of each value times the corresponding y value (Σvy), and the quantity $n\Sigma v^2 - (\Sigma v)^2$ needed in subsequent calculations.

	y	x	x²	log x	√x	1/x
	0	1	1	0	1.0000	1.0000
	0.3	2	4	0.3010	1.4142	0.5000
	0.5	3	9	0.4771	1.7321	0.3333
	0.6	4	16	0.6021	2.0000	0.2500
	0.7	5	25	0.6990	2.2361	0.2000
	0.9	8	64	0.9031	2.8284	0.1250
Σv	3	23	119	2.9823	11.2108	2.4083
Σv²	2	119	5075	1.9849	23.0000	1.4792
Σvy		15.2	90.4	1.9922	6.6011	0.7192
$n\Sigma v^2 - (\Sigma v)^2$	3	185	16289	3.0153	12.3189	3.0753

In general, $r = [n(\Sigma vy) - (\Sigma v)(\Sigma y)]/[\sqrt{n(\Sigma v^2) - (\Sigma v)^2}\sqrt{n(\Sigma y^2) - (\Sigma y)^2}]$

a. For v = x,

$r = [6(15.2) - (23)(3)]/[\sqrt{185}\sqrt{3}] = 0.9423$

b. For v = x²,

$r = [6(90.4) - (119)(3)]/[\sqrt{16289}\sqrt{3}] = 0.8387$

c. For v = log x,

$r = [6(1.9922) - (2.9823)(3)]/[\sqrt{3.0153}\sqrt{3}] = 0.9996$

d. For v = √x ,

$r = [6(6.6011) - (11.2108)(3)]/[\sqrt{12.3189}\sqrt{3}] = 0.9827$

e. For v = 1/x,

$r = [6(0.7192) - (2.4083)(3)]/[\sqrt{3.0753}\sqrt{3}] = -0.9580$

In each case the critical values from Table A-6 for testing significance at the 0.05 level are ±0.811. While all the correlations except for (b) are significant, the largest value for r occurs in part (c).

10-3 Regression

1. The symbol ŷ represents the predicted cholesterol level. The predictor variable x represents weight. The response variable y represents cholesterol level.

3. Since s_y and s_x must be non-negative, the regression line has a slope (which is equal to $r \cdot s_y/s_x$) with the same sign as r. If r is positive, the slope of the regression line is positive and the regression line rises as it goes from left to right. If r is negative, the slope of the regression line is negative and the regression line fall as it goes from left to right.

5. For n=62, C.V. = ±0.254. Since r = 0.759 > 0.254, use the regression line for prediction.

$\hat{y}$ = 0.445 + 0.119x

$\hat{y}_{50}$ = 0.455 + 0.119(50) = 6.4 people

7. For n=40, C.V. = ±0.312. Since r = 0.202 < 0.312, use the mean for prediction.

$\hat{y} = \bar{y}$

$\hat{y}_{70}$ = 76.3 beats/minute

9. Excel produces the following scatterplot.

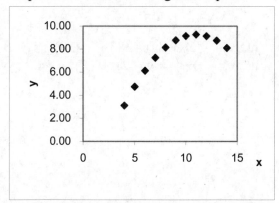

x	y	xy	x^2	y^2
10	9.14	91.40	100	83.5396
8	8.14	65.12	64	66.2596
13	8.74	113.62	169	76.3876
9	8.77	78.93	81	76.9129
11	9.26	101.86	121	85.7476
14	8.10	113.40	196	65.61
6	6.13	36.78	36	37.5769
4	3.10	12.40	16	9.61
12	9.13	109.56	144	83.3569
7	7.26	50.82	49	52.7076
5	4.74	23.70	25	22.4676
99	82.51	797.59	1001	660.1763

See the chart above at the right, where n = 11.

$\bar{x} = (\Sigma x)/n = 99/11 = 9.0$ $n(\Sigma xy) - (\Sigma x)(\Sigma y) = 11(797.59) - (99)(82.51) = 605.00$

$\bar{y} = (\Sigma y)/n = 82.52/11 = 7.50$ $n(\Sigma x^2) - (\Sigma x)^2 = 11(1001) - (99)^2 = 1210$

$b_1 = [n(\Sigma xy) - (\Sigma x)(\Sigma y)]/[n(\Sigma x^2) - (\Sigma x)^2]$

    = 605.00/1210 = 0.500

$b_o = \bar{y} - b_1\bar{x}$

    = 7.50 - 0.500(9.0) = 3.00

$\hat{y} = b_o + b_1 x$

$\hat{y}$ = 3.00 + 0.500x

The scatterplot indicates that the relationship between the variables is quadratic, not linear.

NOTE: In addition to the value of n, calculations associated with regression involve five sums: Σx, Σy, Σx^2, Σy^2 and Σxy. As the sums can usually be found conveniently using a calculator, the remaining exercises give only the values of the sums without constructing a chart as in exercises 9 and 10,. In addition, the calculations typically involve the following subcalculations.

(1) $n(\Sigma xy) - (\Sigma x)(\Sigma y)$ determines the sign of the slope of the regression line. If large values of x are associated with large values of y, it will be positive. If large values of x are associated with small values of y, it will be negative. If not, a mistake has been made.

(2) $n(\Sigma x^2) - (\Sigma x)^2$ cannot be negative. If it is, a mistake had been made.

(3) $n(\Sigma y^2) - (\Sigma y)^2$ cannot be negative. If it is, a mistake had been made.

If any of these mistakes occurs, stop immediately and find the error – continuing is wasted effort.

11. a. using all the points: $n = 10$ $\Sigma x = 28$ $\Sigma y = 28$ $\Sigma xy = 136$ $\Sigma x^2 = 142$ $\Sigma y^2 = 142$

$\bar{x} = (\Sigma x)/n = 28/10 = 2.8$ $n(\Sigma xy) - (\Sigma x)(\Sigma y) = 10(136) - (28)(28) = 576$

$\bar{y} = (\Sigma y)/n = 28/10 = 2.8$ $n(\Sigma x^2) - (\Sigma x)^2 = 10(142) - (28)^2 = 636$

$b_1 = [n(\Sigma xy) - (\Sigma x)(\Sigma y)]/[n(\Sigma x^2) - (\Sigma x)^2]$

   $= 576/636 = 0.906$

$b_0 = \bar{y} - b_1 \bar{x}$

   $= 2.8 - 0.906(2.8) = 0.264$

$\hat{y} = b_0 + b_1 x$

$\hat{y} = 0.264 + 0.906x$

b. without the outlier: $n = 9$ $\Sigma x = 18$ $\Sigma y = 18$ $\Sigma xy = 36$ $\Sigma x^2 = 42$ $\Sigma y^2 = 42$

$\bar{x} = (\Sigma x)/n = 18/9 = 2.0$ $n(\Sigma xy) - (\Sigma x)(\Sigma y) = 9(36) - (18)(18) = 0$

$\bar{y} = (\Sigma y)/n = 18/9 = 2.0$ $n(\Sigma x^2) - (\Sigma x)^2 = 9(42) - (18)^2 = 54$

$b_1 = [n(\Sigma xy) - (\Sigma x)(\Sigma y)]/[n(\Sigma x^2) - (\Sigma x)^2]$

   $= 0/54 = 0$

$b_0 = \bar{y} - b_1 \bar{x}$

   $= 2.0 - 0(2.0) = 2.0$

$\hat{y} = b_0 + b_1 x$

$\hat{y} = 2.0 + 0x$ [or simply $\hat{y} = 2.0$, for any x]

c. The results are very different – without the outlier, x has no predictive value for y. A single outlier can have a dramatic effect on the regression equation.

NOTE: For exercises 13-26, the exact summary statistics (i.e., without any rounding) are given on the right. While the intermediate calculations on the left are presented rounded to various degrees of accuracy, the entire unrounded values were preserved in the calculator until the end. When finding a predicted value, always verify that it is reasonable for the story problem and consistent with the given data points used to find the regression equation. The final prediction is made either using the regression equation $\hat{y} = b_0 + b_1 x$ or the sample mean $\bar{y}$. Refer back to the corresponding test for a significant linear correlation in the previous section (the exercise numbers are the same), and use $\hat{y} = b_0 + b_1 x$ only if there is a significant linear correlation.

13. $\bar{x} = 123.78$ $n = 6$

   $\bar{y} = 1.08$ $\Sigma x = 742.7$

   $b_1 = [n(\Sigma xy) - (\Sigma x)(\Sigma y)]/[n(\Sigma x^2) - (\Sigma x)^2]$ $\Sigma y = 6.50$

      $= 1579.910/157,089.77 = 0.0101$ $\Sigma x^2 = 118115.51$

   $b_0 = \bar{y} - b_1 \bar{x}$ $\Sigma y^2 = 9.7700$

      $= 1.08 - 0.0101(123.78) = -0.162$ $\Sigma xy = 1067.910$

   $\hat{y} = b_0 + b_1 x = -0.162 + 0.0101x$

   $\hat{y}_{182.5} = -0.162 + 0.0101(182.5) = \$1.67$ [$\$1.68$ using rounded values]

15. $\bar{x} = 91.0$ $n = 5$

   $\bar{y} = 163.2$ $\Sigma x = 455$

   $b_1 = [n(\Sigma xy) - (\Sigma x)(\Sigma y)]/[n(\Sigma x^2) - (\Sigma x)^2]$ $\Sigma y = 816$

      $= 3405/2590 = 1.315$ $\Sigma x^2 = 41923$

   $b_0 = \bar{y} - b_1 \bar{x}$ $\Sigma y^2 = 134362$

      $= 163.2 - 1.315(91.0) = 43.56$ $\Sigma xy = 74937$

   $\hat{y} = b_0 + b_1 x = 43.6 + 1.31x$

   $\hat{y}_{100} = \bar{y} = 163.2$ mm Hg [no significant correlation]

17. $\bar{x} = 8.50$ $n = 6$

    $\bar{y} = 184.67$ $\Sigma x = 51.0$

    $b_1 = [n(\Sigma xy) - (\Sigma x)(\Sigma y)]/[n(\Sigma x^2) - (\Sigma x)^2]$ $\Sigma y = 1108$

       $= 1326.0/33.00 = 40.18$ $\Sigma x^2 = 439.00$

    $b_o = \bar{y} - b_1\bar{x}$ $\Sigma y^2 = 214482$

       $= 184.67 - 40.18(8.50) = -156.87$ $\Sigma xy = 9639.0$

    $\hat{y} = b_o + b_1 x = -156.9 + 40.2x$

    $\hat{y}_{9.0} = -156.9 + 40.2(9.0) = 204.8$ kg

19. $\bar{x} = 272.57$ $n = 7$

    $\bar{y} = 690.29$ $\Sigma x = 1908$

    $b_1 = [n(\Sigma xy) - (\Sigma x)(\Sigma y)]/[n(\Sigma x^2) - (\Sigma x)^2]$ $\Sigma y = 4832$

       $= 161,888/22,888 = 7.07$ $\Sigma x^2 = 523336$

    $b_o = \bar{y} - b_1\bar{x}$ $\Sigma y^2 = 3661094$

       $= 690.29 - 7.07(272.57) = -1237.62$ $\Sigma xy = 1340192$

    $\hat{y} = b_o + b_1 x = -1237.6 + 7.07x$

    $\hat{y}_{300} = \bar{y} = \$690.3$ [no significant correlation]

21. $\bar{x} = 2412.86$ $n = 7$

    $\bar{y} = 1614.71$ $\Sigma x = 16890$

    $b_1 = [n(\Sigma xy) - (\Sigma x)(\Sigma y)]/[n(\Sigma x^2) - (\Sigma x)^2]$ $\Sigma y = 11303$

       $= -17,073,275/91,974,238 = -0.186$ $\Sigma x^2 = 53892334$

    $b_o = \bar{y} - b_1\bar{x}$ $\Sigma y^2 = 23922183$

       $= 1614.71 - (-0.186)(2412.86) = 2062.62$ $\Sigma xy = 24833485$

    $\hat{y} = b_o + b_1 x = 2062.6 - 0.186x$

    $\hat{y}_{4594} = \bar{y} = \$1614.7$ [no significant correlation]

The result does not compare very well to the actual repair cost of $982.

23. $\bar{x} = 337.70$ $n = 10$

    $\bar{y} = 14.17$ $\Sigma x = 3377$

    $b_1 = [n(\Sigma xy) - (\Sigma x)(\Sigma y)]/[n(\Sigma x^2) - (\Sigma x)^2]$ $\Sigma y = 141.7$

       $= 365.1/33,441 = 0.0109$ $\Sigma x^2 = 1143757$

    $b_o = \bar{y} - b_1\bar{x}$ $\Sigma y^2 = 2008.39$

       $= 14.17 - 0.0109(337.70) = 10.48$ $\Sigma xy - 47888.6$

    $\hat{y} = b_o + b_1 x = 10.5 + 0.0109x$

    $\hat{y}_{370.9} = 10.5 + 0.0109(370.9) = 14.5$ °C

Yes; in this instance the predicted temperature is equal to the actual temperature of 14.5 °C.

25. $\bar{x} = 22.00$ $n = 7$

    $\bar{y} = 0.504$ $\Sigma x = 154$

    $b_1 = [n(\Sigma xy) - (\Sigma x)(\Sigma y)]/[n(\Sigma x^2) - (\Sigma x)^2]$ $\Sigma y = 3.531$

       $= 283.437/578,396 = 0.000490$ $\Sigma x^2 = 86016$

    $b_o = \bar{y} - b_1\bar{x}$ $\Sigma y^2 = 1.807253$

       $= 0.504 - 0.000490(22.00) = 0.494$ $\Sigma xy = 118.173$

    $\hat{y} = b_o + b_1 x = 0.494 + 0.000490x$

    $\hat{y}_{52} = 0.494 + 0.000490(52) = 0.519$

Yes; the predicted proportion is reasonably close to the actual proportion of 0.543.

27. $\overline{x} = 1082.10$ $n = 10$
    $\overline{y} = 102.80$ $\Sigma x = 10821$
    $b_1 = [n(\Sigma xy) - (\Sigma x)(\Sigma y)]/[n(\Sigma x^2) - (\Sigma x)^2]$ $\Sigma y = 1028$
    $ = 20{,}922/731{,}109 = 0.0286$ $\Sigma x^2 = 11782515$
    $b_o = \overline{y} - b_1\overline{x}$ $\Sigma y^2 = 107544$
    $ = 102.80 - 0.0286(1082.10) = 71.83$ $\Sigma xy = 1114491$
    $\hat{y} = b_o + b_1 x = 71.8 - 0.0286x$
    $\hat{y}_{1275} = \overline{y} = 102.8$ [no significant correlation]

NOTE: Exercises 29-32 involve large data sets from Appendix B. Use statistical software to find the regression equation. When finding a predicted value, always verify that it is reasonable for the story problem and consistent with the given data points used to find the regression equation. The final prediction is made either using the regression equation $\hat{y} = b_o + b_1 x$ or the sample mean $\overline{y}$.
Refer back to the corresponding test for a significant linear correlation in the previous section (the exercise numbers are the same), and use $\hat{y} = b_o + b_1 x$ only if there is a significant linear correlation. If there is no significant linear correlation, use statistical software to find the mean of the response variable (i.e., the y variable) and use that for the predicted value.

29. For the n=35 paired sample values, the Minitab regression of c4 on c3 yields
    gross = 20.6 + 1.38 budget
    $\hat{y} = 20.6 + 1.38x$
    $\hat{y}_{120} = 20.6 + 1.38(120) = 186.2$ million \$

31. For the n=56 paired sample values, the Minitab regression of c2 on c1 yields
    1F = 13439 + 0.302 1M
    $\hat{y} = 13439 + 0.302x$
    $\hat{y}_{6000} = 13439 + 0.302(6000) = 15{,}248$ words per day

33. If $H_o: \rho = 0$ is true, there is no linear correlation between x and y and $\hat{y} = \overline{y}$ is the appropriate
    prediction for y for any x.
    If $Ho: \beta_1 = 0$ is true, then the true regression line is $y = \beta_o + 0x = \beta_o$ and the best estimate for β_o
    is $b_o = \overline{y} - 0\overline{x} = \overline{y}$, producing the line $\hat{y} = \overline{y}$.
    Since both hypotheses imply precisely the same result, they are equivalent.

35. Refer to the table at the right, where
       x = the pulse rate
       y = the systolic blood pressure
       $\hat{y} = 71.68 + 0.5956x$
          = the value predicted by the regression equation
    $y - \hat{y}$ = the residuals for the regression line

x	y	$\hat{y}$	$y-\hat{y}$
68	125	112.181	12.819
64	107	109.798	-2.798
88	126	124.093	1.907
72	110	114.563	-4.563
64	110	109.798	0.202
72	107	114.563	-7.563
428	685	684.997	0.003

The residual plot on the following page was obtained by plotting the predictor variable (pulse rate) on the horizontal axis and the corresponding residual from the table on the vertical axis. The scatterplot shows the original (x,y) = (pulse,systolic) pairs.
    The residual plot seems to suggest that the regression equation is a good model – because the residuals are randomly scattered around the zero line, with no obvious pattern or change in

variability. The scatterplot suggests that the regression equation is not a good model – because the points do not appear to fit a straight line pattern.

10-4 Variation and Prediction Intervals

1. In general, s measures the spread of the data around some reference. For a set of y values in one dimension, s_y measures the spread of the y values around $\overline{y}$. For ordered pairs (x,y) in two dimensions, s_y measures the spread of the points around the line $y = \overline{y}$. For ordered pairs (x,y), s_e measures the spread of the points around the regression line $\hat{y} = b_o + b_1x$.

3. By providing a range of values instead of a single point, a prediction interval gives an indication of the accuracy of the prediction. A confidence interval is an interval estimate of a parameter – i.e., of a conceptually fixed, although unknown, value. A prediction interval is an interval estimate of a random variable – i.e., of a value from a distribution of values.

5. The coefficient of determination is $r^2 = (0.873)^2 = 0.762$.
   The portion of the total variation in y explained by the regression is $r^2 = 0.762 = 76.2\%$

7. The coefficient of determination is $r^2 = (-0.865)^2 = 0.748$.
   The portion of the total variation in y explained by the regression is $r^2 = 0.748 = 74.8\%$.

9. Since the slope of the regression line $b_1 = r \cdot (s_y/s_x)$ is negative, r must be negative.
   Since $r^2 = 65.0\% = 0.650$, $r = -\sqrt{0.650} = -0.806$.
   For n=32 [closest entry is n=30], Table A-6 gives C.V. = ±0.361.
   Since -0.806 < -0.361, there is sufficient evidence to support the claim of a linear correlation between the weights of cars and their highway fuel consumption amounts.

11. The given point estimate is $\hat{y} = 27.028$ mpg.

NOTE: The following summary statistics apply to exercises 13-16 and 17-20. They are all that is necessary to use the chapter formulas to work the problems.

exercise #13	exercise #14	exercise #15	exercise #16
n = 6	n = 6	n = 6	n = 10
$\Sigma x = 742.7$	$\Sigma x = 742.77$	$\Sigma x = 51.0$	$\Sigma x = 3377$
$\Sigma y = 6.50$	$\Sigma y = 6.35$	$\Sigma y = 1108$	$\Sigma y = 141.7$
$\Sigma x^2 = 118115.51$	$\Sigma x^2 = 118115.51$	$\Sigma x^2 = 439.00$	$\Sigma x^2 = 1143757$
$\Sigma y^2 = 9.7700$	$\Sigma y^2 = 9.2175$	$\Sigma y^2 = 214482$	$\Sigma y^2 = 2008.39$
$\Sigma xy = 1067.910$	$\Sigma xy = 1036.155$	$\Sigma xy = 9639.0$	$\Sigma xy = 47888.6$
see also 10.2-3 #13	see also 10.2-3 #14	see also 10.2-3 #17	see also 10.2-3 #23

13. The predicted values were calculated using the regression line $\hat{y} = -0.161601 + 0.0100574x$.

x	y	$\hat{y}$	$\overline{y}$	$\hat{y}-\overline{y}$	$(\hat{y}-\overline{y})^2$	$y-\hat{y}$	$(y-\hat{y})^2$	$y-\overline{y}$	$(y-\overline{y})^2$
30.2	0.15	0.142	1.083	-0.940	0.886	0.008	0.000	-0.930	0.871
48.3	0.35	0.324	1.083	-0.760	0.576	0.026	0.001	-0.730	0.538
112.3	1.00	0.968	1.083	-0.120	0.013	0.032	0.001	-0.080	0.007
162.2	1.25	1.470	1.083	0.386	0.149	-0.220	0.048	0.167	0.028
191.9	1.75	1.768	1.083	0.685	0.469	-0.018	0.000	0.667	0.444
197.8	2.00	1.828	1.083	0.744	0.554	0.172	0.030	0.917	0.840
742.7	6.50	6.500	6.500	0.000	2.648	0.000	0.080	0.000	2.728

a. The explained variation is $\Sigma(\hat{y}-\overline{y})^2 = 2.648$
b. The unexplained variation is $\Sigma(y-\hat{y})^2 = 0.080$
c. The total variation is $\Sigma(y-\overline{y})^2 = 2.728$
d. $r^2 = \Sigma(\hat{y}-\overline{y})^2 / \Sigma(y-\overline{y})^2$
   $= 2.648/2.728 = 0.971$
e. $s_e^2 = \Sigma(y-\hat{y})^2/(n-2)$
   $= 0.080/4 = 0.020$
   $s_e = \sqrt{0.020} = 0.141$

NOTE: A table such as the one in the preceding exercise organizes the work and provides all the values needed to discuss variation. In such a table, the following must always be true (except for minor discrepancies due to rounding) and can be used as a check before proceeding.
   (1) $\Sigma y = \Sigma \hat{y} = \Sigma \overline{y}$
   (2) $\Sigma(\hat{y}-\overline{y}) = \Sigma(y-\hat{y}) = \Sigma(y-\overline{y}) = 0$
   (3) $\Sigma(y-\hat{y})^2 + \Sigma(\hat{y}-\overline{y})^2 = \Sigma(y-\overline{y})^2$

15. The predicted values were calculated using the regression line $\hat{y} = -156.879 + 40.1818x$.

x	y	$\hat{y}$	$\overline{y}$	$\hat{y}-\overline{y}$	$(\hat{y}-\overline{y})^2$	$y-\hat{y}$	$(y-\hat{y})^2$	$y-\overline{y}$	$(y-\overline{y})^2$
7.2	116	132.43	184.67	-52.20	2728.67	-16.43	269.94	-68.70	4715.11
7.4	154	140.47	184.67	-44.20	1953.67	13.53	183.16	-30.70	940.44
9.8	245	236.9	184.67	52.24	2728.60	8.097	65.567	60.33	3640.11
9.4	202	220.83	184.67	36.16	1307.78	-18.83	354.57	17.33	300.44
8.8	200	196.72	184.67	12.05	145.303	3.279	10.753	15.33	235.11
8.4	191	180.65	184.67	-4.02	16.15	10.35	107.16	6.33	40.11
51.0	1108	1108.00	1108.00	0.00	8880.17	0.00	991.15	0.00	9871.33

a. The explained variation is $\Sigma(\hat{y}-\overline{y})^2 = 8880.17$
b. The unexplained variation is $\Sigma(y-\hat{y})^2 = 991.15$
c. The total variation is $\Sigma(y-\overline{y})^2 = 9871.33$
d. $r^2 = \Sigma(\hat{y}-\overline{y})^2 / \Sigma(y-\overline{y})^2$
   $= 8880.17/9871.33 = 0.900$
e. $s_e^2 = \Sigma(y-\hat{y})^2/(n-2)$
   $= 991.15/4 = 247.7875$
   $s_e = \sqrt{247.7875} = 15.74$

17. a. $\hat{y} = -0.161601 + 0.0100574x$

    $\hat{y}_{187.1} = -0.161601 + 0.0100574(187.1) = 1.7201$, rounded to \$1.72

   b. preliminary calculations for $n = 6$

    $\overline{x} = (\Sigma x)/n = 742.7/6 = 123.783$

    $n\Sigma x^2 - (\Sigma x)^2 = 6(118115.51) - (742.7)^2 = 157,089.77$

    $\alpha = 0.05$ and $df = n-2 = 4$

$$\hat{y} \pm t_{\alpha/2}s_e \sqrt{1 + 1/n + n(x_o - \overline{x})^2/[n\Sigma x^2 - (\Sigma x)^2]}$$

$$\hat{y}_{187.1} \pm t_{0.025}(0.141) \sqrt{1 + 1/6 + 6(187.1-123.783)^2/[157089.77]}$$

    $1.7201 \pm (2.776)(0.141)\sqrt{1.31979}$

    1.7201 ± 0.4450

    $1.27 < y_{187.1} < 2.17$ (dollars)

19. a. $\hat{y} = -156.879 + 40.1818x$

    $\hat{y}_{9.0} = -156.879 + 40.1818(9.0) = 204.757$, rounded to 204.8 kg

   b. preliminary calculations for $n = 6$

    $\overline{x} = (\Sigma x)/n = 51.0/6 = 8.50$

    $n\Sigma x^2 - (\Sigma x)^2 = 6(439.00) - (51.0)^2 = 33.00$

    $\alpha = 0.05$ and $df = n-2 = 4$

$$\hat{y} \pm t_{\alpha/2}s_e \sqrt{1 + 1/n + n(x_o - \overline{x})^2/[n\Sigma x^2 - (\Sigma x)^2]}$$

$$\hat{y}_{187.1} \pm t_{0.025}(15.74) \sqrt{1 + 1/6 + 6(9.0-8.50)^2/[33.00]}$$

    $204.757 \pm (2.776)(15.74)\sqrt{1.17424}$

    204.757 ± 47.348

    $157.4 < y_{9.0} < 252.1$ (kg)

Exercises 21–24 refer to the chapter problem of Table 10-1. Use the following, which are calculated and/or discussed in the text,

    $n = 6$ $\Sigma x = 6.50$ $\Sigma x^2 = 9.7700$ $\hat{y} = 0.034560 + 0.945021x$ $s_e = 0.122987$

and the additional values

    $\overline{x} = (\Sigma x)/n = 6.50/6 = 1.083333$ $n\Sigma x^2 - (\Sigma x)^2 = 6(9.7700) - (6.50)^2 = 16.3700$

NOTE: Using a slightly different regression equation for $\hat{y}$ or a slightly different value for s_e may result in slightly different values in exercises 21-24.

21. $\hat{y}_{2.10} = 0.034560 + 0.945021(2.10) = 2.019$

    $\alpha = 0.01$ and $df = n-2 = 4$

$$\hat{y} \pm t_{\alpha/2}s_e \sqrt{1 + 1/n + n(x_o - \overline{x})^2/[n\Sigma x^2 - (\Sigma x)^2]}$$

$$\hat{y}_{2.10} \pm t_{0.005}(0.122987) \sqrt{1 + 1/6 + 6(2.10-1.083333)^2/[16.3700]}$$

    $2.019 \pm (4.604)(0.122987)\sqrt{1.545510}$

    2.019 ± 0.704

    $1.32 < y_{2.10} < 2.72$ (dollars)

23. $\hat{y}_{0.50} = 0.034560 + 0.945021(0.50) = 0.507$

$\alpha = 0.05$ and $df = n-2 = 4$

$$\hat{y} \pm t_{\alpha/2}s_e \sqrt{1 + 1/n + n(x_o - \overline{x})^2/[n\Sigma x^2 - (\Sigma x)^2]}$$

$$\hat{y}_{0.50} \pm t_{0.25}(0.122987) \sqrt{1 + 1/6 + 6(0.50 - 1.083333)^2/[16.3700]}$$

$0.507 \pm (2.776)(0.122987)\sqrt{1.291387}$

0.507 ± 0.388

$0.12 < y_{0.50} < 0.89$ (dollars)

25. Use the following, which are calculated and/or discussed in the text,

$n = 6$ $\Sigma x = 6.50$ $\Sigma x^2 = 9.7700$ $\hat{y} = 0.034560 + 0.945021x$ $s_e = 0.122987$

and the additional values

$\overline{x} = (\Sigma x)/n = 6.50/6 = 1.083333$ $\Sigma x^2 - (\Sigma x)^2/n = 9.7700 - (6.50)^2/6 = 2.728333$

a. $\alpha = 0.05$ and $df = n-2 = 4$

$$b_o \pm t_{\alpha/2}s_e \sqrt{1/n + (\overline{x})^2/[\Sigma x^2 - (\Sigma x)^2/n]}$$

$$0.034560 \pm t_{0.025}(0.122987) \sqrt{1/6 + (1.083333)^2/[2.728333]}$$

$0.034560 \pm (2.776)(0.122987)\sqrt{0.596823}$

0.034560 ± 0.263755

$-0.229 < \beta_o < 0.298$ (dollars)

b. $\alpha = 0.05$ and $df = n-2 = 4$

$$b_1 \pm t_{\alpha/2}s_e / \sqrt{\Sigma x^2 - (\Sigma x)^2/n}$$

$0.945021 \pm t_{0.025}(0.122987)/\sqrt{2.728333}$

$0.945021 \pm (2.776)(0.122987)/\sqrt{2.728333}$

0.945021 ± 0.206695

$0.738 < \beta_1 < 1.152$ (dollars/dollar)

NOTE: The confidence interval for $\beta_o = y_0$ may also be found as the confidence interval [as distinguished from the prediction interval, see exercise #26] for $x = 0$.

$\hat{y}_0 = 0.034560 + 0.945021(0) = 0.034560$

$\alpha = 0.05$ and $df = n-2 = 4$

$$\hat{y} \pm t_{\alpha/2}s_e \sqrt{1/n + n(x_o - \overline{x})^2/[n\Sigma x^2 - (\Sigma x)^2]} \text{ modifies to become}$$

$$\hat{y}_0 \pm t_{\alpha/2}s_e \sqrt{1/n + n(\overline{x})^2/[n\Sigma x^2 - (\Sigma x)^2]}$$

$$0.034560 \pm t_{0.025}(0.122987) \sqrt{1/6 + (1.083333)^2/[2.728333]}$$

$0.034560 \pm (2.776)(0.122987)\sqrt{0.596823}$

0.034560 ± 0.263755

$-0.229 < \beta_o < 0.298$ (dollars)

10-5 Multiple Regression

1. In multiple regression, b_1 is the coefficient of the variable x_1 in the regression line that best fits the sample data – and it is an estimate of β_1, which is the coefficient of the variable x_1 in the regression line that best fits all of the data in the population. In other words, b_1 is the sample statistic that estimate the population parameter β_1.

3. No; the methods of this section apply to quantitative data, and eye color is qualitative data. While it is possible to model qualitative data having only two categories as binomial quantitative data with values 0 and 1, the variety of possible eye colors eliminate that possibility in this context.

5. Nicotine = 1.59 + 0.0231(Tar) – 0.0525(CO)

   $\hat{y} = 1.59 + 0.0231x_1 – 0.0525x_2$

   NOTE: More accurate values may be obtained from the "Coef" [i.e., coefficient] column of the Minitab table.

7. No. The P-value of 0.317 > 0.05 indicates that it would not be considered unusual to get results like those observed when there is no multiple linear relationship among the variables.

9. The best single predictor for predicting selling price is LP (i.e., list price), which has the lowest P-value of 0.000 and the highest adjusted R^2 of 0.990.

11. Of all the regression equations, the best one for predicting selling price is
    $\hat{y} = 99.2 + 0.979(LP)$. It has the lowest P-value of 0.000 and the highest adjusted R^2 of 0.990.

13. Minitab produces the following regressions for predicting nicotine content.

    (1) nicotine = 0.0800 + 0.0633 tar
        S = 0.0869783   R-Sq = 88.2%   R-Sq(adj) = 87.7%   P = 0.000
    (2) nicotine = 0.328 + 0.0397 CO
        S = 0.185937   R-Sq = 46.0%   R-Sq(adj) = 43.7%   P = 0.000
    (3) nicotine = 0.127 + 0.0878 tar - 0.0250 CO
        S = 0.0671065   R-Sq = 93.3%   R-Sq(adj) = 92.7%   P = 0.000

    The best regression for predicting nicotine content is (3)
        $\hat{y} = 0.127 + 0.0878(tar) – 0.0250(CO)$.

    It has the lowest P-value of 0.000 and the highest adjusted R^2 of 0.927. Its P-value and adjusted R^2 value suggest that it is a good equation for predicting nicotine content.

15. Minitab produces the following regressions for predicting highway mpg.

    (1) hway = 50.5 - 0.00587 weight
        S = 2.19498   R-Sq = 65.0%   R-Sq(adj) = 63.9%   P = 0.000
    (2) hway = 77.3 - 0.250 length
        S = 2.61068   R-Sq = 50.5%   R-Sq(adj) = 48.9%   P = 0.000
    (3) hway = 37.7 - 2.57 disp
        S = 2.46348   R-Sq = 55.9%   R-Sq(adj) = 54.5%   P = 0.000
    (4) hway = 56.3 - 0.00510 weight - 0.0447 length
        S = 2.21668   R-Sq = 65.5%   R-Sq(adj) = 63.1%   P = 0.000
    (5) hway = 47.9 - 0.00440 weight - 0.823 disp
        S = 2.17777   R-Sq = 66.7%   R-Sq(adj) = 64.4%   P = 0.000
    (6) hway = 56.0 - 0.110 length - 1.71 disp
        S = 2.40253   R-Sq = 59.5%   R-Sq(adj) = 56.7%   P = 0.000
    (7) hway = 50.6 - 0.00418 weight - 0.0196 length - 0.759 disp
        S = 2.21351   R-Sq = 66.8%   R-Sq(adj) = 63.2%   P = 0.000

    The best regression for predicting movie gross is (1)
        $\hat{y} = 50.5 – 0.00587(weight)$

    It has the lowest P-value of 0.000 and the second highest adjusted R^2 of 0.639. Its P-value and adjusted R^2 value suggest that it is a good equation for predicting highway mpg. Even though (7) had a slightly higher adjusted R^2, the increase gained from adding a second predictor variable is negligible.

17. a. original claim: $\beta_1 = 0$

H_o: $\beta_1 = 0$ inches/inch

H_1: $\beta_1 \neq 0$ inches/inch

$\alpha = 0.05$ [assumed] and df = 17

C.V. $t = \pm t_{\alpha/2} = \pm t_{0.025} = \pm 2.110$

calculations:

$t_{b_1} = (b_1 - \mu_{b_1})/s_{b_1}$

$= (0.7072 - 0)/0.1289$

$= 5.49$ [Minitab]

P-value $= 2 \cdot P(t_{17} > 5.49) = 0.000$ [Minitab]

conclusion:

Reject H_o; there is sufficient evidence to reject the claim that $\beta_1 = 0$ and conclude that $\beta_1 \neq 0$ (in fact, that $\beta_1 > 0$).

b. original claim: $\beta_2 = 0$

H_o: $\beta_2 = 0$ inches/inch

H_1: $\beta_2 \neq 0$ inches/inch

$\alpha = 0.05$ [assumed] and df = 17

C.V. $t = \pm t_{\alpha/2} = \pm t_{0.025} = \pm 2.110$

calculations:

$t_{b_2} = (b_2 - \mu_{b_2})/s_{b_2}$

$= (0.1636 - 0)/0.1266$

$= 1.29$ [Minitab]

P-value $= 2 \cdot P(t_{17} > 1.29) = 0.213$ [Minitab]

conclusion:

Do not reject H_o; there is not sufficient evidence to reject the claim that $\beta_2 = 0$.

The result in (a) implies that β_1 is significantly different from 0 and is appropriate for inclusion in the regression equation. The result in (b), however, implies that β_2 is not significantly different from 0 and should be dropped from the regression equation. It appears that the regression equation should include the height of the mother as a predictor variable, but not the height of the father.

19. The Minitab regression of c9 on c1 and the modified c3 yields the multiple regression equation

WEIGHT = 3.1 + 2.91(AGE) + 82.4(SEX).

$\hat{y} = 3.1 + 2.91x_1 + 82.4x_2$

Yes, but not merely because the coefficient 82.4 is so large. Minitab indicates that for the test H_o: $\beta_2 = 0$, the sample value $b_2 = 82.4$ results in the test statistic $t_{51} = 3.96$ and P-value $= .000$. As suggested by (a) and (b) below, sex does have a significant effect on the weight of a bear.

a. $\hat{y} = 3.1 + 2.91x_1 + 82.4x_2$

$\hat{y}_{20,0} = 3.1 + 2.91(20) + 82.4(0) = 61.3$ lbs

b. $\hat{y} = 3.1 + 2.91x_1 + 82.4x_2$

$\hat{y}_{20,1} = 3.1 + 2.91(20) + 82.4(1) = 143.7$ lbs

10-6 Modeling

1. The value $R^2 = 1$ indicates that the model fits the data perfectly, or at least so closely that the R^2 value rounds to 1.000. Given the fact that the number of vehicles produced in the U.S. does not follow an nice pattern, but fluctuates according to various factors (economic conditions, industry strikes, import regulations, etc.), there are two possible explanations for the claim:
   (1) The analyst was using a large number of predictor variables in the model. With n-1 predictor variables, it is always possible to construct a line (i.e., a curve) that passes through n data points.
   (2) The claim is not correct.

3. The quadratic model relating the year and the number of points scored explains $R^2 = 0.082 = 8.2\%$ of the variation in number of points scored – i.e., there is a lot of variation between the observed and predicted values that the model is not able to account for. This result suggests that the model cannot be expected to make accurate predictions and is not a useful model.

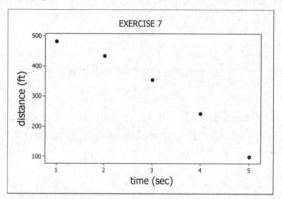

5. The graph appears to be that of a straight line function.
   •Try a linear regression of the form y = ax + b.
   ```
   y = 8.00 + 2.00 x
   S = 0    R-Sq = 100.0%    R-Sq(adj) = 100.0%
   ```
   The $s_e = 0$ and adjusted $R^2 = 100.0\%$ indicate a perfect fit.
   •Choose the linear model y = 8 + 2x

7. The graph appears to be that of a quadratic function.
   •Try a quadratic regression of the form $d = at^2 + bt + c$.
   Let z = t·t. Regress y on t and z.
   ```
   d = 500 + 0.000000 t - 16.0 z
   S = 0    R-Sq = 100.0%    R-Sq(adj) = 100.0%
   ```
   The $s_e = 0$ and adjusted $R^2 = 100.0\%$ indicate a perfect fit.
   •Choose the quadratic model $d = 500 - 16t^2$

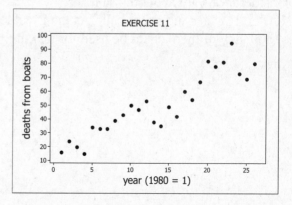

9. The graph appears to be that of a quadratic function or an exponential function.
  • Try a quadratic regression of the form $y = ax^2 + bx + c$.
  Let $z = x \cdot x$. Regress y on x and z.
```
y = 0.109 + 0.0157 x + 0.000516 z
S = 0.193681   R-Sq = 95.5%   R-Sq(adj) = 92.5%
```
  The adjusted $R^2 = 92.5\%$ indicates a very good fit.
  • Try an exponential regression of the form $y = a \cdot b^x$.
$$\ln(y) = \ln(a \cdot b^x) = \ln(a) + x \cdot \ln(b)$$
  Let $z = \ln(y)$. Regress z on x.
```
ln(y) = - 1.8435 + 0.057651 x
S = 0.195222   R-Sq = 97.0%   R-Sq(adj) = 96.2%
```
  The adjusted R^2 rounds to 100%, indicating a nearly perfect fit.
  Solving for the original parameters: $\ln(a) = -1.8435$ $\ln(b) = 0.057651$
$$a = e^{-1.8435} = 0.15826 \qquad b = e^{0.057651} = 1.05935$$
  • Choose the exponential model $y = 0.15826 \cdot (1.05935)^x$
  The year 2020 corresponds to x=61. The predicted subway fare for 2020 is
$$\hat{y}_{61} = 0.15826 \cdot (1.05935)^{61} = \$5.33$$

11. Recode the years, with 1980 = 1. The graph could be that of any of several functions.
  • Try a linear regression of the form $y = ax + b$.
```
y = 14.3 + 2.67 x
S = 9.02645   R-Sq = 84.2%   R-Sq(adj) = 83.6%
```
  The adjusted $R^2 = 83.6\%$ indicates a good fit.
  • Try a quadratic regression of the form $y = ax^2 + bx + c$.
  Let $z = x \cdot x$. Regress y on x and z.
```
y = 15.3 + 2.46 x + 0.0080 z
S = 9.21063   R-Sq = 84.3%   R-Sq(adj) = 82.9%
```
  The adjusted $R^2 = 82.9\%$ indicates a good fit, but not as good as the linear model.
  • Try a power function regression of the form $y = a \cdot x^b$, where b should be close to 3.
$$\ln(y) = \ln(a \cdot x^b) = \ln(a) + b \cdot \ln(x)$$
  Let $z = \ln(y)$ and let $w = \ln(x)$. Regress z on w.
```
ln(y) = 2.53 + 0.545 ln(x)
S = 0.217722   R-Sq = 82.1%   R-Sq(adj) = 81.4%
```
  The adjusted $R^2 = 81.4\%$ is slightly less than the others. The model is not considered further.
  • Choose the linear model $y = 14.3 + 2.67x$
  The year 2006 corresponds to x=27. The predicted number of deaths for 2007 is
$$\hat{y}_{27} = 14.3 + 2.67(27) = 86.4$$

This compares reasonably well to the actual number of 92. In this case the best model was not much better than the others. But not only does the linear model have the highest adjusted R^2, it is also the simplest model. In general, choose the simplest model whenever all other considerations are about the same.

NOTE: This is a judgment call. As the P-value (not shown) for each of the above three models is 0.00, any of them could be used for making predictions.

13. The graph appears to be that of a quadratic function.
   •Try a quadratic regression of the form $y = ax^2 + bx + c$.

   ```
   13y = 0.0048 - 0.0286 13x + 4.90 12y
    S = 0.0308607   R-Sq = 100.0%   R-Sq(adj) = 100.0%
   ```

   The adjusted R^2 rounds to 100.0%, indicating a nearly perfect fit.
   •Choose the quadratic model $y = 4.90x^2 - 0.0286x + 0.0048$.

   $\hat{y}_{12} = 4.90(12)^2 - 0.0286(12) + 0.0048 = 705.3$ meters.

   But if the building from which the ball is dropped is only 50 meters tall, the ball will hit the ground and stop falling long before 12 seconds elapse.

15. Code the years 1950=1, 1955=2, 1960=3, etc.
   The graph appears to be that of a quadratic function or an exponential function.
   •Try a quadratic regression of the form $y = ax^2 + bx + c$.

   Let $z = x \cdot x$. Regress y on x and z.

   ```
   y = 13.82 + 0.01986 x + 0.004471 z
    S = 0.121612   R-Sq = 87.1%   R-Sq(adj) = 84.2%
   ```

   The adjusted $R^2 = 84.2\%$ indicates a very good fit.
   •Try an exponential regression of the form $y = a \cdot b^x$.

   $$\ln(y) = \ln(a \cdot b^x) = \ln(a) + x \cdot \ln(b)$$

   Let $z = \ln(y)$. Regress z on x.

   ```
   ln(y) = 2.62 + 0.00547 x
    S = 0.00876850   R-Sq = 84.8%   R-Sq(adj) = 83.3%
   ```

   The adjusted $R^2 = 83.3\%$ indicates a very good fit, but not quite as good as the quadratic.
   •Choose the quadratic model $y = 0.004471x^2 + 0.01986x + 13.82$
   The year 2010 corresponds to x=13. The predicted temperature for 2010 is

   $\hat{y}_{13} = 0.004471(13)^2 + 0.01986(13) + 13.82 = 14.8\ °C$.

17. NOTE: The following analysis codes the years so that 1971 is x=0 and determines a regression equation. Coding the years so that 1971 is x=1 gives the different equation $y = 1.382 \cdot (1.424)^x$, which considers 1970 as the starting year x=0 but gives the same numerical predictions for each year. In general, the recoding of the years is arbitrary – and while a different equation may result, the individual predictions and other key characteristics will be identical.
   Consider the pattern in the box at the right.
   For a variable that starts with value $y = a$ at year 0 and doubles every 18 months,

   $$y = a \cdot 2^{x/2} = a \cdot 2^{(1/2)x} = a \cdot (2^{1/2})^x = a \cdot (\sqrt{2})^x.$$

   a. If Moore's law applies as indicated, and the years are coded with 1971 = 0, the data should be a good fit to the exponential model
      $y = 2.3 \cdot (1.414)^x$.

year	doubles every 12 months	doubles every 18 months
0	$a = a \cdot 2^0$	$a = a \cdot 2^0$
1	$2a = a \cdot 2^1$	
2	$4a = a \cdot 2^2$	$2a = a \cdot 2^1$
3	$8a = a \cdot 2^3$	
4	$16a = a \cdot 2^4$	$4a = a \cdot 2^2$
…	…	…
x	$a \cdot 2^x$	$a \cdot 2^{x/2}$

b. Try an exponential regression of the form $y = a \cdot b^x$.

$$\ln(y) = \ln(a \cdot b^x) = \ln(a) + x \cdot \ln(b)$$

Let $z = \ln(y)$. Regress z on x.

```
ln(y) = 0.6446 + 0.35792 x
S = 0.506769    R-Sq = 98.8%    R-Sq(adj) = 98.6%
```

The adjusted $R^2 = 98.6\%$ indicates an excellent fit.

Solving for the original parameters: $\ln(a) = 0.6446$ $\qquad\qquad$ $\ln(b) = 0.35792$

$$a = e^{0.6446} = 1.905 \qquad b = e^{-0.35792} = 1.430$$

Choose the exponential model $y = 1.905 \cdot (1.430)^x$

c. Yes. The $1.430 \approx 1.414$ indicates that the y value is doubling approximately every 18 months. In addition, the starting value for 1971 ($x=0$) of 1.9 is close to the actual value of 2.3.

19. The table below was obtained using $\hat{y}_{lin} = -61.93 + 27.20x$ and $\hat{y}_{quad} = 2.77x^2 - 6.00x + 10.01$

year	pop	$\hat{y}_{lin}$	$y - \hat{y}_{lin}$	$(y - \hat{y}_{lin})^2$	$\hat{y}_{quad}$	$y - \hat{y}_{quad}$	$(y - \hat{y}_{quad})^2$
1	5	-34.727	39.7273	1578.26	6.776	-1.77622	3.1550
2	10	-7.527	17.5273	307.21	9.074	0.92587	0.8572
3	17	19.673	-2.6727	7.14	16.906	0.09417	0.0089
4	31	46.873	-15.8727	251.94	30.271	0.72867	0.5310
5	50	74.073	-24.0727	579.50	49.171	0.82937	0.6879
6	76	101.273	-25.2727	638.71	73.604	2.39627	5.7421
7	106	128.473	-22.4727	505.02	103.571	2.42937	5.9018
8	132	155.673	-23.6727	560.40	139.071	-7.07133	50.0037
9	179	182.873	-3.8727	15.00	180.106	-1.10583	1.2229
10	227	210.073	16.9273	286.53	226.674	0.32587	0.1062
11	281	237.273	43.7273	1912.07	278.776	2.22378	4.9452
68	1114	1114.000	0.0000	6641.78	1114.000	0.00000	73.1618

a. $\Sigma(y - \hat{y})^2 = 6641.78$ for the linear model

b. $\Sigma(y - \hat{y})^2 = 73.16$ for the quadratic model

c. Since $73.16 < 6641.78$, the quadratic model is better – using the sum of squares criterion.

Statistical Literacy and Critical Thinking

1. Section 9-4 deals with making inferences about the mean of the differences between matched pairs and requires that each member of the pair have the same unit of measurement. Section 10-2 deals with making inference about the relationship between the members of the pairs and does not require that each member of the pair have the same unit of measurement.

2. Yes; since $0.963 > 0.279$ (the C.V. from Table A-6), there is sufficient evidence to support the claim of a linear correlation between chest size and weight. No; the conclusion is only that larger chest sizes are associated with larger weights – not that there is a cause and effect relationship, and not that the direction of any cause and effect relationship can be identified.

3. No; a perfect positive correlation means only that larger values of one variable are associated with larger values of the other variable and that the value of one of the variables can be perfectly predicted from the value of the other. A perfect correlation does not imply equality between the paired values, or even that the paired values have the same unit of measurement.

4. No; a value of r=0 suggests only that there is no <u>linear</u> relationship between the two variables, but the two variables may be related in some other manner.

Chapter Quick Quiz

1. If the calculation indicate that r = 2.650, then an error has been made. For any set of data, it must be true that $-1 \le r \le 1$.

2. Since 0.989 > 0.632 (the C.V. from Table A-6), there is sufficient evidence to support the claim of a linear correlation between the two variables.

3. True.

4. Since -0.632 < 0.099 < 0.632 (the C.V.'s from Table A-6), there is not sufficient evidence to support the claim of a linear correlation between the two variables.

5. False; the absence of a <u>linear</u> correlation does not preclude the existence of another type of relationship between the two variables.

6. From Table A-6, C.V. = ±0.514.

7. A perfect straight line pattern that falls from left to right describes a perfect negative correlation with r = -1.

8. $\hat{y}_{10} = 2(10) - 5 = 15$

9. The proportion of the variation in y that is explained by the linear relationship between x and y is $r^2 = (0.400)^2 = 0.160$, or 16%.

10. False; the conclusion is only that larger amounts of salt consumption are associated with higher measures of blood pressure – not that there is a cause and effect relationship, and not that the direction of any cause and effect relationship can be identified.

Review Exercises

1. These are the necessary summary statistics.
   n = 6
   $\Sigma x = 586.4$
   $\Sigma y = 590.7$
   $\Sigma x^2 = 57312.44$
   $\Sigma y^2 = 58156.45$
   $\Sigma xy = 57730.62$
   $n(\Sigma x^2) - (\Sigma x)^2 = 6(57312.44) - (586.4)^2$
   $\qquad\qquad = 9.68$
   $n(\Sigma y^2) - (\Sigma y)^2 = 6(58156.45) - (590.7)^2$
   $\qquad\qquad = 12.21$
   $n(\Sigma xy) - (\Sigma x)(\Sigma y) = 6(57730.62) - (586.4)(590.7)$
   $\qquad\qquad = -2.76$

   a. The scatterplot is given above at the right. The scatterplot suggests that there is not a linear relationship between the two variables.

b. $r = [n(\Sigma xy) - (\Sigma x)(\Sigma y)]/[\sqrt{n(\Sigma x^2) - (\Sigma x)^2}\sqrt{n(\Sigma y^2) - (\Sigma y)^2}]$

$\quad = -2.76/[\sqrt{9.68}\sqrt{12.21}]$

$\quad = -0.254$

H_o: $\rho = 0$

H_1: $\rho \neq 0$

$\alpha = 0.05$ [assumed] and df $= 4$

C.V. $t = \pm t_{\alpha/2} = \pm t_{0.025} = \pm 2.776$ [or $r = \pm 0.811$]

calculations:

$\quad t_r = (r - \mu_r)/s_r$

$\quad\quad = (-0.254 - 0)/\sqrt{[1 - (-0.254)^2]/4}$

$\quad\quad = -0.254/0.4836$

$\quad\quad = -0.525$

$\quad$ P-value $= 2 \cdot \text{tcdf}(-99, -0.525, 4) = 0.6274$

conclusion:

Do not reject H_o; there is not sufficient evidence to conclude that $\rho \neq 0$. No; there is not sufficient evidence to support the claim of a linear correlation between the 8 am and midnight temperatures.

c. $\overline{x} = (\Sigma x)/n = 586.4/6 = 97.73$ $\qquad\qquad \overline{y} = (\Sigma y)/n = 590.7/6 = 98.45$

$\quad b_1 = [n(\Sigma xy) - (\Sigma x)(\Sigma y)]/[n(\Sigma x^2) - (\Sigma x)^2]$

$\quad\quad = -2.76/9.68$

$\quad\quad = -0.2851$

$\quad b_o = \overline{y} - b_1 \overline{x}$

$\quad\quad = 98.45 - (-0.2851)(97.73)$

$\quad\quad = 126.32$

$\quad \hat{y} = b_o + b_1 x$

$\quad\quad = 126.32 - 0.2851x$

d. $\hat{y}_{98.3} = \overline{y} = 98.45\ °F$ [no significant correlation]

2. a. Yes. Assuming $\alpha = 0.05$, Table A-6 indicates C.V. $= \pm 0.312$. Since $0.522 > 0.312$, there is sufficient evidence to support a claim of a linear correlation between heights and weights of males.

b. $r^2 = (0.522)^2 = 0.272$, or 27.2%

c. $\hat{y} = -139 + 4.55x$

d. $\hat{y}_{72} = -139 + 4.55(72) = 188.6$ lbs

3. These are the necessary summary statistics.

$n = 5$

$\Sigma x = 265$

$\Sigma y = 917$

$\Sigma x^2 = 14531$

$\Sigma y^2 = 247049$

$\Sigma xy = 54572$

$n(\Sigma x^2) - (\Sigma x)^2 = 5(14531) - (265)^2$

$\qquad = 2430$

$n(\Sigma y^2) - (\Sigma y)^2 = 5(247049) - (917)^2$

$\qquad = 394356$

$n(\Sigma xy) - (\Sigma x)(\Sigma y) = 5(54572) - (265)(917)$

$\qquad\qquad = 29855$

a. The scatterplot is given above at the right. The scatterplot suggests that there is a linear relationship between the two variables.

b. $r = [n(\Sigma xy) - (\Sigma x)(\Sigma y)]/[\sqrt{n(\Sigma x^2) - (\Sigma x)^2}\sqrt{n(\Sigma y^2) - (\Sigma y)^2}]$

$\quad = 29855/[\sqrt{2430}\sqrt{394356}]$

$\quad = 0.964$

$H_o: \rho = 0$

$H_1: \rho \neq 0$

$\alpha = 0.05$ [assumed] and df $= 3$

C.V. $t = \pm t_{\alpha/2} = \pm t_{0.025} = \pm 3.182$ [or $r = \pm 0.878$]

calculations:

$\quad t_r = (r - \mu_r)/s_r$

$\qquad = (0.964 - 0)/\sqrt{[1 - (0.964)^2]/3}$

$\qquad = 0.964/0.1526$

$\qquad = 6.319$

P-value $= 2 \cdot \text{tcdf}(6.319,99,3) = 0.0080$

conclusion:

Reject H_o; there is sufficient evidence to conclude that $\rho \neq 0$ (in fact, that $\rho > 0$).   Yes; there is sufficient evidence to support the claim of a linear correlation between the lengths and weights of bears.

c. $\overline{x} = (\Sigma x)/n = 265/5 = 53.0$ $\overline{y} = (\Sigma y)/n = 917/5 = 183.4$

$\quad b_1 = [n(\Sigma xy) - (\Sigma x)(\Sigma y)]/[n(\Sigma x^2) - (\Sigma x)^2]$

$\qquad = 29855/2430$

$\qquad = 12.286$

$\quad b_o = \overline{y} - b_1\overline{x}$

$\qquad = 183.4 - (12.286)(53.0)$

$\qquad = -467.8$

$\quad \hat{y} = b_o + b_1 x$

$\qquad = -467.8 + 12.286x$

d. $\hat{y}_{72} = -467.8 + 12.286(72) = 416.8$ lbs

4. These are the necessary summary statistics,
   where x = leg and y = height.
   n = 5
   $\Sigma x = 209.0$
   $\Sigma y = 851$
   $\Sigma x^2 = 8771.42$
   $\Sigma y^2 = 145045$
   $\Sigma xy = 35633.2$
   $n(\Sigma x^2) - (\Sigma x)^2 = 5(8771.42) - (209.0)^2 = 176.10$
   $n(\Sigma y^2) - (\Sigma y)^2 = 5(145045) - (851)^2 = 1024$
   $n(\Sigma xy) - (\Sigma x)(\Sigma y) = 5(35633.2) - (209.0)(851) = 307.0$

a. The scatterplot is given above at the right. The scatterplot suggests that there may be a
   linear relationship between the two variables, but only a formal test determine that with any
   degree of conficence.

b. $r = [n(\Sigma xy) - (\Sigma x)(\Sigma y)]/[\sqrt{n(\Sigma x^2) - (\Sigma x)^2}\sqrt{n(\Sigma y^2) - (\Sigma y)^2}]$
   $= 307.0/[\sqrt{176.10}\sqrt{1024}]$
   $= 0.723$
   $H_o: \rho = 0$
   $H_1: \rho \neq 0$
   $\alpha = 0.05$ [assumed] and df = 3
   C.V. $t = \pm t_{\alpha/2} = \pm t_{0.025} = \pm 3.182$ [or $r = \pm 0.878$]
   calculations:

   $t_r = (r - \mu_r)/s_r$
   $= (0.723 - 0)/\sqrt{[1 - (0.723)^2]/3}$
   $= 0.723/0.3989$
   $= 1.812$
   P-value = 2·tcdf(1.812,99,3) = 0.1676
   conclusion:
   Do not reject H_o; there is not sufficient evidence to conclude that $\rho \neq 0$. No; there is not
   sufficient evidence to support a claim of a linear correlation between upper leg length
   and height of males.

c. $\overline{x} = (\Sigma x)/n = 209.0/5 = 41.80$ $\overline{y} = (\Sigma y)/n = 170.2$
   $b_1 = [n(\Sigma xy) - (\Sigma x)(\Sigma y)]/[n(\Sigma x^2) - (\Sigma x)^2] = 307.0/176.10 = 1.743$
   $b_0 = \overline{y} - b_1\overline{x} = 170.2 - (1.743)(41.80) = 97.329$
   $\hat{y} = b_0 + b_1x = 97.33 + 1.743x$

d. $\hat{y}_{45} = \overline{y} = 170.2$ cm [no significant correlation]

5. Minitab produces the following regression for predicting height as a function of leg and arm.
   ```
   height = 140.44 + 2.4961 leg - 2.2738 arm
    S = 1.53317   R-Sq = 97.7%   R-Sq(adj) = 95.4%      P = 0.023
   ```
   $\hat{y} = 140.44 + 2.4961x_1 - 2.2738x_2$
   $R^2 = 0.977$
   adjusted $R^2 = 0.954$
   P-value = 0.023
   Yes; since 0.023 < 0.05, the multiple regression equation can be used to predict the height of a
   male when given his upper leg length and arm circumference.

Cumulative Review Exercises

The following summary statistics apply to exercises 1-6. The ordered heights are as follows.
   1877: 62 64 65 65 66 66 67 68 68 71
   recent: 62 63 66 68 68 69 69 71 72 73

Let the 1877 heights be group 1.

<u>group 1: 1877</u> $(n=10)$	<u>group 2: recent</u> $(n=10)$
$\Sigma x = 662$	$\Sigma x = 681$
$\Sigma x^2 = 43{,}880$	$\Sigma x^2 = 46{,}493$
$\overline{x} = 66.2$	$\overline{x} = 68.1$
$s^2 = 6.178$ (s=2.486)	$s^2 = 12.989$ (s=3.604)

$\overline{x}_1 - \overline{x}_2 = 66.2 - 68.1 = -1.9$

1. For 1877: $\overline{x} = 66.2$ inches $\tilde{x} = (66+66)/2 = 66.0$ inches $s = 2.5$ inches
   For recent: $\overline{x} = 68.1$ inches $\tilde{x} = (68+69)/2 = 68.5$ inches $s = 3.6$ inches

2. original claim: $\mu_1 - \mu_2 < 0$
   H_o: $\mu_1 - \mu_2 = 0$
   H_1: $\mu_1 - \mu_2 < 0$
   $\alpha = 0.05$ and df = 9
   C.V. $t = -t_\alpha = -t_{0.05} = -1.833$
   calculations:

   $t_{\overline{x}_1 - \overline{x}_2} = (\overline{x}_1 - \overline{x}_2 - \mu_{\overline{x}_1 - \overline{x}_2})/s_{\overline{x}_1 - \overline{x}_2}$
   $= (-1.9 - 0)/\sqrt{6.178/10 + 12.989/10}$
   $= -1.9/1.3844$
   $= -1.372$
   P-value = tcdf(-99,-1.372,9) = 0.1016
   conclusion:
   Do not reject H_o; there is not sufficient evidence to conclude that $\mu_1 - \mu_2 < 0$. There is not sufficient evidence to support the claim that the males in 1877 had a mean height that is less than the mean height of males today.

3. original claim: $\mu < 69.1$
   H_o: $\mu = 69.1$
   H_1: $\mu < 69.1$
   $\alpha = 0.05$ and df = 9
   C.V. $t = -t_\alpha = -t_{0.05} = -1.833$
   calculations:

   $t_{\overline{x}} = (\overline{x} - \mu)/s_{\overline{x}}$
   $= (66.2 - 69.1)/(2.486/\sqrt{10})$
   $= -2.9/0.7860$
   $= -3.690$
   P-value = $P(t_9 < -3.690)$ = tcdf(-99,-3.690,9) = 0.0025
   conclusion:
   Reject H_o; there is sufficient evidence to conclude that $\mu < 69.1$. There is sufficient evidence to support the claim that the men from 1877 have a mean height that is less than 69.1 inches.

4. σ unknown (and assuming the distribution is approximately normal), use t with df=9

$\alpha = 0.05$, $t_{df,\alpha/2} = t_{9,0.05} = 2.262$

$$\bar{x} \pm t_{\alpha/2} \cdot s / \sqrt{n}$$

$66.2 \pm 2.262(2.486)/\sqrt{10}$

66.2 ± 1.8

$64.4 < \mu < 68.0$ (inches)

5. $\alpha = 0.05$ and df $= 9$

$(\bar{x}_1 - \bar{x}_2) \pm t_{\alpha/2} s_{\bar{x}_1 - \bar{x}_2}$

$-1.9 \pm 2.262 \sqrt{6.178/10 + 12.989/10}$

-1.9 ± 3.1

$-5.0 < \mu_1 - \mu_2 < 1.2$ (inches)

Yes; the confidence interval includes the value 0. Since the confidence interval includes the value 0, we cannot reject the notion that the two populations may have the same mean.

6. It would not be appropriate to test for a linear correlation between heights from 1877 and current heights because the sample data are not matched pairs, as required for that test.

7. a. A statistic is a numerical value, calculated from sample data, that describes a characteristic of the sample. A parameter is a numerical value that describes a characteristic of the population.

   b. A simple random sample of size n is one chosen in such a way that every group of n members of the population has the same chance of being selected as the sample from that population.

   c. A voluntary response sample is one in which the respondents themselves decide whether or not to be included. Such samples are generally unsuited for making inferences about populations because they are not likely to be representative of the population. In general, those with a strong interest in the topic are more likely to make the effort to include themselves in the sample – and the sample will contain an over-representation of persons with a strong interest in the topic, and an under-representation of persons with little or no interest in the topic.

8. Yes; since 40 is $(40-26)/5 = 2.8$ standard deviation from the mean, it is considered an outlier. In general, any observation more than 2 standard deviations from the mean (which typically accounts for the most extreme 5% of the observations) is considered an outlier.

9. a. Use $\mu = 26$ and $\sigma = 5$.

   $P(x>28) = P(z>0.40) = 1 - 0.6554 = 0.3446$

   b. Use $\mu_{\bar{x}} = \mu = 26$ and $\sigma_{\bar{x}} = \sigma/\sqrt{n} = 5/\sqrt{16} = 1.25$.

   $P(\bar{x}>28) = P(z>1.60) = 1 - 0.9452 = 0.0548$

10. For independent events, $P(G_1 \text{ and } G_2 \text{ and } G_3 \text{ and } G_4) = P(G_1) \cdot P(G_2) \cdot P(G_3) \cdot P(G_4)$

$$= (0.12)(0.12)(0.12)(0.12)$$
$$= 0.000207$$

Because the probably of getting four green-eyed persons by random selection is so small, it appears that the researcher (either knowingly or unknowingly) did not make the selections at random from the population.

Chapter 11

Goodness-of-Fit and Contingency Tables

11-2 Goodness-of-Fit

1. When digits are randomly generated, they should form a uniform distribution – i.e., a distribution in which each of the digits is equally likely. The test for goodness-of-fit is a test of the hypothesis that the sample data fit the uniform distribution.

3. O represents the observed frequencies. The twelve values for O are 5, 8, 7, 9, 13, 17, 11, 10, 10, 12, 8, 10. E represents the expected frequencies. The twelve values for E dependent on the expected (i.e., the hypothesized) distribution. If the hypothesized distribution is that weddings occur in different months with the same frequency, we expect the $\Sigma O = 120$ observations to be equally distributed among the 12 months so that each of the twelve values for E is $120/12 = 12$.

NOTE FOR EXERCISES 5 AND 6: The principle that the null hypothesis must contain the equality still applies. When the original claim is given in words, it will reduce either to a statement that the observed results fit or match (i.e., are equal to) some specified distribution or to a statement that the observed results are different from some specified distribution. In the first case the original claim is the null hypothesis, in the second case the original claim is the alternative hypothesis.

5. original claim: the actual outcomes agree with the expected frequencies
   H_o: the actual outcomes agree with the expected frequencies
   H_1: at least one outcome is not as expected
   $\alpha = 0.05$ and df = 9
   C.V. $\chi^2 = \chi_\alpha^2 = \chi_{0.05}^2 = 16.919$
   calculations:
   $\quad \chi^2 = \Sigma[(O - E)^2/E]$
   $\quad\quad = 8.185$
   $\quad$ P-value $= \chi^2$cdf $(8.815,99,5) = 0.5156$
   conclusion:
   $\quad$ Do not reject Ho; there is not sufficient evidence to reject the claim that the actual outcomes agree with the expected frequencies. There is no reason to say the slot machine is not functioning as expected.

NOTES FOR THE REMAINING EXERCISES IN THIS SECTION:
   (1) In goodness-of-fitl problems, always verify that $\Sigma E = \Sigma O$ before proceeding. If these sums are not equal, then an error has been made and further calculations have no meaning.
   (2) As in the previous uses of the χ^2 distribution, the accompanying illustrations follow the "usual" shape – even though that shape is not correct for df = 1 and df = 2.
   (3) As in the previous sections, a subscript (the df) may be used to identify which χ^2 distribution to use in the tables.
   (4) Exact P-values are found as in previous sections using the χ^2cdf command on the TI-83/84[+] calculator.

7. original claim: the four categories are equally likely

H_o: $p_1 = p_2 = p_3 = p_4 = 0.25$

H_1: at least one $p_i \neq 0.25$

$\alpha = 0.05$ and df = 3

C.V. $\chi^2 = \chi_\alpha^2 = \chi_{0.05}^2 = 7.815$

calculations:

outcome	O	E	$(O-E)^2/E$
0-24	61	25	51.840
25-49	17	25	2.560
50-74	10	25	9.000
75-99	12	25	6.760
	100	100	70.160

$\chi^2 = \Sigma[(O - E)^2/E]$

   $= 70.160$

P-value $= \chi^2 cdf\,(70.160, 99, 3) = 3.9E-15 \approx 0$

conclusion:

Reject Ho; there is sufficient evidence to reject the claim that $p_1 = p_2 = p_3 = p_4 = 0.25$. There is sufficient evidence to reject the claim that the four categories are equally likely. Yes; informally speaking, the results appear to support the statement that the frequency for the first category is disproportionately high. [See the following NOTE.]

NOTE: Formally, the goodness-of-fit test can only consider two cases: either the observed frequencies fit the expected frequencies (which must be the null hypothesis), or the observed frequencies do not fit the expected frequencies (which must be the alternative hypothesis). The test cannot address specific variations of the alternative hypothesis – i.e., it cannot address specific manners in which the observed frequencies do not fit the expected frequencies. When the null hypothesis of a goodness-of-fit is rejected, the conclusion can be only that at least one of the observed frequencies (or proportions) is not as expected – and the test cannot identify in which one(s) of the categories that discrepancy occurs.

9. original claim: the four categories are equally likely

H_o: $p_1 = p_2 = p_3 = p_4 = 0.25$

H_1: at least one $p_i \neq 0.25$

$\alpha = 0.05$ and df = 3

C.V. $\chi^2 = \chi_\alpha^2 = \chi_{0.05}^2 = 7.815$

calculations:

outcome	O	E	$(O-E)^2/E$
0-24	33	25	2.560
25-49	16	25	3.240
50-74	23	25	0.160
75-99	28	25	0.360
	100	100	6.320

$\chi^2 = \Sigma[(O - E)^2/E]$

   $= 6.320$

P-value $= \chi^2 cdf\,(6.320, 99, 3) = 0.0970$

conclusion:

Do not reject Ho; there is not sufficient evidence to reject the claim that $p_i = 0.25$ for each category. There is not sufficient evidence to reject the claim that the four categories are equally likely. No; the results appear to support the statement that the frequency for the first category is disproportionately high. [See the NOTE following Exercise 7.]

11. original claim: the outcomes are not equally likely

H_o: $p_1 = p_2 = p_3 = p_4 = p_5 = p_6 = 1/6$

H_1: at least one $p_i \neq 1/6$

$\alpha = 0.05$ and df = 5

C.V. $\chi^2 = \chi^2_\alpha = \chi^2_{0.05} = 11.071$

calculations:

outcome	O	E	$(O-E)^2/E$
1	27	33.33	1.2033
2	31	33.33	0.1633
3	42	33.33	2.2533
4	40	33.33	1.3333
5	28	33.33	0.8533
6	32	33.33	0.0533
	200	200.00	5.8600

$\chi^2 = \Sigma[(O - E)^2/E]$
$\quad = 5.860$

P-value = χ^2cdf (5.86,99,5) = 0.3201

conclusion:

Do not reject Ho; there is not sufficient evidence to support the claim that $p_i \neq 1/6$ for at least one outcome. There is not sufficient evidence to support the claim that the outcomes of the loaded die are not equally likely. No; it does not appear that the loaded die behaves differently than a fair die.

13. original claim: the likelihood of winning is the same for all post positions

H_o: $p_1 = p_2 = p_3 = \ldots = p_{10} = 1/10$

H_1: at least one $p_i \neq 1/10$

$\alpha = 0.05$ and df = 9

C.V. $\chi^2 = \chi^2_\alpha = \chi^2_{0.05} = 16.919$

calculations:

position	O	E	$(O-E)^2/E$
1	19	11.4	5.0667
2	14	11.4	0.5930
3	11	11.4	0.0140
4	14	11.4	0.5930
5	14	11.4	0.5930
6	7	11.4	1.6982
7	8	11.4	1.0140
8	11	11.4	0.0140
9	5	11.4	3.5930
10	11	11.4	0.0140
	114	114.0	13.1930

$\chi^2 = \Sigma[(O - E)^2/E]$
$\quad = 13.193$

P-value = χ^2cdf (13.193,99,9) = 0.1541

conclusion:

Do not reject Ho; there is not sufficient evidence to reject the claim that $p_i = 1/10$ for each position. There is not sufficient evidence to reject the claim that the likelihood of winning is the same for all post positions. Based on these results, post position is not a significant consideration when betting on the Kentucky Derby.

15. original claim: UFO sightings occur in the different months with equal frequency

H_o: $p_{Jan} = p_{Feb} = p_{Mar} = \ldots = p_{Dec} = 1/12$

H_1: at least one $p_i \neq 1/12$

$\alpha = 0.05$ and df $= 11$

C.V. $\chi^2 = \chi^2_\alpha = \chi^2_{0.05} = 19.675$

calculations:

month	O	E	$(O-E)^2/E$
Jan	1239	1506.08	47.3636
Feb	1111	1506.08	103.6402
Mar	1428	1506.08	4.0483
Apr	1276	1506.08	35.1497
May	1101	1506.08	108.4159
Jun	1225	1506.08	52.4591
Jul	2233	1506.08	350.8490
Aug	2012	1506.08	169.9452
Sep	1680	1506.08	20.0832
Oct	1994	1506.08	158.0674
Nov	1648	1506.08	13.3727
Dec	1125	1506.08	96.4253
	18073	18073.00	1159.8196

$\chi^2 = \Sigma[(O - E)^2/E]$

$\quad = 1159.820$

P-value $= \chi^2 cdf\,(1159.820,9999,11) \approx 0$

conclusion:

Reject Ho; there is sufficient evidence to reject the claim that $p_i = 1/12$ for each month. There is sufficient evidence to reject the claim that UFO sightings occur in the different months with equal frequency. July and August likely have the highest frequency because those are summer months when more people are more likely to be outdoors.

17. original claim: the observed frequencies fit the expected proportions

H_o: $p_1 = 9/16$, $p_2 = 3/16$, $p_3 = 3/16$, $p_4 = 1/16$

H_1: at least one p_i is not as claimed

$\alpha = 0.05$ and df $= 3$

C.V. $\chi^2 = \chi^2_\alpha = \chi^2_{0.05} = 7.815$

calculations:

category	O	E	$(O-E)^2/E$
1	59	45	4.3556
2	15	15	0.0000
3	2	15	11.2667
4	4	5	0.2000
	80	80	15.8222

$\chi^2 = \Sigma[(O - E)^2/E]$

$\quad = 15.822$

P-value $= \chi^2 cdf\,(15.822,99,3) = 0.0012$

conclusion:

Reject Ho; there is sufficient evidence to reject the claim that the proportions are as claimed. There is sufficient evidence to reject the claim that the observed frequencies fit the proportions that were expected according to the principles of genetics.

19. original claim: the color distribution is as stated by Mars

H_o: $p_G = 0.16$, $p_O = 0.20$, $p_Y = 0.14$, $p_{Bl} = 0.24$, $p_R = 0.13$, $p_{Br} = 0.13$

H_1: at least one p_i is not as claimed

$\alpha = 0.05$ and df $= 5$

C.V. $\chi^2 = \chi_\alpha^2 = \chi_{0.05}^2 = 11.071$

calculations:

color	O	E	$(O-E)^2/E$
G	19	16	0.5625
O	25	20	1.2500
Y	8	14	2.5714
Bl	27	24	0.3750
R	13	13	0.0000
Br	8	13	1.9231
	100	100	6.6820

$\chi^2 = \Sigma[(O-E)^2/E]$
$= 6.682$

P-value $= \chi^2\text{cdf}\,(6.682,99,5) = 0.2454$

conclusion:

Do not reject Ho; there is not sufficient evidence to reject the claim that the proportions are as stated. There is not sufficient evidence to reject the claim that the color distribution is as stated by Mars.

21. original claim: the leading digits follow Benford's law.

H_o: $p_1 = 0.301$, $p_2 = 0.176$, $p_3 = 0.125$, $p_4 = 0.097,\ldots$, $p_9 = 0.046$

H_1: at least one p_i is not as claimed

$\alpha = 0.01$ and df $= 8$

C.V. $\chi^2 = \chi_\alpha^2 = \chi_{0.01}^2 = 20.090$

calculations:

digit	O	E	$(O-E)^2/E$
1	0	235.98	235.9840
2	15	137.98	101.6146
3	0	98.00	98.0000
3	76	76.05	0.0000
5	479	61.94	2808.4213
6	183	52.53	324.0737
7	8	45.47	30.8795
8	23	39.98	7.2143
9	0	36.06	36.0640
	784	784.00	3650.2514

$\chi^2 = \Sigma[(O-E)^2/E]$
$= 3650.251$

P-value $= \chi^2\text{cdf}\,(3650.251,9999,8) \approx 0$

conclusion:

Reject Ho; there is sufficient evidence to reject the claim that the observed frequencies fit the stated proportions. There is sufficient evidence to reject the claim that there is goodness-of-fit of the leading digits with Benford's law. Yes; it appears that the checks are the result of fraud.

23. original claim: the leading digits follow Benford's law.

H_o: $p_1 = 0.301$, $p_2 = 0.176$, $p_3 = 0.125$, $p_4 = 0.097,\ldots$, $p_9 = 0.046$

H_1: at least one p_i is not as claimed

$\alpha = 0.01$ and df $= 8$

C.V. $\chi^2 = \chi^2_\alpha = \chi^2_{0.01} = 20.090$

calculations:

digit	O	E	$(O-E)^2/E$
1	52	63.82	2.1865
2	40	37.31	0.1936
3	23	26.50	0.4623
3	20	20.56	0.0155
5	21	16.75	1.0795
6	9	14.20	1.9066
7	8	12.30	1.5009
8	9	10.81	0.3037
9	30	9.75	42.0408
	212	212.00	49.6894

$\chi^2 = \Sigma[(O - E)^2/E]$

$\quad = 49.689$

P-value $= \chi^2$cdf $(49.689,99,8) = 4.9E\text{-}8 = 0.00000005$

conclusion:

Reject Ho; there is sufficient evidence to reject the claim that the observed frequencies fit the stated proportions. There is sufficient evidence to reject the claim that there is goodness-of-fit of the leading digits with Benford's law. No; it appears the campaign contributions are not legitimate.

25. An outlier in one of the categories changes the total number of observations and the expected value for each of the other categories. In Example 1 the calculated $\chi^2 = \Sigma[(O - E)^2/E]$ statistic changes from $\Sigma[(O\text{-}8)^2/8] = 11.250$ to $\Sigma[(O\text{-}14.3)^2/14.3] = 247.280$, and the effect of the outlier is dramatic.

11-3 Contingency Tables

1. To move from scientific notation to standard notation, the E-11 indicates that the decimal point should be moved 11 places to the left, which places 10 zeros between the decimal point and the first significant digit, so that $1.73251E\text{-}11 = 0.0000000000173$. Because the P-value is so low, reject the claim that getting paralytic polio is independent of receiving the Salk vaccine. The vaccine appears to be effective.

3. The P-value is the probability of getting results as extreme as or more extreme than the ones obtained whenever the null hypothesis is true. In this context, it is the probability of achieving a success rate at least as effective as the one observed be chance alone – i.e., if the vaccine has no effect, and getting paralytic polio truly is independent of receiving the vaccine.

5. H_o: winning team (home/visitor) and the sport played are independent

H_1: winning team (home/visitor) and the sport played are related

$\alpha = 0.05$ and df $= 3$

C.V. $\chi^2 = \chi^2_\alpha = \chi^2_{0.05} = 7.815$

calculations:
$$\chi^2 = \Sigma[(O - E)^2/E]$$
$$= 4.737 \quad [\text{TI-83/84}^+]$$
P-value $= \chi^2\text{cdf}\,(4.737,99,3) = 0.1921 \quad [\text{TI-83/84}^+]$

conclusion:

Do not reject Ho; there is not sufficient evidence to reject the claim that the winning team (home/visitor) and the sport played are independent. There is not sufficient evidence to reject the claim that the likelihood of the home team winning is independent of the sport.

NOTES FOR THE REMAINING EXERCISES:

(1) For each row and each column it must be true that $\Sigma O = \Sigma E$. After the marginal row and column totals are calculated, both the row totals and column totals must sum to produce the same grand total. If either of the preceding is not true, then an error has been made and further calculations have no meaning.

(2) Rejecting an hypothesis of independence to conclude that two variables are related does not necessarily imply some desired result (e.g., "the drug is effective"). Making such secondary judgments is a legitimate informal procedure, but it requires checking to make sure the frequencies in the table occur in the appropriate direction.

(3) All contingency table analyses in this manual use the following conventions.

• The E values for each cell are given in parentheses below the O values.

• The addends used to calculate the χ^2 test statistic follow the arrangement of the cells in the contingency table – making it easier to monitor the large number of intermediate steps involved and helping to prevent errors caused by missing or double-counting cells.

• As before, P-values are obtained using the $\chi^2\text{cdf}$ command on the TI-83/84$^+$ calculator.

• The accompanying illustrations follow the "usual" chi-square shape, even though that shape is not correct for df=1 or df=2.

7. Ho: challenge success and gender are independent
   H$_1$: challenge success and gender are related
   $\alpha = 0.05$ and df = 1
   C.V. $\chi^2 = \chi^2_\alpha = \chi^2_{0.05} = 3.841$

calculations:

		SUCCESSFUL?		
		yes	no	
GENDER	M	201 (190.59)	288 (298.41)	489
	F	126 (136.41)	224 (213.59)	350
		327	512	839

$$\chi^2 = \Sigma[(O - E)^2/E]$$
$$= 0.569 + 0.363$$
$$\quad 0.795 + 0.598 = 2.235$$
P-value $= \chi^2\text{cdf}\,(2.235,99,1) = 0.1349$

conclusion:

Do not reject Ho; there is not sufficient evidence to reject the claim that challenge success and gender are independent. There is not sufficient evidence to reject the claim that success in challenges is independent of the gender of the player. No; neither gender appears to be more successful than the other.

9. H_o: subject truthfulness and test result are independent
H_1: subject truthfulness and test result are related
$\alpha = 0.05$ and df $= 1$
C.V. $\chi^2 = \chi^2_\alpha = \chi^2_{0.05} = 3.841$
calculations:

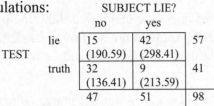

		SUBJECT LIE?		
		no	yes	
	lie	15 (190.59)	42 (298.41)	57
TEST				
	truth	32 (136.41)	9 (213.59)	41
		47	51	98

$\chi^2 = \Sigma[(O - E)^2/E]$
$\quad = 5.567 + 5.131$
$\qquad 7.740 + 7.133 = 25.571$
P-value $= \chi^2 \text{cdf} (25.571, 99, 1) = 4.3\text{E-}7 = 0.0000004$
conclusion:

Reject Ho; there is sufficient evidence to reject the claim that subject truthfulness and test results are independent. There is sufficient evidence to reject the claim that the test result is independent of whether the subject was telling the truth. Yes; the results suggest that polygraphs are effective in distinguishing between truth and lies.

11. H_o: type of filling and adverse health conditions are independent
H_1: type of filling and adverse health conditions are related
$\alpha = 0.05$ and df $= 1$
C.V. $\chi^2 = \chi^2_\alpha = \chi^2_{0.05} = 3.841$
calculations:

		FILLING		
		amalgam	composite	
	yes	135 (140.00)	145 (140.00)	280
ADVERSE?				
	no	132 (127.00)	122 (127.00)	254
		267	267	534

$\chi^2 = \Sigma[(O - E)^2/E]$
$\quad = 0.179 + 0.179$
$\qquad 0.197 + 0.197 = 0.751$
P-value $= \chi^2 \text{cdf} (0.751, 99, 1) = 0.3862$
conclusion:

Do not reject Ho; there is not sufficient evidence to reject the claim that type of filling and adverse health conditions are independent. There is not sufficient evidence to reject the claim that experiencing adverse health conditions is independent of whether the subject received an amalgam or a composite filling. No; amalgam restorations do not appear to affect health conditions.

13. H_o: plea and sentence are independent
    H_1: plea and sentence are related
    $\alpha = 0.05$ and df = 1
    C.V. $\chi^2 = \chi^2_\alpha = \chi^2_{0.05} = 3.841$
    calculations:

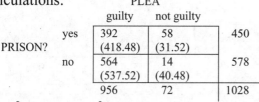

		PLEA		
		guilty	not guilty	
PRISON?	yes	392 (418.48)	58 (31.52)	450
	no	564 (537.52)	14 (40.48)	578
		956	72	1028

$\chi^2 = \Sigma[(O - E)^2/E]$
$= 1.676 + 22.252$
$\quad 1.305 + 17.324 = 42.557$
P-value $= \chi^2$cdf $(42.557,99,1) = 6.9$E-11 ≈ 0

conclusion:

> Reject Ho; there is sufficient evidence to reject the claim that plea and sentence are independent. There is sufficient evidence to reject the claim that whether or not a person is sent to prison is independent of the plea. It appears that those entering a guilty pleas are less likely to be sent to prison. Yes; assuming those who are really guilty will be convicted with a trial, these results suggest that a guilty plea should be encouraged for guilty persons.
> NOTE: The study could be misleading because it appears to compare those who plead guilty only to those who are convicted in trials. Apparently those who plead not guilty and are found not guilty (and therefore receive no prison sentence) are not included – and perhaps a guilty person who has a chance for acquittal should plead not guilty. If 50 guilty people with not guilty pleas were acquitted, as shown at the right, including them in the table in the no prison category changes the above conclusion since
> $\chi^2 = \Sigma[(O - E)^2/E] = 0.125 + 0.982 + 0.090 + 0.704 = 1.901$.

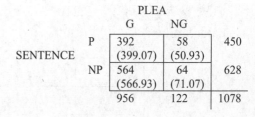

		PLEA		
		G	NG	
SENTENCE	P	392 (399.07)	58 (50.93)	450
	NP	564 (566.93)	64 (71.07)	628
		956	122	1078

15. H_o: success and type of treatment are independent
    H_1: success and type of treatment are related
    $\alpha = 0.01$ and df = 1
    C.V. $\chi^2 = \chi^2_\alpha = \chi^2_{0.01} = 6.635$
    calculations:

		SUCCESS?		
		yes	no	
TYPE	splint	60 (67.57)	23 (15.43)	83
	surgery	67 (59.43)	6 (13.57)	73
		127	29	156

$\chi^2 = \Sigma[(O - E)^2/E]$
$= 0.848 + 3.714$
$\quad 0.964 + 4.223 = 9.750$
P-value $= \chi^2$cdf $(9.750,99,1) = 0.0018$

conclusion:

> Reject Ho; there is sufficient evidence to reject the claim that success and type of treatment are independent. There is sufficient evidence to reject the claim that successfully treating carpal tunnel syndrome is independent of the type of treatment (splint vs. surgery) used. The results suggest that the surgery treatment is associated with a higher rate of success.

17. H_o: gender and opinion are independent
   H_1: gender and opinion are related
   $\alpha = 0.05$ and df = 2
   C.V. $\chi^2 = \chi_\alpha^2 = \chi_{0.05}^2 = 5.991$
   calculations:

		OPINION			
		human	natural	other	
	M	314	146	44	504
GENDER		(307.34)	(152.19)	(44.47)	
	F	308	162	46	516
		(314.66)	(155.81)	(45.53)	
		622	308	90	1020

$\chi^2 = \Sigma[(O - E)^2/E]$
   $= 0.144 + 0.252 + 0.005$
   $0.141 + 0.246 + 0.005 = 0.792$
   P-value $= \chi^2\text{cdf}(0.792,99,2) = 0.6730$
conclusion:

   Do not reject Ho; there is not sufficient evidence to reject the claim that gender and opinion are independent. There is not sufficient evidence to reject the claim that the gender of the respondent is independent of that respondent's opinion about the cause of global warming. Men and women appear to generally agree.

19. H_o: reaction and treatment are independent
   H_1: reaction and treatment are related
   $\alpha = 0.01$ and df = 2
   C.V. $\chi^2 = \chi_\alpha^2 = \chi_{0.01}^2 = 9.210$
   calculations:

		TREATMENT			
		placebo	C-1332	C-1998	
	Y	344	89	8	441
REACT?		(284.12)	(143.72)	(13.16)	
	N	13621	774	71	2207
		(1421.88)	(719.28)	(65.84)	
		1706	863	79	2648

$\chi^2 = \Sigma[(O - E)^2/E]$
   $= 12.621 + 20.837 + 2.021$
   $2.522 + 4.164 + 0.404 = 42.568$
   P-value $= \chi^2\text{cdf}(42.568,99,2) = 5.7E\text{-}10 \approx 0$
conclusion:

   Reject Ho; there is sufficient evidence to reject the claim that reaction and treatment are independent. There is sufficient evidence to reject the claim that having an adverse effect on the digestive system is not independent of the treatment group. Yes; since the observed frequencies in the REACT?-yes row of the two Campral columns are less than the expected frequencies in those cells, Campral appears to be associated with reduced adverse effects on the digestive system.

21. The following table is used for the calculations in exercise #21.

		TREATMENT				
		placebo	A-10	A-40	A-80	
INFECTION?	Y	27 (27.08)	89 (86.56)	8 (7.92)	7 (9.43)	131
	N	243 (242.92)	774 (776.44)	71 (71.08)	87 (84.57)	1175
		270	863	79	94	1306

H_o: getting an infection and the treatment used are independent
H_1: getting an infection and the treatment used are related
$\alpha = 0.05$ [assumed] and df $= 3$
C.V. $\chi^2 = \chi^2_\alpha = \chi^2_{0.05} = 7.815$
calculations:

$\chi^2 = \Sigma[(O - E)^2/E]$
$= 0.000 + 0.069 + 0.001 + 0.626$
$\quad 0.000 + 0.008 + 0.000 + 0.070$
$= 0.773$
P-value $= \chi^2$cdf $(0.773, 99, 3)$
$\quad = 0.8559$

conclusion:

Do not reject Ho; there is not sufficient evidence to reject the claim that getting an infection and the treatment used are independent. There is not sufficient evidence to reject the claim that getting an infection is independent of the atorvastatin treatment used. No; the drug does not seem to have an effect on infection rates.

23. H_o: the proportions of agree/disagree are the same for male and female interviewers
H_1: the proportions of agree/disagree are not the same for male and female interviewers
$\alpha = 0.01$ and df $= 1$
C.V. $\chi^2 = \chi^2_\alpha = \chi^2_{0.01} = 6.635$
calculations:

		INTERVIEWER		
		male	female	
REPLY	agree	512 (565.33)	336 (282.67)	848
	disagree	288 (234.67)	64 (117.33)	352
		800	400	1200

$\chi^2 = \Sigma[(O - E)^2/E]$
$= 5.051 + 10.063$
$\quad 12.121 + 24.242 = 51.458$
P-value $= \chi^2$cdf $(51.458, 99, 1) = 7.3E\text{-}13 \approx 0$

conclusion:

Reject Ho; there is sufficient evidence to reject the claim that the proportions of agree/disagree are the same for male and female interviewers. There is sufficient evidence to reject the claim that women given the proportions of agree/disagree responses to male interviewers as they do to female interviewers. Yes; it appears that the gender of the interviewer affects the responses of women.

25. The test from exercise 7 as a difference between two proportions is as follows.

Let the males be group 1.

$\hat{p}_1 = x_1/n_1 = 201/489 = 0.411$ $\hat{p}_1 - \hat{p}_2 = 0.411 - 0.360 = 0.051$

$\hat{p}_2 = x_2/n_2 = 126/350 = 0.360$

$\bar{p} = (201+126)/(489+350)$

     $= 327/839 = 0.390$

original claim: $p_1 - p_2 \neq 0$

H_o: $p_1 - p_2 = 0$

H_1: $p_1 - p_2 \neq 0$

$\alpha = 0.05$

C.V. $z = \pm z_{\alpha/2} = \pm z_{0.025} = \pm 1.96$

calculations:

$z_{\hat{p}_1 - \hat{p}_2} = (\hat{p}_1 - \hat{p}_2 - \mu_{\hat{p}_1 - \hat{p}_2})/\sigma_{\hat{p}_1 - \hat{p}_2}$

$= (0.051 - 0)/\sqrt{(0.390)(0.610)/489 + (0.390)(0.610)/350}$

$= 0.051/0.03415$

$= 1.4948$

P-value $= 2 \cdot P(z > 1.4948) = 2 \cdot normalcdf(1.4948,99) = 0.1350$

conclusion:

Do not reject H_o; there is not sufficient evidence to conclude that $p_1 - p_2 \neq 0$. There is not sufficient evidence to reject the claim that the percentage of men's challenges that are successful is different from the percentage of women's challenges that are successful. No; neither gender appears to be more successful than the other.

The test as a test of independence is given in Exercise 7. The tests are mathematically equivalent, the relationship being that $(z)^2 = \chi^2$. In particular,

calculated statistic: $(z)^2 = (1.4918)^2 = 2.235 = \chi^2$

critical region: $(z)^2 = (\pm 1.96)^2 = 3.842 = \chi^2$ [$\chi^2 = 3.841$, there is a slight round-off error]

P-value: $0.1350 = 0.1349$ [there is a slight round-off error]

11-4 McNemar's Test for Matched Pairs

1. Neither format appears to be better. Ignore the 74 students for which both methods produced mastery and the 48 students for which neither method produced master, as they do not differentiate between the methods. Consider the discordant cases: 31 times the symbolic format succeeded when the verbal format did not, and 33 times the verbal format succeeded when the symbolic format did not. Since the discordant frequencies of 31 and 33 are nearly identical, the neither format is better.

3. McNemar's test ignores the frequencies of 74 and 48 because they indicate pairs of results in which both formats performed the same. Pairs in which both formats perform the same give no information about which format mat be better.

5. The number of subjects in the experiment is $460 + 4 + 361 + 192 = 1017$.

7. The number of subjects who smoking status did not change is $460 + 192 = 652$. Of those, 460 were smokers both before and after the treatment, and 192 were non-smokers both before and after the treatment.

9. Discordant pairs are pairs for which the BEFORE and AFTER states do not agree.
   a. No. The data from the a=460 subjects who were smokers BEFORE and smokers AFTER do not represent discordant pairs of results.
   b. Yes. The data from the b=361 subjects who were smokers BEFORE and non-smokers AFTER represent discordant pairs of results.
   c. Yes. The data from the c=4 subjects who were non-smokers BEFORE and smokers AFTER represent discordant pairs of results.
   d. No. The data from the d=192 subjects who were non-smokers BEFORE and non-smokers AFTER do not represent discordant pairs of results.

11. For $\alpha = 0.01$ and df = 1, the critical value is $\chi^2 = 6.635$.

NOTE: The null hypotheses of no effect can be stated in a variety of ways: as a difference between states (i.e., using the main labels that distinguish the rows from the columns), a difference between proportions, or as an independence of variables. Each of those statements is equivalent, and in any given problem the one should be used which most clearly conveys the spirit of the experiment. In general, the manual chooses the first option. In the story problem of exercises 5-12, any of the following is valid and communicates the idea that the treatment has no effect.
H_o: there is no difference between the BEFORE and AFTER states
H_o: the proportion of smokers who become non-smokers is equal to the proportion of non-smokers who become smokers
H_o: smoking behavior and taking the treatment are independent

13. H_o: there is no difference between wearing or not wearing the hip protector
    H_1: there is a difference between wearing or not wearing the hip protector
    $\alpha = 0.05$ and df = 1
    C.V. $\chi^2 = \chi^2_\alpha = \chi^2_{0.05} = 3.841$
    calculations:

    $b = 17, c = 21$
    $\chi^2 = (|b-c| - 1)^2/(b+c)$
    $\quad = (4-1)^2/(17+21)$
    $\quad = 9/38$
    $\quad = 0.237$
    P-value = χ^2cdf (0.237,99,1)
    $\quad\quad = 0.6265$

conclusion: Do not reject Ho; there is not sufficient evidence to reject the claim that there is no difference between wearing or not wearing the hip protector. The hip protectors do not appear to be effective in preventing hip fractures.

15. H_o: there is no difference between the treatment and the placebo
    H_1: there is a difference between the treatment and the placebo
    $\alpha = 0.05$ and df = 1
    C.V. $\chi^2 = \chi^2_\alpha = \chi^2_{0.05} = 3.841$
    calculations:

    $b = 12, c = 22$
    $\chi^2 = (|b-c| - 1)^2/(b+c)$
    $\quad = (10-1)^2/(12+22)$
    $\quad = 81/34$
    $\quad = 2.382$
    P-value = χ^2cdf (2.382,99,1)
    $\quad\quad = 0.1227$

conclusion: Do not reject Ho; there is not sufficient evidence to reject the claim that there is no difference between the treatment and the placebo. The treatment does not appear to be effective.

17. H_o: there is no difference between the PET-CT and MRI technologies
H_1: there is a difference between the PET-CT and MRI technologies
$\alpha = 0.05$ [assumed] and df = 1
C.V. $\chi^2 = \chi^2_{\alpha} = \chi^2_{0.05} = 3.841$
calculations:

b = 1, c = 11
$\chi^2 = (|b\text{-}c| - 1)^2/(b+c)$
    $= (10\text{-}1)^2/(1+11)$
    $= 81/12$
    $= 6.750$
P-value $= \chi^2 cdf\,(6.750,99,1)$
    $= 0.0094$

conclusion: Reject Ho; there is sufficient evidence to reject the claim that there is no difference between the PET-CT and MRI technologies. Since the majority of the discordant pairs are within the "PET-CT=correct/MRI=incorrect" cell, conclude that the PET-CT technology is more accurate.

19. b=1, c=14, $\chi^2 = (b\text{-}c)^2/(b+c) = (1\text{-}14)^2/(1+14) = 169/15 = 11.267$
The uncorrected test statistic of 11.267 is slightly larger than the corrected for continuity test statistic of 9.600. In this case the conclusion is the same. Since $(|b\text{-}c|\text{-}1)^2 < (b\text{-}c)^2$ whenever $|b\text{-}c| > 0.5$, the correction for continuity generally lowers the calculated test statistic and could lead to failure to reject the null hypothesis in situations where the uncorrected test statistic would lead to rejection.

21. This exercise follows the usual format for testing hypotheses with two modifications.
   • In order to compare two different techniques, the decision is based on comparing the α level and the P-value (instead of using a C.V.). This approach is valid for any hypothesis test and any test statistic, and it makes comparisons easier.
   • There are two "calculations" sections – one for each of the two techniques being compared.
H_o: there is no difference between the two treatments
H_1: there is a difference between the two treatments
$\alpha = 0.05$
calculations using the exact binomial:

   x = 6 = largest frequency in cells
            representing discordant pairs
   P-value $= 2 \cdot P(x \geq 6)$
            $= 2 \cdot [P(x=6) + P(x=7) + P(x=8)]$
            $= 2 \cdot [28(0.5)^6(0.5)^2 + 8(0.5)^7(0.5)^1 + 1(0.5)^8(0.5)^0]$
            $= 2 \cdot [0.109375 + 0.03125 + 0.00390625]$
            $= 2 \cdot [0.14453125]$
            $= 0.2891$

calculations using the χ^2 approximation:

$b = 2, c = 6$

$\chi^2 = (|b-c| - 1)^2/(b+c)$

$\quad = (4-1)^2/(2+6)$

$\quad = 9/8$

$\quad = 1.125$

P-value $= \chi^2\text{cdf}\,(1.125, 99, 1)$

$\quad\quad = 0.2888$

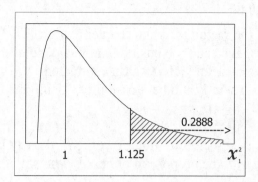

conclusion: Do not reject Ho; there is not sufficient evidence to reject the claim that there is no difference between the two treatments.

Despite the fact that the requirements for McNemar's test are not met, the conclusion is the same for both of the above methods – and the P-values are almost identical.

Statistical Literacy and Critical Thinking

1. The numbers in the table are the frequency counts that summarize the data, and they are not the actual data. The actual data are the (145+23+78+4001+322+1786) = 6355 "nausea" or "no nausea" responses.

2. This is a two-way table because each piece of data is categorized in two dimensions (i.e., according to two different variables): according to which treatment it represents (Celebrex, Ibuprofen, placebo), and according to the response (nausea, no nausea).

3. No. In general, statistics deals with mathematical values and associations – not with cause and effect. The proof of causality requires physical evidence, not just statistical evidence.

4. The expected frequency gives the frequency expected in a particular cell when the two categorizing variable are independent. In this context, if experiencing nausea is independent of the treatment, the expected number of subjects who experience nausea with the Celebrex treatment is 160.490.

Chapter Quick Quiz

1. H_o: $p_{Sun} = p_{Mon} = p_{Tue} = p_{Wed} = p_{Thu} = p_{Fri} = p_{Sat} = 1/7$
H_1: at least one $p_i \neq 1/7$

2. The observed frequency for Sunday is 40. The expected frequency for Sunday is $(\Sigma f)/7 = (40+24+25+28+29+32+38)/7 = 216/7 = 30.857$.

3. The critical value is the value in the chi-square distribution with 6 degrees of freedom that has 0.05 in the upper tail: $\chi^2_\alpha = \chi^2_{0.05} = 12.592$.

4. Since $0.2840 > 0.05$, we fail to reject the null hypothesis. There is not sufficient evidence to reject that claim that fatal DWI crashes occur equally on the different days of the week.

5. The test is right-tailed. The test statistic measures discrepancy between the observed data and the null hypothesis, and we reject only when there is a large discrepancy.

6. chi-square

7. Using the methods of this chapter (section 11-3),
    H_0: response and job position are independent
    H_1: response and job position are related
    NOTE: There is another possibility. Using the methods of section 9-2,
    H_0: $p_1 - p_2 = 0$
    H_1: $p_1 - p_2 \neq 0$
    where p_1 = the proportion of workers who say it is unethical to monitor employee e-mail
          p_2 = the proportion of bosses who say it is unethical to monitor employee e-mail

8. For the test of independence in exercise 7, the critical value is the value in the chi-square distribution with 1 degree of freedom that has 0.05 in the upper tail: $\chi_\alpha^2 = \chi_{0.05}^2 = 3.841$.

9. Since $0.0302 < 0.05$, we reject the null hypothesis. There is sufficient evident to reject the claim that a person's response and job position are independent. There is sufficient evidence to conclude that a person's opinion about monitoring employee e-mail is related to whether that person is a worker or a boss.

10. Since $0.0302 > 0.01$, we fail to reject the null hypothesis. There is not sufficient evidence to reject the claim that a person's response and job status are independent. There is not sufficient evidence to reject the claim that person's opinion about monitoring employee e-mail is independent of whether that person is a worker or a boss.

Review Exercises

1. original claim: experiencing nausea is independent of the treatment
    H_o: nausea and treatment are independent
    H_1: nausea and treatment are related
    $\alpha = 0.05$ and $df = 2$
    C.V. $\chi^2 = \chi_\alpha^2 = \chi_{0.05}^2 = 5.991$
    calculations:

NAUSEA?		celebrex	ibupro	placebo	
	Y	145 (160.45)	23 (13.35)	78 (72.15)	246
	N	4001 (3985.51)	322 (331.65)	1786 (1791.85)	6109
		4146	345	1864	6355

$\chi^2 = \Sigma[(O - E)^2/E]$
$= 1.495 + 6.966 + 0.474$
$0.060 + 0.281 + 0.019 = 9.294$
P-value $= \chi^2 cdf (9.294,99,2) = 0.0096$

conclusion:
    Reject Ho; there is sufficient evidence to reject the claim that nausea and treatment are independent. There is sufficient evidence to reject the claim that a subject's experiencing nausea is independent of the treatment received. No; the adverse reaction of nausea does not appear to be about the same for the different treatments.

2. original claim: deaths by lightning occur on the different days with equal frequency

H_o: $p_{Sun} = p_{Mon} = p_{Tue} = \ldots = p_{Sat} = 1/7$

H_1: at least one $p_i \neq 1/7$

$\alpha = 0.01$ and df $= 6$

C.V. $\chi^2 = \chi^2_\alpha = \chi^2_{0.01} = 16.812$

calculations:

day	O	E	$(O-E)^2/E$
Sun	574	462.57	26.8420
Mon	445	462.57	0.6675
Tue	429	462.57	2.4365
Wed	473	462.57	0.2351
Thu	428	462.57	2.5838
Fri	422	462.57	3.5585
Sat	467	462.57	0.0424
	3238	3238.00	36.3657

$\chi^2 = \Sigma[(O - E)^2/E]$
   $= 36.366$

P-value $= \chi^2$cdf $(36.366,99,6) = 2.34E{-}6 = 0.000002$

conclusion:

Reject Ho; there is sufficient evidence to reject the claim that $p_i = 1/7$ for each day. There is sufficient evidence to reject the claim that lightning deaths occur on the different days with equal frequency. The disproportionate number of deaths on Sunday is due to the fact that more people are involved in outdoor activities on Sundays than on the other days of the week.

3. original claim: the distribution of participants fits the population distribution

H_o: $p_W = 0.757$, $p_H = 0.091$, $p_B = 0.108$, $p_{AP} = 0.038$, $p_{NA} = 0.007$

H_1: at least one p_i is not as claimed

$\alpha = 0.01$ and df $= 4$

C.V. $\chi^2 = \chi^2_\alpha = \chi^2_{0.05} = 9.488$

calculations:

ethnic	O	E	$(O-E)^2/E$
W	644	569.26	9.8117
H	23	68.43	30.1623
B	69	81.22	1.8375
AP	14	28.58	7.4349
NA	2	5.26	2.0239
	752	752.75	51.2703

$\chi^2 = \Sigma[(O - E)^2/E]$
   $= 51.270$

P-value $= \chi^2$cdf $(51.270,99,4) = 1.96E{-}10 \approx 0$

conclusion:

Reject Ho; there is sufficient evidence to reject the claim that the observed frequencies fit the stated proportions. There is sufficient evidence to reject the claim that the distribution of clinical trial participants fits the population distribution. If there is not proportionate representation in clinical trials, then the sample is not representative of the general population and the results should not be applied to the general population.

NOTE: The E's sum to 752.75 because the given probabilities sum to 1.001. In Exercise 20 of section 11-3, a similar problem gives p_W as 0.756 and the given probabilities sum to 1. Using that value gives a calculated $\chi^2 = 51.482$ and the difference is trivial.

4. original claim: continuing to smoke is independent of the treatment received

H_0: smoking and treatment are independent

H_1: smoking and treatment are related

$\alpha = 0.05$ and df = 1

C.V. $\chi^2 = \chi_\alpha^2 = \chi_{0.05}^2 = 3.841$

calculations:

		TREATMENT		
		Bu-Hy	placebo	
SMOKING?	yes	299 (314.33)	167 (151.67)	466
	no	101 (85.67)	26 (41.33)	127
		400	193	593

$\chi^2 = \Sigma[(O - E)^2/E]$

$= 0.748 + 1.550$

$2.745 + 5.689 = 10.732$

P-value $= \chi^2$cdf (10.732,99,1) = 0.0011

conclusion:

Reject Ho; there is sufficient evidence to reject the claim that smoking and treatment are independent. There is sufficient evidence to reject the claim that whether or not a person continues smoking is independent of the treatment received vaccinated. Yes; since the "Bu-Hy/no" cell contains fewer than expected subjects, the vaccine appears to be effective.

5. H_0: there is no difference between the child and parent responses

H_1: there is a difference between the child and parent responses

$\alpha = 0.05$ and df = 1

C.V. $\chi^2 = \chi_\alpha^2 = \chi_{0.05}^2 = 3.841$

calculations:

b = 104, c = 172

$\chi^2 = (|b-c| - 1)^2/(b+c)$

$= (68-1)^2/(104+172)$

$= 4489/276$

$= 16.264$

P-value $= \chi^2$cdf (16.264,99,1)

$= 5.51E-5 = 0.00006$

conclusion: Reject Ho; there is sufficient evidence to reject the claim that there is no difference between child and parent responses. The results indicate significant disagreement between the child's assessment and the parent's assessment of early morning coughing. Since the majority of the discordant pairs are in the "child-cough/parent-no cough" cell, it appears that children report early morning coughing more than the parents – who may not yet be awake, or nearby enough to hear, when the early morning coughing occurs.

Cumulative Review Exercises

1. a. This is an observational study, since the researched merely watched the subjects without applying a treatment or otherwise introducing a factor that would modify the results.

   b. The given numbers are discrete, since they must be whole numbers.

   c. The given numbers are statistics, since they describe the samples and not the populations.

d. If the subjects could tell that they were being observed, they might modify their behavior. Since the study was sponsored by an organization with a vested interest in the results, it is possible that those interests may have (consciously or subconsciously) biased in some way the conducting of the study.

2. Let the males be group 1.

$\hat{p}_1 = x_1/n_1 = 2023/3065 = 0.660$ $\hat{p}_1 - \hat{p}_2 = 0.660 - 0.880 = -0.220$

$\hat{p}_2 = x_2/n_2 = 2650/3011 = 0.880$

$\overline{p} = (2023+2650)/(3065+3011)$

    $= 4673/6076 = 0.769$

original claim: $p_1-p_2 \neq 0$

H_o: $p_1-p_2 = 0$

H_1: $p_1-p_2 \neq 0$

$\alpha = 0.05$

C.V. $z = \pm z_{\alpha/2} = \pm z_{0.025} = \pm 1.96$

calculations:

$z_{\hat{p}_1-\hat{p}_2} = (\hat{p}_1-\hat{p}_2 - \mu_{\hat{p}_1-\hat{p}_2})/\sigma_{\hat{p}_1-\hat{p}_2}$

    $= (-0.220 - 0)/\sqrt{(0.769)(0.231)/3065 + (0.769)(0.231)/3011}$

    $= -0.220/0.01081$

    $= -20.35$

P-value $= 2 \cdot P(z<-20.35) = 2 \cdot (0.0001) = 0.0002$

conclusion:

Reject H_o; there is sufficient evidence to reject the claim that $p_1-p_2 = 0$ and conclude that $p_1-p_2 \neq 0$ (in fact, that $p_1-p_2 < 0$). There is sufficient evidence to reject the claim that the proportion of men who wash their hands is equal to the proportion of women who wash their hands Yes; there is a significant difference.

3. original claim: hand washing is independent of gender

H_o: gender and hand washing are independent

H_1: gender and hand washing are related

$\alpha = 0.05$ and df = 1

C.V. $\chi^2 = \chi^2_{\alpha} = \chi^2_{0.05} = 3.841$

calculations:

		GENDER		
		male	female	
	yes	2023	2650	4673
		(2357.27)	(2315.73)	
WASH?				
	no	1042	361	1403
		(707.73)	(695.27)	
		3065	3011	6076

$\chi^2 = \Sigma[(O - E)^2/E]$

    $= 47.400 + 48.250$

        $157.875 + 160.706 = 414.230$

P-value $= \chi^2\text{cdf}(414.230,999,1) \approx 0$

conclusion:

Reject Ho; there is sufficient evidence to reject the claim that gender and hand washing are independent. There is sufficient evidence to reject the claim that hand washing is independent of gender. Since the "male/yes" cell contains fewer than expected subjects, it appears that the proportion of males who wash their hands is lower than the proportion of females who do so.

The following summary statistics apply to exercises 4-6.

$n = 6$

$\Sigma x = 427$

$\Sigma y = 422$

$\Sigma xy = 30021$ $n(\Sigma xy) - (\Sigma x)(\Sigma y) = 6(30021) - (427)(422) = -68$

$\Sigma x^2 = 30439$ $n(\Sigma x^2) - (\Sigma x)^2 = 6(30439) - (427)^2 = 305$

$\Sigma y^2 = 29696$ $n(\Sigma y^2) - (\Sigma y)^2 = 6(29696) - (422)^2 = 92$

4. first round

$\overline{x} = (\Sigma x)/n = (427)/6 = 71.2$

$\tilde{x} = (71+72)/2 = 71.5$ (use ordered scores)

$R = 75 - 67 = 8$ (use ordered scores)

$s^2 = [n(\Sigma x^2) - (\Sigma x)^2]/[n(n-1)]$

   $= [305]/[6(5) = 305/30 = 10.167$

$s = \sqrt{10.167} = 3.189$, rounded to 3.2

fourth round

$\overline{x} = (\Sigma x)/n = (422)/6 = 70.3$

$\tilde{x} = (69+70)/2 = 69.5$ (use ordered scores)

$R = 73 - 69 = 4$ (use ordered scores)

$s^2 = [n(\Sigma x^2) - (\Sigma x)^2]/[n(n-1)]$

   $= [92]/[6(5) = 92/30 = 3.067$

$s = \sqrt{3.067} = 1.751$, rounded to 1.8

The fourth round scores are slightly lower and closer together.

5. $r = [n(\Sigma xy) - (\Sigma x)(\Sigma y)]/[\sqrt{n(\Sigma x^2) - (\Sigma x)^2} \sqrt{n(\Sigma y^2) - (\Sigma y)^2}]$

   $= -68/[\sqrt{305} \sqrt{92}]$

   $= -0.406$

original claim: there is a linear correlation

H_o: $\rho = 0$

H_1: $\rho \neq 0$

$\alpha = 0.05$ and $df = 4$

C.V. $t = \pm t_{\alpha/2} = \pm t_{0.025} = \pm 2.776$ [or $r = \pm 0.811$]

calculations:

   $t_r = (r - \mu_r)/s_r$

      $= (-0.406 - 0)/\sqrt{[1 - (-0.406)^2]/4}$

      $= -0.406/0.4569$

      $= -0.888$

   P-value $= 2 \cdot tcdf(-99,-0.888,4) = 0.4245$

conclusion:

   Do not reject H_o; there is not sufficient evidence to conclude that $\rho \neq 0$. There is not sufficient evidence to support the claim of a linear correlation between the first round scores and the fourth round scores.

6. σ unknown (and assuming the distribution is approximately normal), use t with df=5

   $\alpha = 0.05$, $t_{df,\alpha/2} = t_{5,0.025} = 2.571$

      $\overline{x} \pm t_{\alpha/2} \cdot s/\sqrt{n}$

   $71.167 \pm 2.571(3.1885)/\sqrt{6}$

   71.167 ± 3.347

   $67.8 < \mu < 74.5$

   We have 95% confidence that the limits of 67.8 and 74.5 contain the true mean first round golf score of all golfers in the tournament.

7. $\alpha = 0.05$, $z_{\alpha/2} = z_{0.025} = 1.96$ and $\hat{p} = x/n = 132/150 = 0.88$

$$\hat{p} \pm z_{\alpha/2}\sqrt{\hat{p}\hat{q}/n}$$
$$0.8800 \pm 1.96\sqrt{(0.88)(0.12)/150}$$

0.8800 ± 0.0520

$0.828 < p < 0.932$

To increase the chances of being hired, it appears that an applicant would be wise to send a thank-you note after every job interview for a position that he desired.

8. original claim: $p > 0.75$

$\hat{p} = x/n = 132/150 = 0.880$

Ho: $p = 0.75$

H$_1$: $p > 0.75$

$\alpha = 0.01$

C.V. $z = z_\alpha = z_{0.01} = 2.326$

calculations:

$$z\hat{p} = (\hat{p} - \mu\hat{p})/\sigma\hat{p}$$
$$= (0.880 - 0.750)/\sqrt{(0.750)(0.250)/150}$$
$$= 0.130/0.03536$$
$$= 3.68$$

P-value = $P(z>3.68) = 1 - 0.9999 = 0.0001$

conclusion:

Reject H$_o$; there is sufficient evidence to conclude that $p > 0.75$. There is sufficient evidence to support the claim that more than 75% of all senior executives believe that a thank-you note following an interview increases the chances of an applicant being hired.

9. a. normal distribution

$\mu = 42.6$

$\sigma = 2.9$

$P(x>45)$

$= P(z>0.83)$

$= 1 - 0.7967$

$= 0.2033$

b. normal distribution,

since the original distribution is so

$\mu_{\bar{x}} = \mu = 42.6$

$\sigma_{\bar{x}} = \sigma/\sqrt{n} = 2.9/\sqrt{16} = 0.725$

$P(\bar{x}>45)$

$= P(z>3.31)$

$= 1 - 0.9995$

$= 0.0005$

c. The result from (a) is more meaningful, since the cockpit will be occupied by one person at a time.

10. Since repeated random selections are independent,
$$P(T_1 \text{ and } T_2 \text{ and } T_3 \text{ and } T_4 \text{ and } T_5) = P(T_1) \cdot P(T_2) \cdot P(T_3) \cdot P(T_4) \cdot P(T_5)$$
$$= (0.925)(0.925)(0.925)(0.925)(0.925)$$
$$= (0.925)^5 = 0.6772$$

No; since $0.6772 > 0.05$, randomly selecting five women that are all over 5 feet tall would not be considered unusual.

Chapter 12

Analysis of Variance

12-2 One-Way ANOVA

1. a. One-way analysis of variance is appropriate for these data because they represent three or more populations categorized by a single characteristic that distinguishes the populations from each other. The distinguishing characteristic in this case is epoch.
   b. One-way analysis of variance tests the equality of two or more population means by analyzing sample variances. It finds a difference in the population means if the variance between the sample means is larger than can be expected considering the variance within the samples.

3. We should reject the hypothesis that the three epochs have the same mean skull breadth. There is sufficient evidence to conclude that at least one of the means is different from the others.

NOTE: When testing the hypothesis that three or more groups have the same mean, the test statistic is F, where F is the ratio is the ratio of the variance between the groups to the variance within the groups as defined in the text. This manual generally uses the generic notation $F = s_B^2/s_p^2$. As in previous chapters, the superscripts and subscripts (the numerator df and denominator df) may be used to identify which F distribution to look up in the tables.

5. H_o: $\mu_1 = \mu_2 = \mu_3$
   H_1: at least one μ_i is different
   $\alpha = 0.05$ and $df_{num} = 2$, $df_{den} = 33$
   C.V. $F = F_\alpha = F_{0.05} = 3.3158$
   calculations:

   $$F = s_B^2/s_p^2$$
   $$= 669.0011/70.6481$$
   $$= 9.4695 \text{ [TI-83/84+]}$$
   P-value = $P(F_{33}^2 > 9.4695) = 0.0006$ [TI-83/84+]

   conclusion:
   Reject Ho; there is sufficient evidence to reject the claim that $\mu_1 = \mu_2 = \mu_3$. There is sufficient evidence to reject the claim that the three books have the same mean Flesch Reading Ease score.

7. H_o: $\mu_1 = \mu_2 = \mu_3 = \mu_4$
   H_1: at least one μ_i is different
   $\alpha = 0.05$ and $df_{num} = 3$, $df_{den} = 156$
   C.V. $F = F_\alpha = F_{0.05} = 2.6626$ [Excel]
   calculations:

   $$F = s_B^2/s_p^2$$
   $$= 11.99995/34.08497$$
   $$= 0.35206 \text{ [Excel]}$$
   P-value = $P(F_{156}^3 > 0.3521) = 0.7877$ [Excel]

conclusion:

Do not reject Ho; there is not sufficient evidence to reject the claim that $\mu_1 = \mu_2 = \mu_3 = \mu_4$. There is not sufficient evidence to reject the claim that the mean weight loss is the same for all four diets. Given the four mean losses that range from 2.1 lbs to 3.3 lbs, it appears that one year of following the diet does not result in a weight loss worth the effort. The effort may be justified, however, when other potential benefits are considered – viz., maintaining one's weight (i.e., not gaining weight), enjoying a healthy lifestyle, etc.

9. H_o: $\mu_1 = \mu_2 = \mu_3$
H_1: at least one μ_i is different
$\alpha = 0.05$ and $df_{num} = 2$, $df_{den} = 32$
C.V. F = $F_\alpha = F_{0.05} = 3.3158$
calculations:

$F = s_B^2/s_p^2$

$= 0.42400.933/7981.129$

$= 5.313$ [SPSS]

P-value = $P(F_{32}^2 > 5.313) = 0.010$ [SPSS]

conclusion:

Reject Ho; there is sufficient evidence to reject the claim that $\mu_1 = \mu_2 = \mu_3$. There is sufficient evidence to reject the claim that PG, PG-13 and R movies have the same gross amount.

NOTE FOR THE REMAINING EXERCISES IN THIS SECTION: This section is calculation-oriented. Do not get so involved with the formulas that you miss the concepts. This manual arranges the calculations to promote both computational efficiency and understanding of the underlying principles. The following notation is used.

k = the number of groups

n_i = the number of scores in group i (where i = 1,2,…,k)

$\overline{x}_i$ = the mean of group i

s_i^2 = the variance of group i

$\overline{\overline{x}}$ = the overall mean of all the scores in all the groups

$= (\Sigma n_i \overline{x}_i)/\Sigma n_i$ = the (weighted) mean of the group means

$= \Sigma \overline{x}_i /k$ = simplified form when each group has equal size n

s_B^2 = the variance between the groups

$= \Sigma n_i(\overline{x}_i - \overline{\overline{x}})^2/(k-1)$

$= n \Sigma(\overline{x}_i - \overline{\overline{x}})^2/(k-1) = ns_{\overline{x}}^2$ = simplified form when each group has equal size n

s_p^2 = the variance within the groups

$= (\Sigma df_i s_i^2)/\Sigma df_i$

$= \Sigma s_i^2 /k$ = simplified form when each group has equal size n

numerator df = k-1

denominator df = Σdf_i

$= k(n-1)$ = simplified form when each group has equal size n

$F = s_B^2/s_p^2$ = (variance between groups)/(variance within groups)

P-value = Fcdf(F, 99, numerator df, denominator df) from the TI-83/84$^+$ calculator

As a crude check against errors, and to help get a feeling for the problem, always verify that the following "overall" values are realistic in that

$\overline{\overline{x}}$ is a value between the lowest and the highest of the $\overline{x}_i$ values.

s_p^2 is a value between the lowest and the highest of the s_i^2 values.

11. Since each group has equal size n=10, use the simplified form of the calculations. The following preliminary values are identified.

	small	medium	large
n	10	10	10
Σx	4315	3906	3858
Σx^2	1918833	1620816	1764878
$\overline{x}$	431.5	390.6	385.8
s^2	6323.389	10570.267	30717.956

$k = 3$ $\qquad \overline{\overline{x}} = \Sigma \overline{x}_i / k$

$n = 10$ $\qquad\qquad = 402.633$

$s_{\overline{x}}^2 = \Sigma(\overline{x}_i - \overline{\overline{x}})^2/(k-1) \qquad s_p^2 = \Sigma s_i^2 / k$

$\qquad = 630.723 \qquad\qquad\qquad = 15870.537$

$H_o: \mu_1 = \mu_2 = \mu_3$

$H_1:$ at least one μ_i is different

$\alpha = 0.05$ and $df_{num} = 2$, $df_{den} = 27$

C.V. $F = F_\alpha = F_{0.05} = 3.3541$

calculations:

$\quad F = ns_{\overline{x}}^2/s_p^2$

$\qquad = 10(630.723)/15870.537$

$\qquad = 0.3974$

P-value $= Fcdf(0.3974,99,2,27) = 0.6759$

conclusion:

Do not reject Ho; there is not sufficient evidence to reject the claim that $\mu_1 = \mu_2 = \mu_3$. There is not sufficient evidence to reject the claim that the different car categories have the same mean head injury values. No, these data do not suggest that larger cars are safer.

NOTE: Since subtracting the same value from each will not affect the differences between the group means or any variances in the problem, that strategy may be used to reduce the mathematical magnitude of each of the following problems. In Exercise 13, for example, one could subtract 3 minutes from each time and deal only with the excess seconds over 3 minutes as follows.

mile 1: 15 24 23 22 21

mile 2: 19 22 21 17 19

mile 3: 34 31 29 31 29

13. Since each group has equal size n=5, use the simplified form of the calculations. The following preliminary values are identified (all measurements in seconds).

	mile 1	mile 2	mile 3
n	5	5	5
Σx	1005	998	1054
Σx^2	202055	199216	222200
$\overline{x}$	201.0	199.6	210.8
s^2	12.50	3.80	4.20

$k = 3$ $\qquad \overline{\overline{x}} = \Sigma\overline{x}_i/k$

$\qquad\qquad = 203.8$

$n = 5$

$s_{\overline{x}}^2 = \Sigma(\overline{x}_i - \overline{\overline{x}})^2/(k-1)$ $\qquad s_p^2 = \Sigma s_i^2/k$

$\qquad = 37.24$ $\qquad\qquad\qquad = 6.8333$

$H_o: \mu_1 = \mu_2 = \mu_3$

$H_1:$ at least one μ_i is different

$\alpha = 0.05$ and $df_{num} = 2$, $df_{den} = 12$

C.V. $F = F_\alpha = F_{0.05} = 3.8853$

calculations:

$\qquad F = ns_{\overline{x}}^2/s_p^2$

$\qquad\quad = 5(37.24)/6.8333$

$\qquad\quad = 27.2488$

$\qquad$ P-value = Fcdf(27.2488,99,2,12) = 3.45E-5 = 0.00003

conclusion:

$\qquad$ Reject Ho; there is sufficient evidence to reject the claim that $\mu_1 = \mu_2 = \mu_3$. There is sufficient evidence to reject the claim that it takes the same time to ride each of the miles. Yes; the data suggest that there may be a hill on mile 3.

15. The Minitab output is as follows

```
Level      N    Mean    StDev          Source   DF     SS      MS      F      P
king      25  1.2560  0.2329          Factor    2   2.2083  1.1041  18.99  0.000
menthol   25  0.8720  0.2424          Error    72   4.1856  0.0581
filter    25  0.9160  0.2478          Total    74   6.3939
```

$H_o: \mu_1 = \mu_2 = \mu_3$

$H_1:$ at least one μ_i is different

$\alpha = 0.05$ and $df_{num} = 2$, $df_{den} = 72$

C.V. $F = F_\alpha = F_{0.05} = 3.1504$

calculations:

$\qquad F = s_B^2/s_p^2$

$\qquad\quad = 1.1041/0.0581$

$\qquad\quad = 18.99$ [Minitab]

$\qquad$ P-value = $P(F_{72}^2 > 18.99)$ = 0.000 [Minitab]

conclusion:

$\qquad$ Reject Ho; there is sufficient evidence to reject the claim that $\mu_1 = \mu_2 = \mu_3$. There is sufficient evidence to reject the claim that the three types of cigarettes yield the mean amount of nicotine. Yes; given that only the king size cigarettes are not filtered, it appears (although not formally justifiable by the ANOVA) that the filters do make a difference.

17. The Tukey results indicate that the only significant difference (P-value = 0.021) is between μ_1 the mean for the small cars) and μ_3 (the mean for the large cars). The conclusion is the same as that obtained with the Bonferroni results. The slightly larger Bonferroni P-value =0.024 indicates that, at least in this case, the Bonferroni method is more conservative (i.e., requires a larger difference before declaring significance).

12-3 Two-Way ANOVA

1. Two-way analysis of variance is appropriate for these data because they are categorized by two characteristics that distinguish groups of data from each other. The distinguishing characteristics in this case are gender and age bracket.

3. Because each cell contains the same number of observations (viz., 5), this is a balanced design.

5. H_o: there is no gender-age interaction
   H_1: there is gender-age interaction
   $\alpha = 0.05$ and $df_{num} = 1$, $df_{den} = 60$
   C.V. $F = F_\alpha = F_{0.05} = 4.0012$
   calculations:
   $F = MS_{Int}/MS_{Err}$
   $= 1.995/8.275$
   $= 0.24$ [Minitab]
   P-value $= P(F_{60}^1 > 0.24) = 0.625$ [Minitab]

   conclusion:
   Do not reject H_o; there is not sufficient evidence to reject the claim that there is no interaction between gender and age bracket. There is not sufficient evidence to reject the claim that heights are not affected by an interaction between a person's gender and that person's age bracket.

7. H_o: $\mu_1 = \mu_2$ [there is no age effect]
   H_1: at least one age μ_i is different
   $\alpha = 0.05$ and $df_{num} = 1$, $df_{den} = 36$
   C.V. $F = F_\alpha = F_{0.05} = 4.0012$
   calculations:
   $F = MS_{Age}/MS_{Err}$
   $= 24.379/8.275$
   $= 2.95$ [Minitab]
   P-value $= P(F_{60}^1 > 2.95) = 0.625$ [Minitab]

   conclusion:
   Do not reject H_o; there is not sufficient evidence to support the claim that age bracket has an effect on height.

9. H_o: $\mu_1 = \mu_2$ [there is no type effect]
   H_1: at least one type μ_i is different
   $\alpha = 0.05$ and $df_{num} = 1$, $df_{den} = 12$
   C.V. $F = F_\alpha = F_{0.05} = 4.7472$
   calculations:
   $F = MS_{Type}/MS_{Err}$
   $= 34060.5/15133.6$
   $= 2.25$ [Minitab]
   P-value $= P(F_{12}^1 > 2.25) = 0.159$ [Minitab]

   conclusion:
   Do not reject H_o; there is not sufficient evidence to support the claim that type has an effect on head injury measurements. There is not enough evidence to support the claim that the type of car (foreign or domestic) has an effect on head injury measurements.

NOTE: In exercises 11-13 the n=12 values in each of the six cells corresponds to df=11 for each of the six cells and df = 6(11) for the pooled estimate of the mean square error.

11. H_o: there is no subject-target interaction
H_1: there is subject-target interaction
$\alpha = 0.05$ and $df_{num} = 2$, $df_{den} = 66$
C.V. $F = F_\alpha = F_{0.05} = 3.1359$ [Statdisk]
calculations:
$\quad F = MS_{Int}/MS_{Err}$
$\quad\quad = 3.375/[\text{not given}]$
$\quad\quad = 3.73$ [Statdisk]
$\quad\quad$ P-value = $P(F_{66}^2 > 3.73) = 0.0291$ [Statdisk]

conclusion:
Reject H_o; there is sufficient evidence to reject the claim that there is no subject-target interaction. There is sufficient evidence to reject the claim that perceived self-esteem in other target people is unaffected by an interaction between the subject self-esteem and the target's self-esteem.

13. H_o: $\mu_1 = \mu_2 = \mu_3$ [there is no subject effect]
H_1: at least one size μ_i is different
$\alpha = 0.05$ and $df_{num} = 2$, $df_{den} = 66$
C.V. $F = F_\alpha = F_{0.05} = 3.1359$ [Statdisk]
calculations:
$\quad F - MS_{Subject}/MS_{Err}$
$\quad\quad = 1.4306/[\text{not given}]$
$\quad\quad = 1.5824$ [Statdisk]
$\quad\quad$ P-value = $P(F_{66}^2 > 1.5824) = 0.2132$ [Statdisk]]

conclusion:
Do not reject H_o; there is not sufficient evidence to support the claim that there is a subject effect. There is not sufficient evidence to support the claim that the perceived self-esteem in other target people is affected by the subject's own measure (low, medium or high) of self-esteem.

NOTE FOR HAND CALCULATIONS: The formulas and principles in this section are logical extensions of those in the previous section. In particular,

$\quad SS_{Row} = \Sigma n_i (\overline{x}_i - \overline{\overline{x}})^2$ for i = 1,2,3… [for each row]

$\quad SS_{Col} = \Sigma n_j (\overline{x}_j - \overline{\overline{x}})^2$ for j = 1,2,3… [for each column]

$\quad SS_{Tot} = \Sigma (x - \overline{\overline{x}})^2$ for all the x's

When there is only one observation per cell, the unexplained variation is
$\quad SS_{Err} = SS_{Tot} - SS_{Row} - SS_{Col}$, and there is not enough data to measure interaction.
When there is more than one observation per cell, the unexplained variation (i.e., the failure of experimental units in the same cell to respond the same) is
$\quad SS_{Err} = \Sigma (x - \overline{x}_{ij})^2 = \Sigma df_{ij} s_{ij}^2$ [for each cell – i.e., for each i,j (row,col) combination]

and the interaction sum of squares is
$\quad SS_{Int} = SS_{Tot} - SS_{Row} - SS_{Col} - SS_{Err}$.

15. There are n=3 observations in each of the indicated whey-supplement cells.

WHEY

S	0%	10%	20%	30%	
U	$\Sigma x = 13.2$	$\Sigma x = 13.9$	$\Sigma x = 14.1$	$\Sigma x = 14.4$	
P	$\Sigma x^2 = 58.10$	$\Sigma x^2 = 64.45$	$\Sigma x^2 = 66.33$	$\Sigma x^2 = 69.26$	
P no	$\overline{x} = 4.4000$	$\overline{x} = 4.6333$	$\overline{x} = 4.7000$	$\overline{x} = 4.8000$	$\overline{x} = 4.6333$
L	$s^2 = 0.0100$	$s^2 = 0.0233$	$s^2 = 0.0300$	$s^2 = 0.0700$	
E	$\Sigma x = 9.6$	$\Sigma x = 11.1$	$\Sigma x = 15.1$	$\Sigma x = 16.3$	
M	$\Sigma x^2 = 30.74$	$\Sigma x^2 = 41.09$	$\Sigma x^2 = 76.13$	$\Sigma x^2 = 88.61$	
E yes	$\overline{x} = 3.2000$	$\overline{x} = 3.7000$	$\overline{x} = 5.0333$	$\overline{x} = 5.4333$	$\overline{x} = 4.3417$
N	$s^2 = 0.0100$	$s^2 = 0.0100$	$s^2 = 0.0633$	$s^2 = 0.0233$	
T	$\overline{x} = 3.8000$	$\overline{x} = 4.1667$	$\overline{x} = 4.8667$	$\overline{x} = 5.1167$	$\Sigma x = 107.7$ $\overline{\overline{x}} = 4.4875$
					$\Sigma x^2 = 494.71$ $s^2 = 0.4959$

$SS_{Supp} = \Sigma n_i (\overline{x}_i - \overline{\overline{x}})^2 = 12(4.6333-4.4875)^2 + 12(4.3417-4.4875)^2 = 0.51041667$

$SS_{Whey} = \Sigma n_j (\overline{x}_j - \overline{\overline{x}})^2 = 6(3.8-4.4875)^2 + 6(4.1667-4.4875)^2 + 6(4.8667-4.4875)^2 + 6(6.1167-4.4875)^2 =$

$SS_{Tot} = \Sigma (x - \overline{\overline{x}})^2 = df \cdot s^2 = 23 \cdot (0.4959) = 11.40625$ 6.69125

$SS_{Err} = \Sigma (x - \overline{x}_{ij})^2 = \Sigma df_{ij} s^2_{ij} = 2(0.0100) + 2(0.0233) + 2(0.0300) + 2(0.0700)$

$2(0.0100) + 2(0.0100) + 2(0.0633) + 2(0.0233) = 0.4800$

$SS_{Int} = SS_{Tot} - SS_{Supp} - SS_{Whey} - SS_{Err} = 11.40625 - 0.51041 - 6.69125 - 0.48000 = 3.72459$

This is the ANOVA table for the tests of hypotheses in this exercise.

Source	df	SS	MS	F
Supplement	1	0.51041	0.51041	17.0137
Whey	3	6.69125	2.23042	74.3472
Interaction	3	3.72459	1.24153	41.3843
Error	16	0.48000	0.03000	
Total	23	11.40625		

H_o: there is no whey-supplement interaction
H_1: there is whey-supplement interaction
$\alpha = 0.05$ [assumed] and $df_{num} = 3$, $df_{den} = 16$
C.V. $F = F_\alpha = F_{0.05} = 3.2389$
calculations:

$F = MS_{Int}/MS_{Err}$
   $= 1.24153/0.03000$
   $= 41.3843$
P-value = Fcdf(41.3843,99,3,16)
       $= 9.11E-8 = 0.00000009$

conclusion:

Reject H_o; there is sufficient evidence to reject the claim that there is no interaction between the whey and the supplement. There is sufficient evidence to conclude that the ratings are affected by an interaction between the use of the supplement and the amount of whey.

Because there is interaction, further testing for whey or supplement effects are not appropriate. The presence of interaction means that the effect of the supplement, for example, depends upon the amount of whey and cannot be determined in a single test that combines the different whey amounts.

17. The table and preliminary calculations are as follows.

WHEY

S		0%	10%	20%	30%		
U	no	4.4	4.6	4.5	4.6	$\overline{x}$ = 4.525	
P	yes	3.3	3.8	5.0	5.4	$\overline{x}$ = 4.375	
P		$\overline{x}$ = 3.85	$\overline{x}$ = 4.20	$\overline{x}$ = 4.75	$\overline{x}$ = 5.00	Σx = 35.6	$\overline{\overline{x}}$ = 4.45
						Σx^2 = 161.42	s^2 = 0.4286

$SS_{Supp} = \Sigma n_i (\overline{x}_i - \overline{\overline{x}})^2 = 4(4.525 - 4.45)^2 + 4(4.375 - 4.45)^2 = 0.0450$

$SS_{Whey} = \Sigma n_j (\overline{x}_j - \overline{\overline{x}})^2 = 2(3.85 - 4.45)^2 + 2(4.20 - 4.45)^2 + 2(4.75 - 4.45)^2 + 2(5.00 - 4.45)^2 = 1.6300$

$SS_{Tot} = \Sigma (x - \overline{\overline{x}})^2 = df \cdot s^2 = 7 \cdot (0.4286) = 3.000$

$SS_{Err} = SS_{Tot} - SS_{Supp} - SS_{Whey} = 3.0000 - 0.0450 - 1.6300 = 1.3250$

This is the ANOVA table for the tests of hypotheses in this exercise.

Source	df	SS	MS	F
Supplement	1	0.0450	0.04500	0.1019
Whey	3	1.6300	0.54333	1.2302
Error	3	1.3250	0.44167	
Total	7	3.0000		

H_o: $\mu_1 = \mu_2$ [there is no supplement effect]
H_1: at least one supplement μ_i is different
$\alpha = 0.05$ [assumed] and $df_{num} = 1$, $df_{den} = 3$
C.V. $F = F_\alpha = F_{0.05} = 10.128$
calculations:

$F = MS_{Supp}/MS_{Err}$
$\quad = 0.04500/0.44167$
$\quad = 0.1019$

P-value = Fcdf(0.1019,99,1,3) = 0.7684

conclusion:

Do not reject H_o; there is not sufficient evidence to support the claim that the supplement has an effect on the rating. There is not enough evidence to support the claim that use of the supplement affects the ratings of the pancakes.

H_o: $\mu_1 = \mu_2 = \mu_3 = \mu_4$ [there is no whey effect]
H_1: at least one age μ_i is different
$\alpha = 0.05$ [assumed] and $df_{num} = 3$, $df_{den} = 3$
C.V. $F = F_\alpha = F_{0.05} = 9.2766$
calculations:

$F = MS_{Whey}/MS_{Err}$
$\quad = 0.54333/0.44167$
$\quad = 1.2302$

P-value = Fcdf(1.2302,99,3,3) = 0.4327

conclusion:

Do not reject H_o; there is not sufficient evidence to support the claim that whey has an effect on the rating. There is not enough evidence to support the claim that the amount of whey used affects the rating of the pancakes.

Statistical Literacy and Critical Thinking

1. Since two of the three samples are from the same power source, the three sets of sample data are not independent. Since one-way analysis of variance requires independent sets of sample data, the one-way analysis of variance should not be used. Even if the three data sets came from three independent power sources, if all the measurements were taken at precisely the same time the data would consist of matched triples and the one-way analysis of variance (since it ignores the time characteristic) would not be the optimal technique to use.

2. One-way analysis of variance is used when the data are categorized according to one factor, while two-way analysis of variance is used when the data are categorized according to two factors.

3. While the factors used to categorize the data may originally be measured at any level, the data to be analyzed must be quantitative data for which meaningful means and standard deviations can be calculated. MPAA rating is not a quantitative variable. Running time could be analyzed using the factors of budget bracket and MPAA rating, but MPAA rating could not be analyzed using the factors of budget bracket and running time bracket.

4. No. Samples must be representative of the population to which the inference is to be made. If only less expensive cars are used in the crash tests, then the results may be used to make inferences only about the population of less expensive cars and not the population of all cars.

Chapter Quick Quiz

1. The one-way analysis of variance is used to test the null hypothesis that three or more samples are from populations with equal means.

2. One-way analysis of variance tests are right tailed. The test statistic is the ratio of the variance between groups to the variance within groups, and we reject the hypothesis that the groups have equal means in favor of them having means that are not all equal only if the variance between the groups is significantly larger than the variance between groups.

3. In one-way analysis of variance tests, the P-value is the probability of chance alone producing a test statistic as large as or larger than the calculated test statistic. The larger the calculated test statistic, therefore, the smaller the P-value.

4. The value of test statistic is $F = 8.98$.

5. The null hypothesis is that the three books (by Rowling, Clancy and Tolstoy) have the same mean grade level reading scores. Since P-value = 0.001 < 0.05, we reject the null hypothesis and conclude that the three books do not have the same mean grade level reading score.

6. The final conclusion is that at least one of the three books has a mean grade level reading score that is different from the others.

7. One-way analysis of variance is used when data are categorized according to one factor, while two-way analysis of variance is used when data are categorized according to two factors.

8. There is not sufficient evidence to reject the claim that a student's estimate is unaffected by interaction between that student's gender and that student's major.

9. No. Since P-value = 0.395 > 0.05 for testing for a gender effect, conclude that there is not sufficient evidence to support a claim that students' estimates of length are affected by their genders.

10. No. Since P-value = 0.876 > 0.05 for testing for a major effect, conclude that there is not sufficient evidence to support a claim that students' estimates of length are affected by their majors.

Review Exercises

1. a. The age data from the sample was categorized according to one variable, diet. Specifically, it was noted whether the subject with each age in the age participated in the Atkins, Ornish, Weight Watchers or Zones diet.
   b. The data was analyzed using a one-way analysis of variance.
   c. Since P-value = 0.41 > 0.05, we fail to reject the claim that the subjects come from populations with the same mean age. It appears that the mean ages of the four treatment groups are about the same.
   d. A small P-value like 0.001 < 0.05 would lead to rejection of the claim that the subjects come from populations with the same mean age. The conclusion would be that the four treatment groups did not have the same mean age. If such were the case, it would not be possible to declare whether any observed differences between the groups were due to the different treatments or to the difference in ages.

2. H_o: $\mu_1 = \mu_2 = \mu_3$
   H_1: at least one μ_i is different
   $\alpha = 0.05$ and $df_{num} = 2$, $df_{den} = 29$
   C.V. $F = F_\alpha = F_{0.05} = 3.3277$
   calculations:

   $F = s_B^2/s_p^2$

   $= 3083363/56373$

   $= 54.70$ [Minitab]

   P-value $= P(F_{29}^2 > 54.70) = 0.000$ [Minitab]

conclusion:
   Reject Ho; there is sufficient evidence to reject the claim that $\mu_1 = \mu_2 = \mu_3$. There is sufficient evidence to reject the claim that the different car categories (4 cylinder, 6 cylinder, 8 cylinder) have different weights. Yes, it does appear that cars with more cylinders tend to weigh more – but a formal conclusion about such a relationship would require additional statistical methodology.

3. H_o: there is no type-size interaction
   H_1: there is type-size interaction
   $\alpha = 0.05$ and $df_{num} = 2$, $df_{den} = 12$
   C.V. $F = F_\alpha = F_{0.05} = 3.8853$
   calculations:
   $\quad F = MS_{Int}/MS_{Err}$
   $\quad\quad = 57385/155086$
   $\quad\quad = 0.37$ [Minitab]
   $\quad$ P-value $= P(F_{12}^2 > 0.37) = 0.698$ [Minitab]

conclusion:
   Do not reject H_o; there is not sufficient evidence to reject the claim that there is no
   interaction between type and size. There is not sufficient evidence to reject the claim that
   left femur crash loads are unaffected by an interaction between a car's type (foreign or
   domestic) and its size (small, medium or large).

4. H_o: $\mu_1 = \mu_2$ [there is no type effect]
   H_1: at least one type μ_i is different
   $\alpha = 0.05$ and $df_{num} = 1$, $df_{den} = 12$
   C.V. $F = F_\alpha = F_{0.05} = 4.7472$
   calculations:
   $\quad F = MS_{Type}/MS_{Err}$
   $\quad\quad = 282752/155086$
   $\quad\quad = 1.82$ [Minitab]
   $\quad$ P-value $= P(F_{12}^1 > 1.82) = 0.202$ [Minitab]

conclusion:
   Do not reject H_o; there is not sufficient evidence to support the claim that type has an effect
   on left femur crash loads. There is not enough evidence to support the claim that the
   type of car (foreign or domestic) has an effect on left femur crash loads..

5. H_o: $\mu_1 = \mu_2 = \mu_3$ [there is no size effect]
   H_1: at least one size μ_i is different
   $\alpha = 0.05$ and $df_{num} = 2$, $df_{den} = 12$
   C.V. $F = F_\alpha = F_{0.05} = 3.8853$
   calculations:
   $\quad F = MS_{Size}/MS_{Err}$
   $\quad\quad = 74067/155086$
   $\quad\quad = 0.48$ [Minitab]
   $\quad$ P-value $= P(F_{12}^2 > 0.48) = 0.632$ [Minitab]

conclusion:
   Do not reject H_o; there is not sufficient evidence to support the claim that size has an effect
   on left femur crash loads. There is not sufficient evidence to support the claim that the
   size of a car (small, medium or large) has an effect on left femur crash loads.

6. These are precisely the values from Data Set 4. The Minitab output is as follows.

Level	N	Mean	StDev		Source	DF	SS	MS	F	P
king	25	15.720	0.936		Factor	2	12.1	6.0	0.50	0.608
menthol	25	14.960	4.168		Error	72	868.0	12.1		
non-menth	25	14.800	4.233		Total	74	880.1			

H_o: $\mu_1 = \mu_2 = \mu_3$

H_1: at least one μ_i is different

$\alpha = 0.05$ and $df_{num} = 2$, $df_{den} = 72$

C.V. $F = F_\alpha = F_{0.05} = 3.1504$

calculations:

$$F = s_B^2/s_p^2$$

$$= 6.0/12.1$$

$$= 0.50 \text{ [Minitab]}$$

P-value = $P(F_{72}^2 > 0.50) = 0.608$ [Minitab]

conclusion:

Do not reject Ho; there is not sufficient evidence to reject the claim that $\mu_1 = \mu_2 = \mu_3$. There is not sufficient evidence to reject the claim that the three types of cigarettes yield the same mean amount of carbon monoxide. No; it appears that filters do not make a difference in the amount of carbon monoxide – but a formal test using only two groups (filtered and non-filtered) would provide a better test for answering that question.

7. There are n=4 observations in each of the indicated gender-smoking cells.

		SMOKE?	
		yes	no
GENDER		$\Sigma x = 394.8$	$\Sigma x = 391.8$
		$\Sigma x^2 = 38967.44$	$\Sigma x^2 = 38378.44$
male		$\overline{x} = 98.700$	$\overline{x} = 97.950$	$\overline{x} = 98.3250$
		$s^2 = 0.22667$	$s^2 = 0.54333$
		$\Sigma x = 393.9$	$\Sigma x = 393.0$
		$\Sigma x^2 = 38789.69$	$\Sigma x^2 = 38613.34$
female		$\overline{x} = 98.475$	$\overline{x} = 98.250$	$\overline{x} = 98.3625$
		$s^2 = 0.12917$	$s^2 = 0.36333$

$\overline{x} = 98.5875$ $\overline{x} = 98.1000$ $\Sigma x = 1573.5$ $\overline{\overline{x}} = 98.34375$

$\Sigma x^2 = 154748.91$ $s^2 = 0.334625$

$SS_{Gen} = \Sigma n_i (\overline{x}_i - \overline{\overline{x}})^2 = 8(98.3250 - 98.34375)^2 + 8(98.3625 - 98.34375)^2 = 0.005625$

$SS_{Smo} = \Sigma n_j (\overline{x}_j - \overline{\overline{x}})^2 = 8(98.5875 - 98.34375)^2 + 8(98.1000 - 98.34375)^2 = 0.950625$

$SS_{Tot} = \Sigma (x - \overline{\overline{x}})^2 = df \cdot s^2 = 15 \cdot (0.334625) = 5.019375$

$SS_{Err} = \Sigma (x - \overline{x}_{ij})^2 = \Sigma df_{ij} s_{ij}^2 = 3(0.22667) + 3(0.54333)$

$$3(0.12917) + 3(0.36333) = 3.78750$$

$SS_{Int} = SS_{Tot} - SS_{Supp} - SS_{Whey} - SS_{Err} = 5.019375 - 0.005625 - 0.950625 - 3.78750 = 0.275625$

This is the ANOVA table for the tests of hypotheses in this exercise.

Source	df	SS	MS	F
Gender	1	0.005625	0.005625	0.0178
Smoke	1	0.950625	0.950625	3.0119
Interaction	1	0.275625	0.275625	0.8733
Error	12	3.787500	0.315625	
Total	15	5.019375		

H_o: there is no gender-smoking interaction
H_1: there is gender-smoking interaction
$\alpha = 0.05$ and $df_{num} = 1$, $df_{den} = 12$
C.V. $F = F_\alpha = F_{0.05} = 4.7472$
calculations:
    $F = MS_{Int}/MS_{Err}$
      $= 0.275625/0.315625$
      $= 0.8733$
    P-value = Fcdf(0.8733,99,1,12) = 0.3685
conclusion:

    Do not reject H_o; there is not sufficient evidence to reject the claim that there is no interaction between gender and smoking. There is not sufficient evidence to reject the claim that a person's body temperature is not affected by an interaction between gender and smoking habits.

H_o: $\mu_1 = \mu_2$ [there is no gender effect]
H_1: at least one μ_i is different
$\alpha = 0.05$ and $df_{num} = 1$, $df_{den} = 12$
C.V. $F = F_\alpha = F_{0.05} = 4.7472$
calculations:
    $F = MS_{Gen}/MS_{Err}$
      $= 0.005625/0.315625$
      $= 0.0178$
    P-value = Fcdf(0.0178,99,1,12) = 0.8960
conclusion:

    Do not reject H_o; there is not sufficient evidence to reject the claim that there is no gender effect. There is not sufficient evidence to reject the claim that a person's body temperature is unaffected by that person's gender.

H_o: $\mu_1 = \mu_2$ [there is no smoking effect]
H_1: at least one μ_i is different
$\alpha = 0.05$ and $df_{num} = 1$, $df_{den} = 12$
C.V. $F = F_\alpha = F_{0.05} = 4.7472$
calculations:
    $F = MS_{Smo}/MS_{Err}$
      $= 0.950625/0.315625$
      $= 3.0119$
    P-value = Fcdf(3.0119,99,1,12) = 0.1082
conclusion:

    Do not reject H_o; there is not sufficient evidence to reject the claim that there is no smoking effect. There is not sufficient evidence to reject the claim that a person's body temperature is unaffected by that person's smoking habits.

8. The following preliminary values are identified.

	presidents	popes	monarchs
n	38	24	14
Σx	589	315	318
Σx^2	12627	5981	11722
$\overline{x}$	15.500	13.125	22.714
s^2	94.5270	80.2880	346.0659

$\overline{\overline{x}} = (\Sigma n_i \overline{x}_i)/\Sigma n_i = [38(15.500) + 24(13.125) + 14(22.714)]/[38+24+14] = 1222/76 = 16.079$

$\Sigma n_i(\overline{x}_i - \overline{\overline{x}})^2 = 38(15.500-16.079)^2 + 24(13.125-16.079)^2 + 14(22.714-16.079)^2 = 838.54$

$\Sigma df_i s_i^2 = 37(94.5270) + 23(80.2880) + 13(346.0659) = 9842.98$

$s_B^2 = \Sigma n_i(\overline{x}_i - \overline{\overline{x}})^2/(k-1) = 838.54/2 = 419.27$

$s_p^2 = (\Sigma df_i s_i^2)/\Sigma df_i = 9842.98/73 = 134.84$

H_o: $\mu_1 = \mu_2 = \mu_3$
H_1: at least one μ_i is different
$\alpha = 0.05$ [assumed] and $df_{num} = 2$, $df_{den} = 73$
C.V. $F = F_\alpha = 3.1504$
calculations:

$\quad F = s_B^2/s_p^2$

$\qquad = 419.27/134.84$

$\qquad = 3.1095$

$\quad$ P-value = Fcdf(3.1095,99,2,73) = 0.0506
conclusion:

Do not reject Ho; there is not sufficient evidence to reject the claim that $\mu_1 = \mu_2 = \mu_3$. There is not sufficient evidence to reject the claim that the three groups have the same mean longevity. No, the survival times for the three groups do not appear to differ significantly.

Cumulative Review Exercises

The following summary information applies to exercises 1-4.

	presidents	popes	monarchs
n	38	24	14
Σx	589	315	318
Σx^2	12627	5981	11722
$\overline{x}$	15.500	13.125	22.714
s^2	94.5270	80.2880	346.0659

1. a. The means for the three groups are also follows.
   pres: 15.5 years $\qquad$ popes: 13.1 years $\qquad$ mon: 22.7 years
   b. The standard deviations for the three groups are as follows.
   pres: $\sqrt{94.5270} = 9.7$ years $\quad$ popes: $\sqrt{80.2880} = 9.0$ years $\quad$ mon: $\sqrt{346.0659} = 18.6$ years

2. Let the presidents be group 1.
   $\overline{x}_1 - \overline{x}_2 = 15.500 - 22.714 =$
   original claim: $\mu_1 - \mu_2 \neq 0$ mg
   H_o: $\mu_1 - \mu_2 = 0$ years
   H_1: $\mu_1 - \mu_2 \neq 0$ years
   $\alpha = 0.05$ [assumed] and df = 13
   C.V. $t = \pm t_{\alpha/2} = t_{0.025} = \pm 2.160$
   calculations:

$\quad t_{\overline{x}_1 - \overline{x}_2} = (\overline{x}_1 - \overline{x}_2 - \mu_{\overline{x}_1 - \overline{x}_2})/s_{\overline{x}_1 - \overline{x}_2}$

$\qquad = (-7.214 - 0)/\sqrt{94.5270/38 + 346.0659/14}$

$\qquad = -7.214/5.2160 = -1.383$

$\quad$ P-value = 2·tcdf(-99,-1.383,13) = 0.1899

conclusion:

Do not reject H_o; there is not sufficient evidence to conclude that $\mu_1 - \mu_2 \neq 0$. There is not sufficient evidence to support a claim that there is a difference between the mean for presidents and the mean for British monarchs.

3. Yes, the longevity times for presidents appear to come from a population that is approximately normal. The histogram is approximately bell-shaped and the points on the normal probability plot approximate a straight line. Because of the truncation of the times at 0, there is not the opportunity for unusually low scores to symmetrically balance unusually high scores – and this slight departure form normality is evident in both the histogram and the normal probability plot.

4. $n = 39$ and $\bar{x} = 15.000$ and $s = \sqrt{94.5270} = 9.7224$

σ unknown, $n > 30$: use t with df=37

$\alpha = 0.05$, $t_{df,\,\alpha/2} = t_{37,0.025} = 2.026$

$\bar{x} \pm t_{\alpha/2} \cdot s/\sqrt{n}$

$15.500 \pm 2.026(9.7224)/\sqrt{38}$

15.500 ± 3.195

$12.3 < \mu < 18.7$ (years)

5. Listed in order, the n=9 values are: 96.5   97.3   97.6   97.6   98.0   98.2   98.3   98.4   98.9

$\Sigma x = 880.8$ $\Sigma x^2 = 86204.96$

a. Temperatures are measurements at the interval level, because differences are meaningful but ratios are not (since there is not a meaningful zero).

b. Body temperatures are continuous data. Even though they arc typically reported to the nearest tenth of a degree, they actually can take on any values on a continuum.

c. $\bar{x} = (\Sigma x)/n = 880.8/9 = 97.8667$, rounded to 97.87 °F

d. $\tilde{x} = x_5 = 98.0$ °F

e. $R = 98.9 - 96.5 = 2.4$ °F

f. $s^2 = [n(\Sigma x^2) - (\Sigma x)^2]/[n(n-1)]$

$= [9(86204.96) - (880.8)^2]/[9(8)] = 36.00/72 = 0.5000$

$s = \sqrt{0.5000} = 0.7071$, rounded to 0.71 °F

g. $s^2 = 0.5000$ $(°F)^2$

6. The requested frequency distribution is given
In the box at the right

rating	frequency
1.0 – 1.9	1
2.0 – 2.9	4
3.0 – 3.9	0
4.0 – 4.9	1
5.0 – 5.9	4
6.0 – 6.9	9
7.0 – 7.9	9
8.0 – 8.9	7
	35

7. The histogram is given below. The actual cut points for the bars are 0.95, 1.95, 2.95, etc.

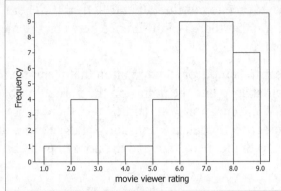

No; it does not appear that the ratings are from a population with a normal distribution. The distribution appears to be negatively skewed, and not bell-shaped.

8. This exercise requires the following summary statistics.

	PG	PG-13	R
n	3	20	12
Σx	19.9	128.6	74.1
Σx^2	133.89	904.88	510.05
$\overline{x}$	6.633	6.430	6.175
s^2	0.9433	4.1043	4.7711

informal comparison of the three sample means:

The means of 6.63, 6.43, and 6.18 do not appear to be very different – especially considering the fairly wide range of ratings that extend from 1.9 to 8.6.

$\overline{\overline{x}} = (\Sigma n_i \overline{x}_i)/\Sigma n_i = [3(6.633) + 20(6.430) + 12(6.175)]/[3+20+12] = 222.6/35 = 6.360$

$\Sigma n_i(\overline{x}_i - \overline{\overline{x}})^2 = 3(6.633-6.360)^2 + 20(6.430-6.360)^2 + 12(6.175-6.360)^2 = 0.7328$

$\Sigma df_i s_i^2 = 2(0.9433) + 19(4.1043) + 11(4.7711) = 132.3504$

$s_B^2 = \Sigma n_i(\overline{x}_i - \overline{\overline{x}})^2/(k-1) = 0.7328/2 = 0.3664$

$s_p^2 = (\Sigma df_i s_i^2)/\Sigma df_i = 132.3504/32 = 4.13595$

H_o: $\mu_1 = \mu_2 = \mu_3$
H_1: at least one μ_i is different
$\alpha = 0.05$ and $df_{num} = 2$, $df_{den} = 32$
C.V. $F = F_\alpha = F_{0.05} = 3.3158$
calculations:

$F = s_B^2/s_p^2$

$= 0.3664 4.13595$

$= 0.0886$

P-value = Fcdf(0.0886,99,2,32) = 0.9154

conclusion:

Do not reject Ho; there is not sufficient evidence to reject the claim that $\mu_1 = \mu_2 = \mu_3$. There is not sufficient evidence to reject the claim that PG, PG-13 and R movies have the same mean viewer rating.

9. These are the necessary summary statistics.

n = 6	$\Sigma xy = 110.9247$	$n(\Sigma xy) - (\Sigma x)(\Sigma y) = 5(110.9247) - (52.41)(10.40) = 9.5595$
$\Sigma x = 52.41$	$\Sigma x^2 = 554.0767$	$n(\Sigma x^2) - (\Sigma x)^2 = 5(554.0767) - (52.41)^2 = 23.5754$
$\Sigma y = 10.40$	$\Sigma y^2 = 23.7642$	$n(\Sigma y^2) - (\Sigma y)^2 = 5(23.7642) - (10.40)^2 = 10.6610$

a. $r = [n(\Sigma xy) - (\Sigma x)(\Sigma y)]/[\sqrt{n(\Sigma x^2) - (\Sigma x)^2}\sqrt{n(\Sigma y^2) - (\Sigma y)^2}]$

   $= 9.5595/[\sqrt{23.5754}\sqrt{10.6610}]$

   $= 0.6030$

H_o: $\rho = 0$

H_1: $\rho \neq 0$

$\alpha = 0.05$ [assumed] and df = 3

C.V. $t = \pm t_{\alpha/2} = \pm t_{0.025} = \pm 3.182$ [or r = ±0.878]

calculations:

   $t_r = (r - \mu_r)/s_r$

     $= (0.6030 - 0)/\sqrt{[1 - (0.6030)^2]/3}$

     $= 0.6030/0.4605$

     $= 1.309$

   P-value = 2·tcdf(1.309,99,3) = 0.2817

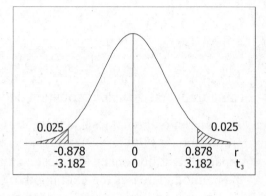

conclusion:

   Do not reject H_o; there is not sufficient evidence to conclude that $\rho \neq 0$. No; there is not sufficient evidence to support the claim of a linear correlation between the amounts of discarded paper and discarded plastic.

b. $\bar{x} = (\Sigma x)/n = 52.41/5 = 10.482$ $\bar{y} = (\Sigma y)/n = 10.40/5 = 2.080$

  $b_1 = [n(\Sigma xy) - (\Sigma x)(\Sigma y)]/[n(\Sigma x^2) - (\Sigma x)^2]$

    $= 9.5595/23.5754 = 0.4055$

  $b_o = \bar{y} - b_1\bar{x}$

    $= 2.080 - (0.4055)(10.482) = -2.1703$

  $\hat{y} = b_o + b_1 x$

    $= -2.1703 + 0.4055x$

c. No. Since the correlation in part (a) was not significant, the regression line in part (b) should not be used for making predictions. The best predictive equation is simply $\hat{y} = \bar{y}$ for all values of x.

10. H_o: gun type and circumstances are independent

    H_1: gun type and circumstances are related

    $\alpha = 0.05$ and df = 2

    C.V. $\chi^2 = \chi^2_\alpha = \chi^2_{0.05} = 5.991$

    calculations:

		CIRCUMSTANCES			
		accident	self-inflict	assault	
TYPE	hand-gun	31 (31.85)	35 (30.40)	162 (165.75)	228
	rifle	13 (12.15)	7 (11.60)	67 (63.25)	87
		44	42	229	315

    $\chi^2 = \Sigma[(O - E)^2/E]$

      $= 0.023 + 0.696 + 0.085$

        $0.059 + 1.824 + 0.223 = 2.909$

    P-value = χ^2cdf (2.909,99,2) = 0.2335

conclusion:

   Do not reject Ho; there is not sufficient evidence to reject the claim that gun type and circumstances are independent. For firearm injuries, there is not sufficient evidence to reject the claim that the type (handgun or rifle/shotgun) of gun involved and the circumstances of the event (unintentional, self-inflicted or assault) are independent.

Chapter 13

Nonparametric Statistics

13-2 Sign Test

1. The sign test is "nonparametric" or "distribution-free" because it does not require the data to come from a particular distribution defined in terms of certain parameters.

3. The alternative hypothesis is that the proportion of girls is greater than 0.5, but the sample proportion of $39/(172+39) = 39/211$ is less than 0.5. Without doing any mathematics, it is apparent that the conclusion must be that there is not enough evidence to support the claim that the method increases the likelihood of girl births. In order for the data to support the claim the number of girl births would have to be significantly greater than $(\frac{1}{2}) \cdot (211) = 105.5$.

5. 13 +'s and 1 −'s
   $n = 14 \leq 25$; use C.V. = 2 from Table A-7
   Since $\min(13,1) = 1 \leq 2 = $ C.V., reject the hypothesis of no difference.
   Since there were more +'s, conclude that the first variable has the larger scores.

7. 360 +'s and 374 −'s
   $n = 734 > 25$; use C.V. = -1.96 from the z table
   $z = [(x+0.5) - n/2]/[\sqrt{n}/2] = [360.5 - 367]/[\sqrt{734}/2] = -6.5/13.546 = -0.480$
   Since $-0.480 > -1.96$, do not reject the hypothesis of no difference.
   There is not sufficient evidence to reject the hypothesis of no difference.

NOTES FOR THE REMAINING EXERCISES IN THIS SECTION.
(1) FOR $n \leq 25$. Table A-7 gives only x_L, the <u>lower</u> critical value for the sign test. And so the text lets x be the <u>smaller</u> of the number of +'s or the number of −'s, and warns the reader to use common sense to avoid concluding the reverse of what the data indicates. But the problem's symmetry means that the upper critical value is $x_U = n - x_L$ and that $\mu_x = n/2$, the natural expected value of x when H_o is true. For completeness, this manual indicates those values whenever using the sign test – and uses a normal curve for illustration, even though the distribution of x is discrete.

Letting x always be the number of +'s is an alternative approach that maintains the natural agreement between the alternative hypothesis and the critical region – and is consistent with the logic and notation of parametric tests. Many sign test software programs use this approach.
(2) FOR $n > 25$. The correction for continuity is a conservative adjustment (intending to make less likely a false rejection of H_o) by shifting the x value 0.5 units toward the middle. When x is the smaller of the number of +'s or the number of −'s, this always involves replacing x with $(x + 0.5)$. The alternate approach suggested above (of letting x always be the number of +'s) maintains the natural agreement between the alternative hypothesis and the critical region – and is consistent with the logic and notation of parametric tests. In this approach, x is replaced with either $(x + 0.5)$ or $(x - 0.5)$ according to which one shifts the value toward the middle – i.e., with $(x + 0.5)$ when $x < \mu_x = n/2$, and with $(x - 0.5)$ when $x > \mu_x = n/2$.
The formula given is the usual one for converting a score into its standard score, using the correction for continuity: $\quad z_x = \dfrac{x - \mu_x}{\sigma_x} = \dfrac{(x \pm 0.5) - (n/2)}{\sqrt{n}/2}$.

9. Let the actresses (females) be group 1.

pair	1	2	3	4	5	6	7	8	9	10	11	12	13	14	15
f − m	−	−	−	−	−	−	−	−	+	−	−	−	−	−	+

n = 15: 2 +'s and 13 −'s

H_o: median difference = 0 years

H_1: median difference ≠ 0 years

α = 0.05

C.V. $x = x_{L,α/2} = 3$

$x = x_{U,α/2} = 15 - 3 = 12$

calculations:

$x = 2$

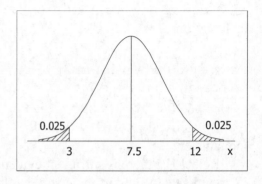

conclusion:

Reject H_o; there is sufficient evidence to reject the claim of no difference. There is sufficient evidence to reject the claim that there is no difference between the ages of the best actress and the best actor of that same year and to conclude that there is difference (in fact, that the actors tend to be older).

NOTE: Remember that the critical values are included in the critical region – so that we reject H_o in exercise #9, for example, if x≤3 or x≥12. Some exercises in this section also appeared previously in a parametric context in chapters 8 and 9. Because comparison of the parametric and non-parametric results are usually informative and provide statistical insights, we list those exercises.

this section	#9	#10	#11	#12	#13	#14	#16	#18	#19	#20
previously	9-4	9-4	9-4	9-4	8-3	8-3	8-3	8-5	8-4	8-5
	#11	#12	#16	#19	#18	#17	#20	#30	#16	#29

11. Let the alcohol times be group 1.

pair	1	2	3	4	5	6	7	8	9	10
a − t	−	−	+	−	−	+	+	−	−	−

n = 10: 3 +'s and 7 −'s

claim: alcohol time less than tobacco time

H_o: median difference = 0 seconds

H_1: median difference < 0 seconds

α = 0.05

C.V. $x = x_{L,α} = 1$

calculations:

$x = 3$

conclusion:

Do not reject H_o; there is not sufficient evidence to reject the claim of no difference. There is not sufficient evidence to support the claim that the time spent depicting alcohol use is less than the time spent depicting tobacco use.

13. Let x = the number of boy births.

claim: p > 0.5

n = 211: 172 +'s and 39 −'s [the data are in the proper direction to proceed]

Since n>25, use z with $μ_x = n/2 = 211/2 = 105.5$ and $σ_x = \sqrt{n}/2 = \sqrt{211}/2 = 7.263$

H_o: p = 0.5
H_1: p > 0.5
α = 0.01
C.V. z = $-z_\alpha$ = $-z_{0.01}$ = -2.326
calculations:
  x = 39
  z = [(x+0.5) – n/2]/[$\sqrt{n}$/2]
   = [39.5 – 105.5]/7.263
   = -66.0/7.263 = -9.09
  P-value = P(z<-9.09) = 0.0001

conclusion:
  Reject H_o; there is sufficient evidence to support the claim that p > 0.5. There is sufficient
  evidence to support the claim that the YSORT method is effective in increasing the
  likelihood of a boy birth.
NOTE: Working the exercise using the alternate method preserves the natural agreement
between the alternative hypothesis and the critical region as follows.
H_o: p = 0.5
H_1: p > 0.5
α = 0.01
C.V. z = z_α = $z_{0.01}$ = 2.326
calculations:
  x = 172
  z = [(x – 0.5) – n/2]/[$\sqrt{n}$/2]
   = [171.5 – 105.5]/7.263 = 9.09

conclusion:
  Reject H_o; conclude the same as above.

15. Let p = P(correctly predicting a baby's gender).
    n = 104: (0.55)(104) = 57 +'s and 47 –'s
    Since n>25, use z with μ_x = n/2 = 104/2 = 52 and σ_x = $\sqrt{n}$/2 = $\sqrt{104}$/2 = 5.099
    NOTE: There are two valid interpretations of the claim in this exercise.
    ***INTERPRETATION #1
    ***claim: p = 0.5 [that women do not have the ability to predict a baby's gender]
    H_o: p = 0.5
    H_1: p ≠ 0.5
    α = 0.05
    C.V. z = $\pm z_{\alpha/2}$ = $\pm z_{0.025}$ = ±1.96
    calculations:
      x = 47
      z = [(x+0.5) – n/2]/[$\sqrt{n}$/2]
       = [47.5 – 52]/5.099
       = -4.5/5.099 = -0.88
      P-value = 2·P(z<-0.88) = 2(0.1894) = 0.3788

conclusion:
      Do not Reject H_o; there is not sufficient evidence to reject the claim that p = 0.5. There is
      not sufficient evidence to reject the claim that women do not have the ability to predict the
      gender of their babies.

NOTE: Using the alternate approach where x = 57, the number of +'s, the direction of the calculated statistic is consistent with the rest of the problem. Since z = [56.5 − 52]/5.099 = 0.88, P-value = 2·P(z>0.88) = 0.3788 and the conclusions are the same.

***INTERPRETATION #2

***claim: $p \leq 0.5$ [that women do not have the ability to predict a baby's gender]

H_o: p = 0.5

H_1: p > 0.5

$\alpha = 0.05$

C.V. $z = z_\alpha = z_{0.05} = 1.645$

calculations and conclusions: same as above, with P-value = ½(0.3788) = 0.1894

17. The Minitab results for the sign test are as follows.

```
Sign test of median =  5.670 versus not = 5.670
                    N  Below  Equal  Above      P  Median
Post-1964 Quarters  40     28      0     12  0.0166   5.636
```

n = 40: 12 +'s and 28 −'s

Since n>25, use z with $\mu_x = n/2 = 40/2 = 20$ and $\sigma_x = \sqrt{n}/2 = \sqrt{40}/2 = 3.162$

H_o: median = 5.670 grams [p = 0.5]

H_1: median ≠ 5.670 grams

$\alpha = 0.01$

C.V. $z = \pm z_{\alpha/2} = \pm z_{0.005} = \pm 2.575$

calculations:

    x = 12

    $z = [(x+0.5) - n/2]/[\sqrt{n}/2]$

      = [12.5 − 20]/3.162

      = -7.5/3.162 = -2.37

    P-value = 2·P(z<-2.37) = 2(0.0089) = 0.0178

conclusion:

    Do not reject H_o; there is not sufficient evidence to reject the claim that median = 5.670. There is not sufficient evidence to reject the claim that the median weight of the post-1964 quarters is 5.670 grams. The quarters appear to be minted according to specifications.

NOTE: The Minitab P-value = 0.0166 is slightly different than the one given above because that software uses an exact test, and not a normal approximation.

19. The Minitab results for the sign test are as follows.

```
Sign test of median =  12.00 versus not = 12.00
              N  Below  Equal  Above      P  Median
CKREGVOL     36      1      2     33  0.0000   12.20
```

n = 34: 33 +'s and 1 −'s

Since n>25, use z with $\mu_x = n/2 = 34/2 = 17$ and $\sigma_x = \sqrt{n}/2 = \sqrt{34}/2 = 2.915$

H_o: median = 12 oz [p = 0.5]

H_1: median ≠ 12 oz

$\alpha = 0.05$

C.V. $z = \pm z_{\alpha/2} = \pm z_{0.025} = \pm 1.96$

calculations:

    x = 0

    $z = [(x+0.5) - n/2]/[\sqrt{n}/2]$

      = [1.5 − 17]/2.915

      = -15.5/2.915 = -5.32

    P-value = 2·P(z<-5.32) = 2(0.0001) = 0.0002

conclusion:

Reject H_o; there is sufficient evidence to reject the claim that median = 12 and to conclude that median ≠ 12 (in fact, that median > 12). There is sufficient evidence to reject the claim that the median volume in cans of regular Coke is 12 ounces. Because the median appears to be greater than 12, consumers are not being cheated.

NOTE: Using the alternate approach where x = 33, the number of +'s, the direction of the calculated statistic is consistent with the rest of the problem. Since z = [32.5 – 17]/2.915 = 5.32, P-value = 2·P(z>5.32) = 0.0002 and the conclusions are the same.

21. Originally there were 23 +'s, 68 –'s and 15 0's.
    a. Use the second method, counting half the 0's each way (and discarding the odd zero).
       n = 105: 30 +'s and 75 –'s
       Since n>25, use z with μ_x = n/2 = 105/2 = 52.5 and σ_x = $\sqrt{n}$/2 = $\sqrt{105}$/2 = 5.123
       H_o: median = 98.6 °F
       H_1: median < 98.6 °F
       α = 0.05
       C.V. z = $-z_\alpha$ = $-z_{0.05}$ = -1.645
       calculations:

       x = 30
       z = [(x+0.5) – n/2]/[$\sqrt{n}$/2]
         = [30.5 – 52.5]/5.123
         = -22.0/5.123 = -4.29
       P-value = P(z<-4.29) = 0.0001
    conclusion:
       Reject H_o; there is sufficient evidence to support the claim that median < 98.6.

    b. Use the third method, counting all the 0's to favor H_o (i.e., against H_1).
       n = 106: 38 +'s and 68 –'s
       Since n>25, use z with μ_x = n/2 = 106/2 = 53 and σ_x = $\sqrt{n}$/2 = $\sqrt{106}$/2 = 5.148
       H_o: median = 98.6 °F
       H_1: median < 98.6 °F
       α = 0.05
       C.V. z = $-z_\alpha$ = $-z_{0.05}$ = -1.645
       calculations:
       x = 38
       z = [(x+0.5) – n/2]/[$\sqrt{n}$/2]
         = [38.5 – 53]/5.148
         = -14.5/5.148 = -2.82
       P-value = P(z<-2.82) = 0.0024
    conclusion:
       Reject H_o; there is sufficient evidence to support the claim that median < 98.6.

    c. In this case the three approaches all lead to the same conclusion. But they illustrate that each approach is more conservative (i.e., requires more evidence to reject the null hypothesis) than the one before it as illustrated by the movement of the calculated z statistics from -4.61 [original Example 4] to -4.29 [part (a) above] to -2.82 [part (b) above]. In some cases the three approaches could lead to very different results.

13-3 Wilcoxon Signed-Ranks Test for Matched Pairs

1. The sign test in Section 13-2 converts differences in sample data pairs to positive and negative signs, thus transforming the values in the sample to nominal data. The Wilcoxon signed-ranks test converts differences in sample data pairs to ranks, thus transforming the values in the sample to ordinal data. Since ordinal data conveys more information than nominal data, the signed-ranks test is using more of the information contained in the original sample values and is to be preferred over the ordinary sign test.

3. Since there are 11 pairs but only 10 non-zero differences between the pairs, the sample size is $n = 10$. Table 8 indicates that the two-tailed critical value at the 0.05 level of significance is $T = 8$.

NOTE FOR THE REMAINING EXERCISES IN THIS SECTION.

Table A-8 gives only T_L, the <u>lower</u> critical value for the signed-ranks test. And so the text lets T be the <u>smaller</u> of the sum of the positive ranks or the sum of the negative ranks, and warns the reader to use common sense to avoid concluding the reverse of what the data indicates. But the problem's symmetry means that the upper critical value is $T_U = \Sigma R - T_L$ and that $\mu_T = \Sigma R/2$, the natural expected value of T when H_o is true. For completeness, this manual indicates those values whenever using the signed-ranks test – and uses a normal curve for illustration, even though the distribution of T is discrete. And remember that the critical values are included in the critical region – so that we reject H_o in exercise #5, for example, if $T \le 1$ or $T \ge 20$.

Letting T always be the sum of the positive ranks is an alternative approach that maintains the natural agreement between the alternative hypothesis and the critical region – and is consistent with the logic and notation of parametric tests. Many software programs use this approach.

Finally, this manual uses a minus sign preceding ranks associated with negative differences. The ranks themselves are not negative, but use of the minus sign helps to organize the information.

5. Let the Friday the 6th values be group 1.

F6-F13	-4	-6	-3	1	-1	-7
R	-4	-5	-3	1.5	-1.5	-6

$\Sigma R- = 19.5$ $n = 6$ non-zero ranks
$\Sigma R+ = 1.5$ check: $\Sigma R = n(n+1)/2 = 6(7)/2 = 21$
$\Sigma R = 21.0$
H_o: median difference = 0 admissions
H_1: median difference $\ne$ 0 admissions
$\alpha = 0.05$
C.V. $T = T_{L,\alpha/2} = 1$
      $T = T_{U,\alpha/2} = 21 - 1 = 20$

calculations:
    $T = 1.5$
conclusion:
    Do not reject H_o; there is not sufficient evidence to reject the claim that
    median difference = 0. There is not sufficient evidence to reject the claim that the numbers
    of hospital admissions are unaffected when the 13th day of the month falls on a Friday.

7. Let the actresses be group 1.

F–M	-34	-9	-25	-14	-8	-16	-31	-12	2	-9	-21	-14	-9	-41	6
R	-14	-5	-12	-8.5	-3	-10	-13	-7	1	-5	-11	-8.5	-5	-15	2

$\Sigma R- = 117$ n = 15 non-zero ranks

$\Sigma R+ = 3$ check: $\Sigma R = n(n+1)/2 = 15(16)/2 = 120$

$\Sigma R = 120$

H_0: median difference = 0 years

H_1: median difference $\neq$ 0 years

$\alpha = 0.05$

C.V. $T = T_{L,\alpha/2} = 25$

     $T = T_{U,\alpha/2} = 120 - 25 = 95$

calculations:

    T = 3

conclusion:

    Reject H_0; there is sufficient evidence to reject the claim of no difference. There is
    sufficient evidence to reject the claim that there is no difference between the ages of the
    best actress and the best actor of that same year and to conclude that there is a difference
    (in fact, that the actors tend to be older).

9. Let the tobacco times be group 1.

A–T	88	18	-113	248	74	-1	-23	132	1	81
R	7	3	-8	10	5	-1.5	-4	9	1.5	6

$\Sigma R- = 13.5$ n = 10 non-zero ranks

$\Sigma R+ = 41.5$ check: $\Sigma R = n(n+1)/2 = 10(11)/2 = 55$

$\Sigma R = 55.0$

H_0: median difference = 0 seconds

H_1: median difference $\neq$ 0 seconds

$\alpha = 0.05$

C.V. $T = T_{L,\alpha/2} = 8$

     $T = T_{U,\alpha/2} = 55 - 8 = 47$

calculations:

    T = 13.5 (using smaller ranks)

    T = 41.5 (using positive ranks)

conclusion:

    Do not reject H_0; there is not sufficient evidence to reject the claim of no difference. There
    is not sufficient evidence to reject the claim that the time spent depicting alcohol use is
    the same as the time spent depicting tobacco use.

NOTE: The directions for exercises 7-10 call for a two tailed test. When the data was given in
the previous section, the directions called for a one-tailed test. In that case the C.V. is T = 11,
and the conclusion is the same – but the final conclusion is worded: There is not sufficient
evidence to support the claim that the time spent depicting alcohol use is less than the time
spent depicting tobacco use.

NOTE FOR EXERCISES 11-14. Minitab follows the pattern of most statistical software programs
and uses the alternate method of reporting $T = \Sigma R+$, the sum of the positive ranks. Since the sum
of the ranks is $\Sigma R = n(n+1)/2$, $\Sigma R-$ can be found by subtraction and the pattern in the text can be
followed using $T = \min(\Sigma R+, \Sigma R-)$.

11. The Minitab output for the Wilcoxon signed-rank test is as follows.

```
Test of median = 5.670 versus median not = 5.670
                        N for   Wilcoxon           Estimated
                    N   Test   Statistic     P      Median
Post-1964 Quarters  40   40      196.0     0.004     5.638
```

$\Sigma R- = 624$ n = 20 non-zero ranks

$\Sigma R+ = 196$ $\Sigma R = n(n+1)/2 = 40(41)/2 = 820$

$\Sigma R = 820$ $\Sigma R- = \Sigma R - \Sigma R+ = 820 - 196 = 624$

For n=40 ranks, use the z approximation with

$\mu_T = n(n+1)/4 = 40(41)/4 = 410$ and $\sigma_T = \sqrt{[n(n+1)(2n+1)]/24} = \sqrt{[40(41)(81)]/24} = 74.398$

H_o: median = 5.670 grams

H_1: median $\neq$ 5.670 grams

$\alpha = 0.01$

C.V. $z = \pm z_{\alpha/2} = \pm z_{0.005} = \pm 2.575$

calculations:

   T = 196

   $z = (T - \mu_T)/\sigma_T$

     $= (196 - 410)/74.398$

     $= -214/74.398 = -2.88$

   P-value $= 2 \cdot P(z < -2.88) = 2(0.0020) = 0.0040$

conclusion:

Reject H_o; there is sufficient evidence to reject the claim that median = 5.670 and conclude that median $\neq$ 5.670 (in fact, that median < 5.670). There is sufficient evidence to reject the claim that the median weight of the post-1964 quarters is 5.670 grams. If the quarters were taken from circulation, however, they would be expected to weigh less than when they were minted.

13. The Minitab output for the Wilcoxon signed-rank test is as follows.

```
Test of median = 12.00 versus median not = 12.00
                     N for   Wilcoxon           Estimated
                 N   Test   Statistic     P      Median
CKREGVOL        36    34      579.5     0.000     12.20
```

$\Sigma R- = 15.5$ n = 34 non-zero ranks

$\Sigma R+ = 579.5$ $\Sigma R = n(n+1)/2 = 34(35)/2 = 595$

$\Sigma R = 595$ $\Sigma R- = \Sigma R - \Sigma R+ = 595 - 579.5 = 15.5$

For n=34 ranks, use the z approximation with

$\mu_T = n(n+1)/4 = 34(35)/4 = 297.5$ and $\sigma_T = \sqrt{[n(n+1)(2n+1)]/24} = \sqrt{[34(35)(69)]/24} = 58.491$

H_o: median = 12 oz

H_1: median $\neq$ 12 oz

$\alpha = 0.05$

C.V. $z = \pm z_{\alpha/2} = \pm z_{0.025} = \pm 1.96$

calculations:

   T = 15.5

   $z = (T - \mu_T)/\sigma_T$

     $= (15.5 - 297.5)/58.491$

     $= -282/58.491 = -4.82$

   P-value $= 2 \cdot P(z < -4.82) = 2(0.0001) = 0.0002$

conclusion:

　　Reject H_o; there is sufficient evidence to reject the claim that median = 12 and to conclude
　　that median ≠ 12 (in fact, that median > 12). There is sufficient evidence to reject the
　　claim that the median volume in cans of regular Coke is 12 ounces. Because the median
　　appears to be greater than 12, consumers are not being cheated.

　NOTE: Using the alternate approach, T = 579.5 and z = +4.82.

15. For any set of ranks, $(\Sigma R+) + (\Sigma R-) = \Sigma R = n(n+1)/2$.
　　a. For n = 50, $\Sigma R = 50(51)/2 = 1275$. If $(\Sigma R+) = 300$, then $(\Sigma R-) = 1275 - 300 = 975$.
　　b. If one of $(\Sigma R+)$ or $(\Sigma R-)$ is equal to k, then the other must equal $\Sigma R - k = n(n+1)/2 - k$.

13-4 Wilcoxon Rank-Sum Test for Two Independent Samples

1. Because the samples are small and are from populations not having normal distributions, the
　parametric t test should not be used.

3. a. The theoretical median of each population is $(1+1000)/2 = 500.5$.
　　b. No. The rank-sum test procedure tests only whether the two populations have the same
　　　median, but it does not specify or test for the value of that common median.

NOTE: On the following page are the ordered scores and the accompanying rank for exercises 5-8.
All the values necessary to work the exercises are found on that page. Make certain that you
understand the format of that page, and how each of the ranks was obtained.

　　Always letting $R = \Sigma R_1$ (i.e., the sum of the ranks from the scores designated as group 1)
guarantees that there will be agreement between H_1 and the C.R. (regardless of whether the test is
two-tailed or one-tailed in either direction) and that the sign of the test statistic indicates in the
natural manner which group has the larger median.

　　When working the problems by hand, it is strongly recommended that you calculate both
ΣR_1 and ΣR_2 (even though only the former is necessary for the test) and that you verify that
$\Sigma R_1 + \Sigma R_2 = n(n+1)/2$. This simple check can identify the presence of errors that must be
corrected before proceeding with the calculations.

5. Let the unsuccessful candidates be group 1.

$n_1 = 23$	$R = \Sigma R_1 = 692$	chk: $\Sigma R_i = n(n+1)/2$	$\mu_R = n_1(n+1)/2$	$\sigma^2_R = n_1 n_2 (n+1)/12$
$\underline{n_2 = 30}$	$\underline{\Sigma R_2 = \ 739}$	$= 53(54)/2$	$= 23(54)/2$	$= (23)(30)(54)/12$
$n = \Sigma n = 53$	$\Sigma R_i = 1431$	$= 1431$	$= 621$	$= 3105$ $\sigma_R = 55.722$

H_o: $median_1 = median_2$
H_1: $median_1 \neq median_2$
$\alpha = 0.05$
C.V. $z = \pm z_{\alpha/2} = \pm z_{0.025} = \pm 1.960$
calculations:

　　$z = (R - \mu_R)/\sigma_R$

　　　$= (692 - 621)/\sqrt{3105}$

　　　$= 71/55.722 = 1.27$

　　P-value $= 2 \cdot P(z > 1.27) = 2(0.1020) = 0.2040$

conclusion:

　　Do not reject H_o; there is not sufficient evidence to reject the claim that $median_1 = median_2$.
　　There is not enough evidence to reject the claim that unsuccessful applicants are from a
　　population with the same median age as the successful applicants. No; based on this result,
　　there does not appear to be discrimination based on age.

Below are the ordered scores and accompanying ranks for exercises 5-8.

Row	exercise #5 n	R	y	R	exercise #6 PA	R	NY	R	exercise #7 po	R	mo	R	exercise #8 pr	R	po	R
1	34	3.0	27	1.0	130	3.0	128	1.0	0	1.0	6	7.5	0	2.0	0	2.0
2	37	6.0	33	2.0	133	5.5	129	2.0	2	2.5	7	10.0	0	2.0	2	6.0
3	37	6.0	36	4.0	138	9.0	131	4.0	2	2.5	9	12.5	1	4.0	2	6.0
4	38	9.0	37	6.0	142	15.0	133	5.5	3	4.0	10	14.5	2	6.0	3	8.5
5	41	12.0	38	9.0	142	15.0	134	7.0	5	5.0	12	18.0	3	8.5	5	13.0
6	42	14.0	38	9.0	149	18.0	137	8.0	6	7.5	13	19.5	4	11.0	6	15.0
7	43	17.0	39	11.0	151	19.5	140	11.5	6	7.5	13	19.5	4	11.0	6	15.0
8	44	21.0	42	14.0	151	19.5	140	11.5	6	7.5	15	22.0	4	11.0	6	15.0
9	44	21.0	42	14.0	155	21.0	140	11.5	8	11.0	17	24.5	7	17.0	8	18.0
10	45	27.0	43	17.0	156	22.0	140	11.5	9	12.5	25	31.0	9	20.0	9	20.0
11	45	27.0	43	17.0	161	23.0	142	15.0	10	14.5	33	35.0	9	20.0	10	23.0
12	45	27.0	44	21.0	163	24.0	143	17.0	11	16.5	36	36.0	10	23.0	11	26.0
13	46	32.0	44	21.0		194.5		105.5	11	16.5	59	37.0	10	23.0	11	26.0
14	48	37.0	44	21.0					15	22.0	63	38.0	11	26.0	15	32.0
15	49	40.0	45	27.0					15	22.0		325.0	12	29.0	15	32.0
16	53	45.5	45	27.0					17	24.5			12	29.0	17	38.0
17	53	45.5	45	27.0					18	26.0			12	29.0	18	40.5
18	54	48.0	45	27.0					19	27.0			15	32.0	19	42.0
19	54	48.0	46	32.0					21	28.0			16	35.0	21	44.5
20	55	50.0	46	32.0					23	29.0			16	35.0	23	47.0
21	56	51.0	47	34.5					25	31.0			16	35.0	25	52.5
22	57	52.0	47	34.5					25	31.0			17	38.0	25	52.5
23	60	53.0	48	37.0					26	33.0			17	38.0	26	55.5
24		692.0	48	37.0					32	34.0			18	40.5	32	60.5
25			49	40.0						416.0			20	43.0		690.5
26			49	40.0									21	44.5		
27			51	42.5									23	47.0		
28			51	42.5									23	47.0		
29			52	44.0									24	49.5		
30			54	48.0									24	49.5		
31				739.0									25	52.5		
32													25	52.5		
33													26	55.5		
34													28	57.5		
35													28	57.5		
36													29	59.0		
37													32	60.5		
38													36	62.0		
														1262.5		

7. Let the popes be group 1.

$n_1 = 24$	$R = \Sigma R_1 = 416$	chk: $\Sigma R_i = n(n+1)/2$	$\mu_R = n_1(n+1)/2$	$\sigma^2_R = n_1 n_2(n+1)/12$
$\underline{n_2 = 14}$	$\underline{\Sigma R_2 = 325}$	$= 38(39)/2$	$= 24(39)/2$	$= (24)(14)(39)/12$
$n = \Sigma n = 38$	$\Sigma R_i = 741$	$= 741$	$= 468$	$= 1092$ $\sigma_R = 33.045$

H_o: median$_1$ = median$_2$

H_1: median$_1 \neq$ median$_2$

$\alpha = 0.05$

C.V. $z = \pm z_{\alpha/2} = \pm z_{0.025} = \pm 1.960$

calculations:

$z = (R - \mu_R)/\sigma_R$

$= (416 - 468)/\sqrt{1092}$

$= -52/33.045 = -1.57$

P-value $= 2 \cdot P(z < -1.57) = 2(0.0582) = 0.1164$

conclusion:

Do not reject H_o; there is not sufficient evidence to reject the claim that median$_1$ = median$_2$. There is not enough evidence to reject the claim that the two samples of longevity data from popes and monarchs are from populations with the same median.

NOTE FOR EXERCISES 9-12. Minitab uses the name Mann-Whitney U test instead of Wilcoxon rank-sum test, and it uses the Greek symbol eta (which it spells out, since it cannot print Greek letters) for the median. It prints the value of $R = \Sigma R_1$ (which it identifies as W, since it is the Wilcoxon statistic), states the null and alternative hypotheses, and gives the P-value for the test. Since Minitab uses an exact procedure and not a normal approximation, it does not give a z value for the calculated test statistic and the P-values may differ slightly from the ones obtained below.

9. The Minitab output for the Wilcoxon rank-sum test [Mann-Whitney U test] is as follows.

```
W = 863.0
Test of ETA1 = ETA2 vs ETA1 > ETA2 is significant at 0.0000
The test is significant at 0.0000 (adjusted for ties)
```

Let the king size cigarettes be group 1.

$n_1 = 25$	$n_2 = 25$	$n = n_1 + n_2 = 50$
$\mu_R = n_1(n+1)/2$	$\sigma^2_R = n_1 n_2(n+1)/12$	
$= 25(51)/2$	$= (25)(25)(51)/12$	
$= 637.5$	$= 2656.25$	

H_o: median$_1$ = median$_2$

H_1: median$_1 >$ median$_2$

$\alpha = 0.01$

C.V. $z = z_\alpha = z_{0.01} = 2.326$

calculations:

$R = [W] = \Sigma R_1 = 863.0$

$z = (R - \mu_R)/\sigma_R$

$= (863.0 - 637.5)/\sqrt{2656.25}$

$= 225.5/51.5388 = 4.38$

P-value $= P(z > 4.38) = 1 - 0.9999 = 0.0001$

conclusion:

Reject H_o; there is sufficient evidence to reject the claim that median$_1$ = median$_2$ and to conclude that median$_1 >$ median$_2$. There is sufficient evidence to support the claim that the median amount of nicotine in nonfiltered king size cigarettes is greater than the median amount of nicotine in 100mm filtered cigarettes.

11. The Minitab output for the Wilcoxon rank-sum test [Mann-Whitney U test] is as follows.
    NOTE: First place the PG & PG-13 values in one column and the R values in another column.
    ```
    W = 487.5
    Test of ETA1 = ETA2 vs ETA1 > ETA2 is significant at 0.0056
    The test is significant at 0.0056 (adjusted for ties)
    ```
    Let the PG/PG-13 movies be group 1.

    $n_1 = 23$ $n_2 = 12$ $n = n_1 + n_2 = 35$

    $\mu_R = n_1(n+1)/2$ $\sigma^2_R = n_1n_2(n+1)/12$
    $\quad = 23(36)/2$ $\quad = (23)(12)(36)/12$
    $\quad = 414$ $\quad = 828$

    H_o: median$_1$ = median$_2$
    H_1: median$_1$ > median$_2$
    $\alpha = 0.05$
    C.V. $z = z_\alpha = z_{0.05} = 1.645$
    calculations:

    $R = [W] = \Sigma R_1 = 487.5$
    $z = (R - \mu_R)/\sigma_R$
    $\quad = (487.5 - 414)/\sqrt{828}$
    $\quad = 73.5/28.7750$
    P-value = $P(z>2.55) = 1 - 0.9946 = 0.0054$
    conclusion:

    Reject H_o; there is sufficient evidence to reject the claim that median$_1$ = median$_2$ and to conclude that median$_1$ > median$_2$. There is sufficient evidence to support the claim that movies with ratings of PG or PG-13 have a higher median gross than do movies with R ratings.

13. The denominator of the Mann-Whitney and Wilcoxon statistics are the same.
    Assuming $R = \Sigma R_1$, the numerator of the Wilcoxon statistic is
    $R - \mu_R = R - n_1(n+1)/2$
    Assuming $R = \Sigma R_1$, the numerator of the Mann-Whitney statistic is
    $U - n_1n_2/2 = [n_1n_2 + n_1(n_1+1)/2 - R] - n_1n_2/2$
    $\quad = n_1n_2/2 + n_1(n_1+1)/2 - R$
    $\quad = (n_1/2)(n_2 + n_1 + 1) - R$
    $\quad = n_1(n+1)/2 - R = \mu_R - R = $ -[numerator of the Wilcoxon statistic]
    For the braking distances measures listed in Table 13-5,
    $U = n_1n_2 + n_1(n_1+1)/2 - R = 13(12) + 13(14)/2 - 180.5 = 66.5$
    $\mu_U = n_1n_2/2 = 13(12)/2 = 78$
    $\sigma^2_U = n_1n_2(n_1+n_2+1)/12 = 13(12)(26)/12 = 338$
    $z_U = (\mu_U - U)/\sigma^2_U = (66.5 - 78)/\sqrt{338} = -11.5/18.3848 = -0.63$
    As predicted, z_U has the same absolute value but the opposite sign of z_R.

13-5 Kruskal-Wallis Test

1. Yes. Because each sample was obtained without any considerations regarding the other samples, and because the values from the samples are not matched in any way, the samples are independent.

NOTE: The following strategy to label the ranks, used in Exercise 3 below, has proven to be helpful in organizing the information. Show a whole number for the rank only when that rank refers to a value not involved in any ties. When an odd number of values are tied (as in Exercise 3 when there 3 136's that would have ranks 13 and 14 and 15), show the mean rank to one decimal place (i.e., use 14.0). When an even number of values are tied, of course, the rank will end in ".5" out of necessity.

3. Below are the ordered scores and ranks for each era.

4000	R	1850	R	150	R
125	1	129	4.0	128	2
129	4.0	129	4.0	136	14.0
131	6	134	10.0	137	17.5
132	7.5	134	10.0	137	17.5
132	7.5	136	14.0	138	21.5
134	10.0	136	14.0	139	24
135	12	137	17.5	141	25
138	21.5	137	17.5	142	26
138	21.5	138	21.5	145	27
	91.0		112.5		174.5

$n_1 = 9$ $R_1 = 91$
$n_2 = 9$ $R_2 = 112.5$
$n_3 = 9$ $R_3 = 174.5$
$n = \Sigma n = 27$ $\Sigma R = 378.0$

chk: $\Sigma R = n(n+1)/2 = 27(28)/2 = 378$

The value of R_1, the sum of the ranks of the first sample, is 91.0.

NOTE: The critical value and accompanying graph for the exercises in this section are given in terms of χ^2, which approximates the distribution of the calculated statistic H. Some of the exercises in this section also appeared previously in a parametric context in section 12-2. Because comparison of the parametric and non-parametric results are usually informative and provide statistical insights, we list those pairings.

this section	#7	#8	#9	#11	#12
section 12-2	#11	#12	#14	#15	#16

5. The ordered scores and ranks for each era, including the values for R_1 and R_2 and R_3, are given above in Exercise 3.

H_0: the populations have the same median
H_1: at least one median is different
$\alpha = 0.05$ and df = 2
C.V. $H = \chi^2_{\alpha} = \chi^2_{0.05} = 5.991$

calculations:

$$H = \{12/[n(n+1)]\} \cdot [\Sigma (R_i^2/n_i)] - 3(n+1)$$
$$= \{12/[27(28)]\} \cdot [(91)^2/9 + (112.5)^2/9$$
$$+ (174.5)^2/9] - 3(28)$$
$$= \{0.0159\} \cdot [5709.7] - 84 = 6.631$$

P-value = χ^2cdf (6.631,99,2) = 0.0363

conclusion:

Reject H_0; there is sufficient evidence to reject the claim that the three samples come from populations with the same median. Yes, the data suggests interbreeding of cultures.

7. Below are the ordered scores and ranks for each group. The group listed first is group 1, etc.

small	R	medium	R	large	R
290	6	245	4	169	1
371	13	264	5	216	2.5
374	14	296	7	216	2.5
376	15	349	11	332	8
406	17	368	12	335	9
475	20	393	16	342	10
479	21	474	19	432	18
499	22	502	24	510	26.5
501	23	505	25	608	29
544	28	510	26.5	698	30
	179		149.5		136.5

$n_1 = 10$ $R_1 = 179$
$n_2 = 10$ $R_2 = 149.5$
$n_3 = 10$ $R_3 = 136.5$
$n = \Sigma n = 30$ $\Sigma R = 465.0$

chk: $\Sigma R = n(n+1)/2 = 30(31)/2 = 465$

H_o: the populations have the same median
H_1: at least one median is different
$\alpha = 0.05$ and df = 2
C.V. $H = \chi_\alpha^2 = \chi_{0.05}^2 = 5.991$
calculations:

$H = \{12/[n(n+1)]\} \cdot [\Sigma (R_i^2/n_i)] - 3(n+1)$

$= \{12/[30(31)]\} \cdot [(179)^2/10 + (149.5)^2/10$
$\qquad + (136.5)^2/10] - 3(31)$

$= \{0.0129\} \cdot [7302.35] - 93$

$= 1.2239$

P-value $= \chi^2 cdf\ (1.2239, 99, 2) = 0.5423$

conclusion:

Do not reject H_o; there is not sufficient evidence to reject the claim that the three samples come from populations with the same median. There is not sufficient evidence to reject the claim that the three car types have the same median head injury values (hic). No; these data do not provide sufficient evidence to suggest that larger cars are safer.

9. Below are the ordered scores and ranks for each group. The group listed first is group 1, etc.

four	R	six	R	eight	R			
5.5	1	7.0	9	8.6	22		$n_1 = 13$	$R_1 = 106.5$
6.5	3.0	7.2	12.0	8.7	24.0		$n_2 = 12$	$R_2 = 228.0$
6.5	3.0	7.2	12.0	9.3	29.5		$n_3 = 7$	$R_3 = 193.5$
6.5	3.0	7.4	15.5	9.3	29.5		$n = \Sigma n = 32$	$\Sigma R = 528.0$
6.6	5	7.7	17.5	9.3	29.5			
6.7	6.5	7.7	17.5	9.3	29.5		chk: $\Sigma R = n(n+1)/2 = 32(33)/2 = 528$	
6.7	6.5	8.2	20.5	9.3	29.5			
6.8	8	8.2	20.5		193.5			
7.1	10	8.7	24.0					
7.2	12.0	8.7	24.0					
7.3	14	9.0	26					
7.4	15.5	9.3	29.5					
7.9	19		228.0					
	106.5							

H_o: the populations have the same median
H_1: at least one median is different
$\alpha = 0.05$ and df = 2
C.V. $H = \chi_\alpha^2 = \chi_{0.05}^2 = 5.991$
calculations:

$H = \{12/[n(n+1)]\} \cdot [\Sigma (R_i^2/n_i)] - 3(n+1)$

$= \{12/[32(33)]\} \cdot [(106.5)^2/13 + (228.0)^2/12$
$\qquad + (193.5)^2/7] - 3(33)$

$= \{0.0114\} \cdot [10553.374] - 99$

$= 20.9247$

P-value $= \chi^2 cdf\ (20.9247, 99, 2) = 2.86E\text{-}5 = 0.00003$

conclusion:

Reject H_o; there is sufficient evidence to reject the claim that the three samples come from populations with the same median. There is sufficient evidence to reject the claim that the three car types have the same median amount of greenhouse gas emissions. Yes; based on these results, the number of cylinders does appear to affect the amount of greenhouse gas emissions.

11. The Minitab output for the Kriskal-Wallis test is as follows.

NOTE: First place all the nicotine values in one column and the type of cigarette from which that value came in another column.

```
Kruskal-Wallis Test on nicotine
     type      N   Median  Ave Rank      Z
     king      25  1.1000     56.5     5.21
     menthol   25  0.8000     26.0    -3.37
     filter    25  1.0000     31.4    -1.84
     Overall   75             38.0
H = 27.91   DF = 2   P = 0.000
H = 29.07   DF = 2   P = 0.000   (adjusted for ties)
```

H_0: the populations have the same median
H_1: at least one median is different
$\alpha = 0.05$ and df $= 2$
C.V. $H = \chi_\alpha^2 = \chi_{0.05}^2 = 5.991$
calculations:

$$H = \{12/[n(n+1)]\} \cdot [\Sigma(R_i^2/n_i)] - 3(n+1)$$

$$= 27.91 \text{ [Minitab]}$$

P-value $= P(\chi_2^2 > 27.91)$

$$= 0.000 \text{ [Minitab]}$$

conclusion:

Reject H_0; there is sufficient evidence to reject the claim that the three samples come from populations with the same median. There is sufficient evidence to reject the claim that the three categories of cigarettes yield the same median amount of nicotine. Yes; given that only the king size cigarettes are not filtered, and that their mean rank (56.5) is considerably higher than the other mean ranks (26.0 & 31.4), the filters do appear to make a difference.

13. For the n=30 observations in Example 1, the correction is determined as follows.

rank	t	$T = t^3 - t$
5	3	24
8	3	24
10.5	2	6
13	3	24
17	5	120
21.5	4	60
25	3	24
		282

correction factor:
$$1 - \Sigma T/(n^3 - n) = 1 - 282(30^3 - 30)$$
$$= 1 - 282/26970$$
$$= 1 - 0.010456062$$
$$= 0.9895439381$$

The original calculated test statistic in Example 1 is H = 5.774

The corrected calculated test statistic is H = 5.774/0.989543939 = 5.835

No, the corrected value of H does not differ substantially from the original one – even in a case such as this when the number of ties seems relatively large

NOTE: Be careful when identifying the tied ranks. In addition to the easily recognized ".5's" that result from an even number of ties, there may be multiple whole number ranks resulting from an odd number of ties. The NOTE following Exercise 3 gives a labeling strategy that helps to identify tied ranks and to bring some order to the process of ranking across the different groups.

13-6 Rank Correlation

1. No. When there is a significant linear correlation between two sets of raw data, the linear regression line can be used for prediction using the raw scores. But a significant linear correlation between two sets of ranks does not necessarily mean that there is a linear correlation between the two sets of raw data that produced the ranks, and so the linear regression line should not be used for prediction using the raw scores. Consider, for example, (x,y) pairs that have a perfect quadratic relationship (x,x^2): there is a perfect correlation $r_s =$ 1.00 between the ranks, but there is not a linear relationship between the raw scores – and the plot of the (x,y) ordered pairs of raw data is not a straight line.

3. In general, r_s refers to the rank correlation in the sample and ρ_s refers the rank correlation in the population. In this context, r_s is the rank correlation for five recent years between the amount spent on new cars and the amount spent on clothing, and ρ_s is that rank correlation for all years in the population of interest – that population being determined by the details and intention of the study. The subscript "s" is necessary to distinguish rank correlation from the linear correlation of section 10-2. The "s" stands for "Spearman" – and rank correlation is sometimes referred to as Spearman's rank correlation. The subscript "s" used to designate Spearman's rank correlation is not related to the lower case "s" used to designate standard deviation.

5. The scatterplot includes n=6 (x,y) points. The critical values are $r_s = \pm 0.886$.
   Since the points can be joined to form a constantly ascending line from right to left, the ranks of both the x variable and the y variable will be in ascending order from 1 to n. This means that $r_s = 1$.
   Since 1 > 0.866, reject Ho:$\rho_s = 0$ and conclude $\rho_s \neq 0$ (in fact, that $\rho_s > 0$). There is sufficient evidence to support the claim that there is a correlation between distance and time.

7. a. From Table A-9, the critical values are ± 0.521.
   b. From Table A-9, the critical values are ± 0.521.
   c. From Formula 13-1, the critical values are $\pm z_{\alpha/2}/\sqrt{n-1} = \pm 1.960/\sqrt{99} = \pm 0.197$.
   d. From Formula 13-1, the critical values are $\pm z_{\alpha/2}/\sqrt{n-1} = \pm 2.575/\sqrt{64} = \pm 0.322$.

NOTE: The rank correlation is correctly calculated using the ranks in the Pearson product moment correlation Formula 10-1 to produce
$$r_s = [n\Sigma R_x R_y - (\Sigma R_x)(\Sigma R_y)]/[\sqrt{n(\Sigma R_x^2) - (\Sigma R_x)^2}\sqrt{n(\Sigma R_y^2) - (\Sigma R_y)^2}]$$
Since $\Sigma Rx = \Sigma Ry = 1 + 2 + \ldots + n = n(n+1)/2$ [always]
 and $\Sigma R_x^2 = \Sigma R_y^2 = 1^2 + 2^2 + \ldots + n^2 = n(n+1)(2n+1)/6$ [when there are no ties in the ranks]
it can be shown by algebra that the above formula can be shortened to
$$r_s = 1 - [6(\Sigma d^2)]/[n(n^2-1)] \underline{\text{ when there are no ties in the ranks}}.$$
This manual calculates $d = R_x - R_y$, thus preserving the sign of d. This convention means that $\Sigma d = 0$ must always be true and provides a check for the assigning and differencing of the ranks – and it must also be true that $\Sigma R_x = \Sigma R_y = n(n+1)/2$.
   Unless there is an excessive number of ties, the difference between using Formula 10-1 and the shortcut formula is usually trivial – in fact, some textbooks give only the shortcut formula and ignore the fact that it does not give the exact answer when there are ties. A related item ignored by virtually every textbook is the fact that the critical values in Table A-9 are not exact when there are ties.

In addition, some exercises in this section appeared previously in a parametric context in section 10-2. Because comparison of the parametric and non-parametric results are usually informative and provide statistical insights, we list those pairings.

this section	#15	#16	#17	#18
section 10-2	#17	#26	#31	#32

9. The following table summarizes the calculations.

R_x	R_y	d	d^2
1	6	-5	25
3	4	-1	1
4	5	-1	1
7	1	6	36
5	3	2	4
6	2	4	16
2	7	-5	25
28	28	0	108

$r_s = 1 - [6(\Sigma d^2)]/[n(n^2-1)]$
$= 1 - [6(108)]/[7(48)]$
$= 1 - 1.929$
$= -0.929$

H_o: $\rho_s = 0$
H_1: $\rho_s \neq 0$
$\alpha = 0.05$
C.V. $r_s = \pm 0.786$
calculations:
   $r_s = -0.929$
conclusion:

Reject H_o; there is sufficient evidence to reject the claim that $\rho_s = 0$ and to conclude that $\rho_s \neq 0$ (in fact, that $\rho_s < 0$). There is sufficient evidence to support the claim that there is a correlation between the rankings of the two judges. In this case the correlation is negative, and so the judges rank very differently – in general, the first judge gave high ranks to those bands that received low ranks from the second judge, and vice-versa.

11. The following table summarizes the calculations.

R_x	R_y	d	d^2
1	6	-5	25
2	2	0	0
3	10	-7	49
4	4	0	0
5	12	-7	49
6	9	-3	9
7	8	-1	1
8	7	1	1
9	1	8	64
10	5	5	25
11	3	8	64
12	11	1	1
78	78	0	288

$r_s = 1 - [6(\Sigma d^2)]/[n(n^2-1)]$
$= 1 - [6(288)]/[12(143)]$
$= 1 - 1.007$
$= -0.007$

H_o: $\rho_s = 0$
H_1: $\rho_s \neq 0$
$\alpha = 0.05$
C.V. $r_s = \pm 0.587$
calculations:
   $r_s = -0.007$
conclusion:

Do not reject H_o; there is not sufficient evidence to reject the claim that $\rho_s = 0$. There is not sufficient evidence to support the claim that there is a correlation between the conviction rate and the recidivism rate. No; recidivism rates do not appear to be related to conviction rates.

13. The following table summarizes the calculations.

Let x be the quality. Give the highest quality rank 1 and give the highest cost rank 1.

x	R_x	y	R_y
74	1	27	3.5
71	2	30	2
68	3	38	1
65	4	23	5.5
63	5	20	7.5
62	6	13	10.5
59	7	27	3.5
57	8.5	23	5.5
57	8.5	14	9
53	10	13	10.5
51	11	20	7.5
	66		66

$\Sigma R_x R_x = 505.50$
$\Sigma R_y R_y = 504.00$
$\Sigma R_x R_y = 468.25$

$H_0: \rho_s = 0$
$H_1: \rho_s \neq 0$
$\alpha = 0.05$
C.V. $r_s = \pm 0.618$
calculations:

$$r_s = [n\Sigma R_x R_y - (\Sigma R_x)(\Sigma R_y)]/ \ [\sqrt{n(\Sigma R_x^2) - (\Sigma R_x)^2} \sqrt{n(\Sigma R_y^2) - (\Sigma R_y)^2}]$$

$$= [11(468.25) - (66)(66)]/ \ [\sqrt{11(505.50) - (66)^2} \sqrt{11(504.00) - (66)^2}]$$

$$= [794.75]/[\sqrt{1204.50}\sqrt{1188.00}] = 0.664$$

conclusion:

Reject H_0; there is sufficient evidence to reject the claim that $\rho_s = 0$ and conclude that $\rho_s \neq 0$ (in fact, that $\rho_s > 0$). There is sufficient evidence to support the claim that there is a correlation between the quality and price of LCD TV's. Yes; based on these results you can expect to get higher quality by purchasing a more expensive LCD TV.

15. The following table summarizes the calculations.

Let x be the overhead width

x	R_x	y	R_y	d	d^2
7.2	6	116	6	0	0
7.4	5	154	5	0	0
9.8	1	245	1	0	0
9.4	2	202	2	0	0
8.8	3	200	3	0	0
8.4	4	191	4	0	0
	21		21	0	0

$r_s = 1 - [6(\Sigma d^2)]/[n(n^2-1)]$
$= 1 - [6(0)]/[6(35)]$
$= 1 - 0$
$= 1$

$H_0: \rho_s = 0$
$H_1: \rho_s \neq 0$
$\alpha = 0.05$
C.V. $r_s = \pm 0.886$
calculations:
$r_s = 1$
conclusion:

Reject H_0; there is sufficient evidence to reject the claim that $\rho_s = 0$ and to conclude that $\rho_s \neq 0$ (in fact, that $\rho_s > 0$). There is sufficient evidence to support the claim that there is a correlation between the overhead width of seals and the weights of the seals.

17. There are n=56 male-female pairs in the first two columns of Data Set 4.

    From Formula 13-1, the critical values are $\pm z_{\alpha/2}/\sqrt{n-1} = \pm 1.960/\sqrt{55} = \pm 0.264$.

    The Minitab correlation command on the ranked data yields the following.

    `Pearson correlation of men and women = 0.231`

    H_o: $\rho_s = 0$

    H_1: $\rho_s \neq 0$

    $\alpha = 0.05$

    C.V. $r_s = \pm 0.264$

    calculations:

        $r_s = 0.231$

conclusion:

    Do not reject H_o; there is not sufficient evidence to reject the claim that $\rho_s = 0$. There is not sufficient evidence to conclude that there is a correlation between the numbers of words spoken by men and women who are in a couple relationship.

19. There are n=52 Monday-Sunday pairs in Data Set 14.

    From Formula 13-1, the critical values for $\alpha=0.05$ are $\pm z_{\alpha/2}/\sqrt{n-1} = \pm 1.960/\sqrt{51} = \pm 0.274$.

    a. Using the shortcut method requires finding the ranks for the Monday values, finding the ranks for the Sunday values, subtracting to find the difference in ranks for each pair, and then summing the squares of the differences. Doing this in Minitab produces the following.

        `Sum of d-squared = 14415`

    Using the shortcut formula

        $r_s = 1 - [6(\Sigma d^2)]/[n(n^2-1)]$

          $= 1 - [6(14425)]/[52(2703)]$

          $= 1 - 0.615$

          $= 0.385$

    H_o: $\rho_s = 0$

    H_1: $\rho_s \neq 0$

    $\alpha = 0.05$ [assumed]

    C.V. $r_s = \pm 0.274$

    calculations:

        $r_s = 0.385$

    conclusion:

        Reject H_o; there is sufficient evidence to reject the claim that $\rho_s = 0$ and conclude that $\rho_s \neq 0$ (in fact, that $\rho_s > 0$). There is sufficient evidence to conclude that there is a correlation in Boston between the amounts of rainfall on Mondays and the following Sundays.

    b. Using Formula 10-1 requires finding the ranks for the Monday values, finding the ranks for the Sunday values, and then using the regular command for Pearson's linear correlation on the ranks. Doing this in Minitab produces the following.

        `Pearson correlation of Monday and Sunday = 0.109`

H_o: $\rho_s = 0$
H_1: $\rho_s \neq 0$
$\alpha = 0.05$ [assumed]
C.V. $r_s = \pm 0.274$
calculations:
    $r_s = 0.109$

conclusion:
    Do not reject H_o; there is not sufficient evidence to reject the claim that $\rho_s = 0$. There is not sufficient evidence to conclude that there is a correlation in Boston between the amounts of rainfall on Mondays and the following Sundays.

c. In this instance there is a substantial difference between the two tests – caused by the high number of individual ties, and especially by the many (0,0) pairs. The two conclusions are different. The second method (using Formula 10-1, the Pearson linear correlation formula, on the ranks) always gives the correct answer and is the preferred method. The first method (using the d^2 values) exists only as a computational shortcut for Formula 10-1 – a shortcut that always gives the same answer when there are no ties, but that fails to give the correct answer when there are ties.

13-7 Runs Test for Randomness

1. No. The runs test can be used only to determine whether the order of fixed numbers of two types of items within a sequence is random The runs test does not address the issue of whether or not those fixed numbers of the two types of items indicate some bias or unbalanced representation in the sample.

3. No. The fact that the ordering of the subjects within the sample is random tells us nothing about how the subjects were selected to be in the sample. The subjects could be a convenience sample, a voluntary response sample, or some other type of sample inappropriate for statistical purposes.

NOTE: In each of the following exercises, the item that appears first in the sequence is considered to be of the first type and its count is indicated by n_1. The bell-shaped curve accompanying each test is given for illustrative purposes only. The actual sampling distribution of G is discrete and not necessarily symmetric – hence no value is given for μ_G. The critical value is included in the critical region, and so in Exercise 5 we reject the null hypothesis if $G \leq 4$ or if $G \geq 14$.

5. Since $n_1 = 9$ and $n_2 = 7$, use Table A-10.
    H_o: the sequence is random
    H_1: the sequence is not random
    $\alpha = 0.05$
    C.V. G = 4, 14
    calculations:
        G = 4

conclusion:

Reject Ho; there is sufficient evidence to reject the claim that the values occur in a random sequence. There is sufficient evidence to reject the claim that the genders of the younger Best Actor/Actress winners occur randomly in consecutive years.

7. Since $n_1 = 19$ and $n_2 = 14$, use Table A-10.

H_o: the sequence is random
H_1: the sequence is not random
$\alpha = 0.05$
C.V. G = 11,23
 calculations:
    G = 15

conclusion:

Do not reject Ho; there is not sufficient evidence to reject the claim that the values occur in a random sequence. There is not sufficient evidence to reject the claim that we elect Democrat and Republican candidates in a random sequence.

9. The sequence is ABABAAAAAAAABAABAABABAABAABBB

Since $n_1 = 20$ and $n_2 = 10$, use Table A-10.
H_o: the sequence is random
H_1: the sequence is not random
$\alpha = 0.05$
C.V. G = 9, 20
 calculations:
    G = 16

conclusion:

Do not reject Ho; there is not sufficient evidence to reject the claim that the values occur in a random sequence. There is not sufficient evidence to reject the claim that the days before and after July 1 occur at random.

11. Since $n_1 = 23$ and $n_2 = 18$, use the normal approximation.

$\mu_G = 2n_1n_2/(n_1+n_2) + 1$
   $= 2(23)(18)/41 + 1 = 21.195$
$\sigma^2_G = [2n_1n_2(2n_1n_2 - n_1 - n_2)]/[(n_1+n_2)^2(n_1 + n_2 - 1)]$
   $= [2(23)(18)(787)]/[(41)^2(40)] = 9.691$
H_o: the sequence is random
H_1: the sequence is not random
$\alpha = 0.05$
C.V. $z = \pm z_{\alpha/2} = \pm z_{0.025} = \pm 1.96$
 calculations:
    G = 16
    $z_G = (G - \mu_G)/ \sigma_G$
       $= (16 - 21.195)/\sqrt{9.691}$
       $= -5.195/3.113$
       $= -1.669$
    P-value $= 2 \cdot P(z < -1.67) = 2(0.0475) = 0.0950$

conclusion:

Do not reject Ho; there is not sufficient evidence to reject the claim that the values occur in a random sequence. There is not sufficient evidence to reject the claim that the sequence of NFC/AFC Super Bowl wins is random. The runs test does not test to see whether one conference wins more often than another, and so the results cannot suggest that either conference is superior.

13. There are 43 values in the sequence. When the values are ordered by magnitude, the median is determined to be $x_{22} = 1956$. When the values are examined in their original chronological order, it is determined (ignoring any values that are equal to the median) that the first 21 values are below the median and the last 21 values are above the median.

Since $n_1 = 21$ and $n_2 = 21$, use the normal approximation.

$\mu_G = 2n_1 n_2/(n_1+n_2) + 1$
$\quad = 2(21)(21)/42 + 1 = 22$
$\sigma^2_G = [2n_1 n_2(2n_1 n_2 - n_1 - n_2)]/[(n_1+n_2)^2(n_1 + n_2 - 1)]$
$\quad = [2(21)(21)(840)]/[(42)^2(41)] = 10.244$

H_o: the sequence is random
H_1: the sequence is not random
$\alpha = 0.05$
C.V. $z = \pm z_{\alpha/2} = \pm z_{0.025} = \pm 1.96$

calculations:

$G = 2$

$z_G = (G - \mu_G)/\sigma_G$
$\quad = (2 - 22)/\sqrt{10.244}$
$\quad = -20/3.201$
$\quad = -6.249$

P-value $= 2 \cdot P(Z < -6.25) = 2(0.0001) = 0.0002$

conclusion:

Reject Ho; there is sufficient evidence to reject the claim that the values occur in a random sequence. There is sufficient evidence to reject the claim that the sequence of annual high values in the DJIA is a random sequence of below the median values and above the median values. The result suggests that the stock market is increasing in value.

15. a. and b. The 84 sequences and the number of runs in each are as follows.

```
AAABBBBBB-2  BAAABBBBB-3  BBAAABBBB-3  BBBAAABBB-3  BBBBAAABB-3  BBBBBAAAB-3  BBBBBBAAA-2
AABABBBBB-4  BAABABBBB-5  BBAABABBB-5  BBBAABABB-5  BBBBAABAB-5  BBBBBAABA-4
AABBABBBB-4  BAABBABBB-5  BBAABBABB-5  BBBAABBAB-5  BBBBAABBA-4
AABBBABBB-4  BAABBBABB-5  BBAABBBAB-5  BBBAABBBA-4
AABBBBABB-4  BAABBBBAB-5  BBAABBBBA-4
AABBBBBAB-4  BAABBBBBA-4
AABBBBBBA-3
ABAABBBBB-4  BABAABBBB-5  BBABAABBB-5  BBBABAABB-5  BBBBABAAB-5  BBBBBABAA-4
ABABABBBB-6  BABABABBB-7  BBABABABB-7  BBBABABAB-7  BBBBABABA-6
ABABBABBB-6  BABABBABB-7  BBABABBAB-7  BBBABABBA-6
ABABBBABB-6  BABABBBAB-7  BBABABBBA-6
ABABBBBAB-6  BABABBBBA-6
ABABBBBBA-5
ABBAABBBB-4  BABBAABBB-5  BBABBAABB-5  BBBABBAAB-5  BBBBABBAA-4
ABBABABBB-6  BABBABABB-7  BBABBABAB-7  BBBABBABA-6
ABBABBABB-6  BABBABBAB-7  BBABBABBA-6
ABBABBBAB-6  BABBABBBA-6
ABBABBBBA-5
ABBBAABBB-4  BABBBAABB-5  BBABBBAAB-5  BBBABBBAA-4
ABBBABABB-6  BABBBABAB-7  BBABBBABA-6
ABBBABBAB-6  BABBBABBA-6
ABBBABBBA-5
```

ABBBBAABB-4   BABBBBAAB-5   BBABBBBAA-4
ABBBBABAB-6   BABBBBABA-6
ABBBBABBA-5
ABBBBBAAB-4   BABBBBBAA-4
ABBBBBABA-5
ABBBBBBAA-3

c. At the right is the distribution for G, the number of runs, found from the above sequences. Based on this distribution, a two-tailed test at the 0.025 level (that places 0.025 or less in each tail) has

C.V. G = 2,8

(i.e., reject Ho for G≤2 or G≥8 – for which it will never be possible to reject by being in the upper tail).

G	P(G)	G·P(G)	G^2·P(G)
2	2/84 = 0.023	4/84	8/84
3	7/87 = 0.083	21/84	63/84
4	20/84 = 0.238	80/84	320/84
5	25/84 = 0.298	125/84	625/84
6	20/84 = 0.238	120/84	720/84
7	10/84 = 0.229	70/84	490/84
	84/84 = 1.000	420/84	2226/84

d. The critical region in part (c) agrees exactly with Table A-10.

NOTE: The distribution of the 84 possible G values also gives insight into the use of the normal approximation for G. Even for the small n_1=3 and n_2=6, the distribution appears to be approximately normal. Furthermore, the mean and variance of the G's calculated from the table agree with those obtained from the formula for the normal approximation. The mean and variance will always be the values obtained from the formula, and the shape of the distribution will grow closer to the normal as the n's increase.

From the table,

$\mu_G = \Sigma G \cdot P(G)$
     $= 420/84 = 5$

$\sigma^2_G = \Sigma G^2 \cdot P(G) - (\mu_G)^2$
     $= 2226/84 - 25 = 126/84 = 3/2$

From the formula,

$\mu_G = 2n_1n_2/(n_1 + n_2) + 1$
     $= 2(3)(6)/9 + 1 = 5$

$\sigma^2_G = [2n_1n_2(2n_1n_2 - n_1 - n_2)]/[(n_1 + n_2)^2(n_1 + n_2 - 1)]$
     $= [2(3)(6)(27)]/[(9)^2(8)] = 972/648 = 3/2$

Statistical Literacy and Critical Thinking

1. A nonparametric test is one that does not require the data to come from a particular distribution defined by one or more parameters. A parametric test is one that requires the data to come from a particular distribution defined by one or more parameters.

2. There is no difference – "distribution-free" is another name for "nonparametric."

3. A rank is a number assigned to an sample item that corresponds to its position in a list of all the sample items ordered according to the their values. When there are ties, each of the tied items is assigned the mean of the ranks for which the items would otherwise be entitled. The given values re-arranged in order of magnitude are 54, 54, 122, 169, 195. The ranks assigned to the values in their original order are 1.5, 1.5, 5 ,4 , 3.

4. The efficiency of a nonparametric test refers to the sample size the corresponding parametric test requires to obtain the same precision that a nonparametric reaches with a sample of size 100 – whenever all the requirements of the parametric test are met. The efficiency of a nonparametric test is a number less than 1.00 and can be interpreted as "percent of effectiveness" in the following sense: *The efficiency of the sign test for detecting the difference between matched pairs is 0.63. This means that when the parametric assumptions are met (i.e., when the original populations have normal distributions), the sign test requires a sample of size 100 to detect the same differences that the parametric test can detect with a sample of size 63. And so there is a sense in which the nonparametric test is using only 63% of the information available in the data.* We should use nonparametric tests when the requirements (usually that the data come from a normal distribution) of the corresponding test are not met.

Chapter Quick Quiz

1. When ordered by magnitude, the values are 65, 75, 88, 88, 95. The ranks corresponding the values in the order in which they were originally given are 2, 1, 3.5, 3.5, 5.

2. The runs test for randomness determines whether data that can be classified into two categories appear in a order that is random so far as the two categories are considered.

3. Rank correlation must be used when the data are ranks, other data in ordinal form, or interval/ratio data that do not follow a normal distribution. In addition, rank correlation can detect some relationships that are not linear.

4. A nonparametric test is less efficient than a corresponding parametric test in the sense that it requires stronger evidence, such as a larger sample size or a greater difference, in order to reject the null hypothesis.

5. Distribution-free. "Nonparametric" and "distribution-free" are different names for the same category of tests.

6. False. The rank correlation test can detect many non-linear patterns, but it cannot detect "any pattern" of the points.

7. Four independent samples. The sign test is limited to information that can be communicated using only two symbols, designated by + and −.

8. Two independent samples. While the test could be applied mathematically to matched pairs, that would be inappropriate and inefficient because it would not use the information about how the data were paired.

9. Matched pairs. The Wilcoxon signed-rank test deals with differences, and it requires a matched design that indicates that pairs the data in order to perform meaningful subtractions.

10. Four independent samples. While the test could be applied mathematically to two independent samples, that would be inappropriate because it would prevent the use of one-tailed tests.

Review Exercises

1. Sign Test: Section 13-2. Let $p = P$(the American League wins)
   claim: $p = 0.5$
   $n = 103$: 61 +'s and 42 −'s
   Since n>25, use z with
   $\mu_x = n/2 = 103/2 = 51.5$
   $\sigma_x = \sqrt{n}/2 = \sqrt{103}/2 = 5.074$
   H_o: $p = 0.5$
   H_1: $p \neq 0.5$
   $\alpha = 0.05$
   C.V. $z = \pm z_{\alpha/2} = \pm z_{0.025} = \pm 1.96$
   calculations:
   $x = 42$
   $z = [(x+0.5) - n/2]/[\sqrt{n}/2]$
   $\quad = [42.5 - 51.5]/5.074$
   $\quad = -9.0/5.074 = -1.77$

P-value = $2 \cdot P(z < -1.77) = 2(0.0384) = 0.0768$

conclusion:

Do not reject H_o; there is not sufficient evidence to reject the claim that $p = 0.5$. There is not sufficient evidence to reject the claim that the American League has a 0.5 probability of winning.

NOTE: Using the alternate approach where $x = 61$, the number of +'s, the direction of the calculated statistic is consistent with the rest of the problem. Since $z = [60.5 - 51.5]/5.074 = 1.77$, P-value = $2 \cdot P(z > 1.77) = 0.0768$ and the conclusion is the same.

2. Sign Test: Section 13-2.

claim: median = 98.6 °F

temp	98.0	98.0	97.0	97.7	98.2	98.4	96.5	98.8	97.4	98.6
sign	–	–	–	–	–	–	–	+	–	0

$n = 9$: 1 + and 8 –'s

H_o: median = 98.6 °F

H_1: median ≠ 98.6 °F

$\alpha = 0.05$

C.V. $x = x_{L,\alpha/2} = 1$

      $x = x_{U,\alpha/2} = 9 - 1 = 8$

calculations:

     $x = 1$

conclusion:

Reject H_o; there is sufficient evidence to reject the claim that median = 98.6 and conclude that median ≠ 98.6 (in fact, that median < 98.6). There is sufficient evidence to reject the claim that body temperatures have a median equal to 98.6 °F.

3. Wilcoxon Signed-Ranks Test: Section 10-3.

claim: median = 98.6 °F

x – 98.6	-0.6	-0.6	-1.6	-0.9	-0.4	-0.2	-2.1	0.2	-1.4	0
R	-4.5	-4.5	-8	-6	-3	-1.5	-9	1.5	-7	–

$\Sigma R- = 43.5$ $n = 9$ non-zero ranks

$\Sigma R+ = 1.5$ check: $\Sigma R = n(n+1)/2 = 9(10)/2 = 45$

$\Sigma R = 45.0$

H_o: median = 98.6 °F

H_1: median ≠ 98.6 °F

$\alpha = 0.05$

C.V. $T = T_{L,\alpha/2} = 6$

      $T = T_{U,\alpha/2} = 45 - 6 = 39$

calculations:

     $T = 1.5$

conclusion:

Reject H_o; there is sufficient evidence to reject the claim that median = 98.6 and conclude that median ≠ 98.6 (in fact, that median < 98.6). There is sufficient evidence to reject the claim that body temperatures have a median equal to 98.6 °F.

4. Runs Test: Section 13-7.
   The sequence is EEEOEEOEEE
   Since $n_1 = 8$ and $n_2 = 2$, use Table A-10.
   H_o: the sequence is random
   H_1: the sequence is not random
   $\alpha = 0.05$
   C.V. G = 1, 6
   calculations:
      G = 5

   conclusion:
      Do not reject Ho; there is not sufficient evidence to reject the claim that the values occur in a random sequence. There is not sufficient evidence to reject the claim that the sequence of odd and even decimal digits is random.

5. Rank Correlation: Section 13-6.
   The following table summarizes the calculations.

R_x	R_y	d	d^2
1	1	0	0
2	2	0	0
3	5	-2	4
4	4	0	0
5	7	-2	4
6	6	0	0
7	3	4	16
8	8	0	0
36	36	0	24

$r_s = 1 - [6(\Sigma d^2)]/[n(n^2-1)]$
$= 1 - [6(24)]/[8(63)]$
$= 1 - 0.286$
$= 0.714$

   H_o: $\rho_s = 0$
   H_1: $\rho_s \neq 0$
   $\alpha = 0.05$
   C.V. $r_s = \pm 0.738$
   calculations:
      $r_s = 0.714$

   conclusion:
      Do not reject H_o; there is not sufficient evidence to reject the claim that $\rho_s = 0$. There is not sufficient evidence to support the claim that there is a correlation between the rankings of the students and the ranking of the magazine.

6. Sign Test: Section 13-2. Let the actual highs be group 1.

pair	1	2	3	4	5	6	7	8	9	10
act. − 5d	−	+	−	−	0	−	−	−	−	+

   n = 9: 2 +'s and 7 −'s
   H_o: median difference = 0 °F
   H_1: median difference $\neq$ 0 °F
   $\alpha = 0.05$
   C.V. $x = x_{L,\alpha/2} = 1$
      $x = x_{U,\alpha/2} = 9 - 1 = 8$
   calculations:
      x = 2

conclusion:

    Do not reject H_o; there is not sufficient evidence to reject the claim of no difference. There is not sufficient evidence to reject the claim that median difference between the actual high temperature and the high temperature predicted 5 days in advance is zero. The results suggest that the five-day predicted high temperatures are on target in the sense that they do not tend to be too high or too low, but the results do not speak to the accuracy of the forecast in the sense of limited variability – i.e., the forecast could be on target in the long run by being 10° too high some days and 10° too low on other days, but that would not be considered being accurate.

7. Wilcoxon Signed-Ranks Test: Section 10-3.

act. – 5d	-1	6	-4	-3	0	-1	-12	-14	-1	3
R	-2	7	-6	-4.5	-	-2	-8	-9	-2	4.5

$\Sigma R- = 33.5$ n = 9 non-zero ranks
$\Sigma R+ = 11.5$ check: $\Sigma R = n(n+1)/2 = 9\,(10)/2 = 45$
$\Sigma R \ = 45.0$
H_o: median difference $= 0\ °F$
H_1: median difference $\neq 0\ °F$
$\alpha = 0.05$
C.V. $T = T_{L,\alpha/2} = 6$
      $T = T_{U,\alpha/2} = 45 - 6 = 39$
calculations:
    $T = 11.5$
conclusion:

    Do not reject H_o; there is not sufficient evidence to reject the claim of no difference. There is not sufficient evidence to reject the claim that median difference between the actual high temperature and the high temperature predicted 5 days in advance is zero. The results suggest that the five-day predicted high temperatures are on target in the sense that they do not tend to be too high or too low, but the results do not speak to the accuracy of the forecast in the sense of limited variability – i.e., the forecast could be on target in the long run by being 10° too high some days and 10° too low on other days, but that would not be considered being accurate.

8. Kruskal-Wallis Test: Section 13-5.

Below are the ordered scores and ranks for each group. The group listed first is group 1, etc.

small	R	medium	R	large	R
63	1	257	2	554	7
307	3	461	4	669	9.5
491	5	547	6	669	9.5
750	14	632	8	683	12
917	20	677	11	740	13
925	21	756	16	752	15
1001	22	787	18	786	17
1008	23	905	19	1023	24.5
1048	26	1023	24.5	1202	27
1261	28	1444	30	1290	29
	163		138.5		163.5

$n_1 = 10$ $R_1 = 163$
$n_2 = 10$ $R_2 = 138.5$
$\underline{n_3 = 10}$ $\underline{R_3 = 163.5}$
$n = \Sigma n = 30$ $\Sigma R = 465.0$

chk: $\Sigma R = n(n+1)/2 = 30(31)/2 = 465$

H_o: the populations have the same median
H_1: at least one median is different
$\alpha = 0.05$ and df = 2
C.V. H = χ_α^2 = $\chi_{0.05}^2$ = 5.991
calculations:

$$H = \{12/[n(n+1)]\}\cdot[\Sigma(R_i^2/n_i)] - 3(n+1)$$
$$= \{12/[30(31)]\}\cdot[(163)^2/10 + (138.5)^2/10$$
$$+ (163.5)^2/10] - 3(31)$$
$$= \{0.0129\}\cdot[7248.35] - 93$$
$$= 0.527$$

P-value = χ^2cdf (0.527,99,2) = 0.7683
conclusion:

Do not reject H_o; there is not sufficient evidence to reject the claim that the three samples come from populations with the same median. There is not sufficient evidence to reject the claim that the three car types have the same median right femur crash load values (lbs). No; these data do not provide sufficient evidence to suggest that larger cars are safer.

9. Wilcoxon Rank-Sum Test: Section 13-4.
   Below are the ordered scores and ranks for each group.

1st	R	2nd	R
51	6.5	40	1
52	8	47	2
53	10.0	48	3.5
54	13.5	48	3.5
54	13.5	49	5
54	13.5	51	6.5
55	16.5	53	10.0
56	18.5	53	10.0
56	18.5	54	13.5
59	22	55	16.5
60	23	57	20
61	24.5	58	21
62	26.5	61	24.5
64	28.5	62	26.5
68	30	64	28.5
	273.0		192.0

$n_1 = 15$ R = $\Sigma R_1 = 273.0$ chk: $\Sigma R_i = n(n+1)/2$
$n_2 = 15$ $\Sigma R_2 = 192.0$ = 30(31)/2
n = Σn= 30 ΣR_i = 465.0   = 465

$\mu_R = n_1(n+1)/2$ $\sigma_R^2 = n_1 n_2(n+1)/12$
$= 15(31)/2$ $= (15)(15)(31)/12$
$= 232.5$ $= 581.25$

$\sigma_R = 24.109$

H_o: median$_1$ = median$_2$
H_1: median$_1 \neq$ median$_2$
$\alpha = 0.05$
C.V. z = $\pm z_{\alpha/2}$ = $\pm z_{0.025}$ = ± 1.960
calculations:

$$z = (R - \mu_R)/\sigma_R$$
$$= (273.0 - 232.5)/\sqrt{581.25}$$
$$= 40.5/24.109$$
$$= 1.680$$

P-value = $2\cdot P(z>1.68)$ = 2(0.0465) = 0.0930
conclusion:

Do not reject H_o; there is not sufficient evidence to reject the claim that median$_1$ = median$_2$. There is not enough evidence to reject the claim that the temperatures in the first half of September and the temperatures in the second half of September have the same median. No; these sample data do not the known trend.

10. Runs Test: Section 13-7.

The sequence is EEOEEOOOEEOEEEEEEOEEOOOOEEEOOO

Since $n_1 = 17$ and $n_2 = 13$, use Table A-10.

H_o: the sequence is random

H_1: the sequence is not random

$\alpha = 0.05$

C.V. G = 10, 22

calculations:

G = 12

conclusion:

Do not reject Ho; there is not sufficient evidence to reject the claim that the values occur in a random sequence. There is not sufficient evidence to reject the claim that the sequence of odd and even daily low temperatures is random.

Cumulative Review Exercises

1. scores in order: 62   89   98   126   130   181   264   267   384   531

n = 10 $\bar{x}$ = 2132/10 = 213.2 mg   R = 531 – 62 = 469 mg

$\Sigma x = 2132$ $\tilde{x}$ = (130+181)/2 = 155.5 mg $s^2 = [10(657308) – (2132)^2]/[10(9)] = 22529.5$ mg^2

$\Sigma x^2 = 657308$ $s = \sqrt{22529.5}$ = 150.1 mg

For exercises 2-5, let x = the cholesterol levels and let y = the weights.

The following table contains all the preliminary calculations for n=10 pairs of scores.

x	R_x	y	R_y	d	other needed calculations
264	7	114.8	4	3	$\Sigma xy = 326173.2$
181	6	149.3	7	-1	
267	8	107.8	1	7	$n\Sigma xy – (\Sigma x)(\Sigma y) = 321277.6$
384	9	160.1	9	0	$n\Sigma x^2 – (\Sigma x)^2 = 2027656$
98	3	127.1	6	-3	$n\Sigma y^2 – (\Sigma y)^2 = 107351.96$
62	1	123.1	5	-4	
126	4	111.7	3	1	$\bar{x} = (\Sigma x)/n = 2132/10 = 213.2$
89	2	156.3	8	-6	$\bar{y} = (\Sigma y)/n = 1379.2/10 = 137.92$
531	10	218.8	10	0	
130	5	110.2	2	3	
Σv 2132	55	1379.2	55	0	
Σv^2 657308		200954.46		130	

2. The scatterplot is given at the right.

3. $r = [n(\Sigma xy) - (\Sigma x)(\Sigma y)]/[\sqrt{n(\Sigma x^2) - (\Sigma x)^2}\sqrt{n(\Sigma y^2) - (\Sigma y)^2}]$

   $= 321277.6/[\sqrt{2027656}\sqrt{107351.96}]$

   $= 0.6886$

H_o: $\rho = 0$

H_1: $\rho \neq 0$

$\alpha = 0.05$ and df = 8

C.V. $t = \pm t_{\alpha/2} = -2.306$ [or $r = \pm 0.632$]

calculations:

   $t_r = (r - \mu_r)/s_r$

   $= (0.689 - 0)/\sqrt{[1 - (0.689)^2]/8}$

   $= 0.689/0.2564 = 2.686$

   P-value $= 2 \cdot tcdf(2.686, 99, 8) = 0.0277$

conclusion:

   Reject H_o; there is sufficient evidence to conclude that $\rho \neq 0$ (in fact, that $\rho > 0$). There is sufficient evidence to support the claim that there is a linear correlation between cholesterol level and weight.

4. $b_1 = [n(\Sigma xy) - (\Sigma x)(\Sigma y)]/[n(\Sigma x^2) - (\Sigma x)^2]$

   $= 321277.6/2027656 = 0.1584$

$b_o = \overline{y} - b_1\overline{x}$

   $= 137.92 - 0.1584(213.2) = 104.14$

$\hat{y} = b_o + b_1 x$

$\hat{y} = 104.14 + 0.1584x$

$\hat{y}_{100} = 104.14 + 0.1584(100) = 120.0$ lbs

5. The summary calculations indicate $\Sigma d^2 = 130$.

H_o: $\rho_s = 0$

H_1: $\rho_s \neq 0$

$\alpha = 0.05$

C.V. $r_s = \pm 0.648$

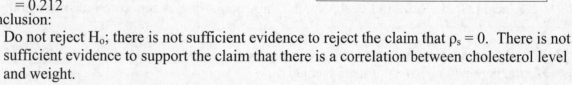

calculations:

   $r_s = 1 - [6(\Sigma d^2)]/[n(n^2-1)]$

   $= 1 - [6(130)]/[10(99)]$

   $= 1 - 0.788$

   $= 0.212$

conclusion:

   Do not reject H_o; there is not sufficient evidence to reject the claim that $\rho_s = 0$. There is not sufficient evidence to support the claim that there is a correlation between cholesterol level and weight.

6. The ordered values for each of the three authors, along with appropriate summary statistics for exercises 6 and 7, are given below.

Clancy R_1		Rowling R_2		Tolstoy R_3		
3.2	1.5	3.2	1.5	5.9	16.5	check: $n_1 = 12$ $\Sigma R_1 = 213.0$
4.6	7	3.7	3	5.9	16.5	$n_2 = 12$ $\Sigma R_2 = 133.0$
4.9	8.5	4.1	4	6.1	19.0	$n_3 = 12$ $\Sigma R_3 = 320.0$
5.0	10.5	4.4	5.5	7.2	24	$n = 36$ $\Sigma R = 666.0$
5.0	10.5	4.4	5.5	7.7	26	
5.4	13	4.9	8.5	8.2	27	$\Sigma R = n(n+1)/2 = 36(37)/2 = 666$
6.1	19.0	5.2	12	8.4	28	
6.4	21	5.6	14	8.6	29	
7.6	25	5.7	15	9.8	32	
8.8	30	6.1	19.0	10.9	33	
9.0	31	6.7	22	11.0	34	
12.0	36	6.9	23	11.5	35	
78.0	213.0	60.9	133.0	101.2	320.0	

Σx^2	572.94	324.07	897.82
$\overline{x}$	6.500	5.075	8.433
s^2	5.9945	1.3639	4.0333

Since each group has equal size n=12, use the simplified form of the calculations.

$k = 3$ $\overline{\overline{x}} = \Sigma\overline{x}_i/k$

$n = 12$ $= 20.008/3 = 6.669$

$s_{\overline{x}}^2 = \Sigma(\overline{x}_i - \overline{\overline{x}})^2/(k-1)$ $s_p^2 = \Sigma s_i^2/k$

$= 5.6823/2 = 2.8411$ $= 11.2918/3 = 3.7973$

$H_o: \mu_1 = \mu_2 = \mu_3$
$H_1:$ at least one μ_i is different
$\alpha = 0.05$ and $df_{num} = 2$, $df_{den} = 33$
C.V. $F = F_\alpha = F_{0.05} = 3.3158$
calculations:

$F = ns_{\overline{x}}^2/s_p^2$

$= 12(2.8411)/3.7973$

$= 8.978$

P-value $= Fcdf(8.978, 99, 2, 33)$

$= 3.45E\text{-}5 = 0.0008$

conclusion:

Reject Ho; there is sufficient evidence to reject the claim that $\mu_1 = \mu_2 = \mu_3$. There is sufficient evidence to reject the claim that the three books have the same mean Flesch-Kincaid Grade Reading Level. No; the three books do not appear to be written at the same reading level.

7. Refer to the summary statistics given in Exercise 6.
   $H_o:$ the populations have the same median
   $H_1:$ at least one median is different
   $\alpha = 0.05$ and $df = 2$
   C.V. $H = \chi_\alpha^2 = \chi_{0.05}^2 = 5.991$
   calculations:

$H = \{12/[n(n+1)]\}\cdot[\Sigma(R_i^2/n_i)] - 3(n+1)$

$= \{12/[36(37)]\}\cdot[(213.0)^2/12 + (133.0)^2/12$
$+ (320.0)^2/12] - 3(37)$

$= \{0.0090\}\cdot[13788.167] - 111$

$= 13.2177$

P-value = χ^2cdf$(13.2177,99,2) = 0.0013$

conclusion:

Reject H_o; there is sufficient evidence to reject the claim that the three samples come from populations with the same median. There is sufficient evidence to reject the claim that the three books have the same mean Flesch-Kincaid Grade Reading Level.

8. The ordered values for each of the three authors, along with appropriate summary statistics for exercises 6 and 7, are given below.

Clancy	R_1	Rowling	R_2
3.2	1.5	3.2	1.5
4.6	7	3.7	3
4.9	8.5	4.1	4
5.0	10.5	4.4	5.5
5.0	10.5	4.4	5.5
5.4	13	4.9	8.5
6.1	16.5	5.2	12
6.4	18	5.6	14
7.6	21	5.7	15
8.8	22	6.1	16.5
9.0	23	6.7	19
12.0	24	6.9	20
78.0	175.5	60.9	124.5

check: $n_1 = 12$ $\Sigma R_1 = 175.5$
$\underline{n_2 = 12}$ $\underline{\Sigma R_2 = 124.5}$
$n = 24$ $\Sigma R = 300.0$

$\Sigma R = n(n+1)/2 = 24(25)/2 = 300$

$R = \Sigma R_1 = 175.5$
$\mu_R = n_1(n+1)/2$ $\sigma^2_R = n_1 n_2(n+1)/12$
    $= 12(25)/2$ $= (12)(12)(25)/12$
    $= 150$ $= 300$

Σx^2	572.94	324.07	
$\overline{x}$	6.500	5.075	
s^2	5.9945	1.3639	

Let the Clancy values be group 1.

$\overline{x}_1 - \overline{x}_2 = 6.500 - 5.075 = 1.425$

original claim: $\mu_1 - \mu_2 = 0$
H_o: $\mu_1 - \mu_2 = 0$
H_1: $\mu_1 - \mu_2 \neq 0$
$\alpha = 0.05$ and df $= 11$
C.V. $t = \pm t_{\alpha/2} = t_{0.025} = \pm 2.201$
calculations:

$t_{\overline{x}_1 - \overline{x}_2} = (\overline{x}_1 - \overline{x}_2 - \mu_{\overline{x}_1 - \overline{x}_2})/s_{\overline{x}_1 - \overline{x}_2}$

$= (1.425 - 0)/\sqrt{5.9945/12 + 1.3639/12}$

$= 1.425/0.7831$

$= 1.820$

P-value = $2 \cdot$tcdf$(1.829,99,11) = 0.0961$

conclusion:

Do not reject H_o; there is not sufficient evidence to reject the claim that conclude that $\mu_1 - \mu_2 = 0$. There is not sufficient evidence reject the claim that the samples of Flesch-Kincaid Grade Level scores from Clancy and Rowling have the same mean.

9. Let the Clancy values be group 1. Refer to the summary statistics given in Exercise 8.

H_o: median$_1$ = median$_2$

H_1: median$_1$ ≠ median$_2$

$\alpha = 0.05$

C.V. $z = \pm z_{\alpha/2} = \pm z_{0.025} = \pm 1.960$

calculations:

$z = (R - \mu_R)/\sigma_R$

$\quad = (175.5 - 150)/\sqrt{300}$

$\quad = 25.5/17.321$

$\quad = 1.47$

P-value $= 2\cdot P(z>1.47) = 2(0.1020) = 0.2040$

conclusion:

Do not reject H_o; there is not sufficient evidence to reject the claim that median$_1$ = median$_2$. There is not enough evidence to reject the claim that the samples of Flesch-Kincaid Grade Level scores from Clancy and Rowling are from populations having the same median.

10. In order for 13.7% of 100 drivers with accidents to own a cell phone, the number of cell phone owners would have to be 13.7. In order for 10.6% of the accident-free drivers to own a cell phone, the number of cell phone owners would have to be 10.6. Assuming the survey cannot include fractions of people, the results as stated are not mathematically possible.

Chapter 14

Statistical Process Control

14-2 Control Charts for Variation and Mean

1. No. Control charts are constructed entirely from the observed data, and they are totally independent of any product specifications. The upper and lower control limits of control charts, for example, are based on the actual behavior of the process under study, and not the desired behavior. It is possible for the process of filling 12 oz cans of Coke to be well within statistical control by consistently filling the cans with, say, 11.5 oz.

3. Random variation is the natural fluctuation due to chance alone that is inherent in every process which does not function with perfect precision. Assignable variation is deviance from perfect precision that can be attributed to an identifiable source – e.g., machinery that needs adjustment.

5. a. Within statistical control.
   b. Not applicable; none of the three criteria applies.
   c. No; the variability is too large. Even though the process may be within control statistically, it is not within control practically. For the amounts of cola in 12 oz cans to vary from 10 oz to 13 oz is clearly unacceptable.

7. a. Not within statistical control.
   b. Criterion 3 applies; there are 8 consecutive points below the centerline. For these data, the criterion applies two times – as there are also 8 consecutive points above the centerline.
   c. No; there appears to have been a shift in the process after observation 8, as all the values before then are below the centerline and all the observations after that are above the centerline. It is usually important to identify the cause for such a shift.

Exercises 9-12 refer to the following table.

decade	00	01	02	03	04	05	06	07	08	09	$\overline{x}$	R
1880	290.7	291.0	291.2	291.4	291.6	291.9	292.1	292.3	292.6	292.9	291.77	2.2
1890	293.2	293.5	293.8	294.1	294.3	294.6	294.9	295.2	295.5	295.8	294.49	2.6
1900	295.6	295.3	295.1	294.8	295.9	296.9	297.5	298.1	298.6	299.2	296.70	4.4
1910	299.4	299.6	299.9	300.1	300.3	300.5	300.7	300.9	301.1	301.2	300.37	1.8
1920	301.4	301.6	302.3	302.9	303.6	304.2	304.9	305.5	305.6	305.8	303.78	4.4
1930	305.9	306.1	306.2	306.3	306.5	306.6	306.8	306.9	307.1	307.3	306.57	1.4
1940	307.4	307.6	307.7	307.9	308.4	308.9	309.3	309.8	310.3	310.8	308.81	3.4
1950	311.3	311.7	312.2	312.7	313.2	313.7	314.3	314.8	315.3	316.0	313.52	4.7
1960	316.9	317.6	318.5	319.0	319.5	320.1	321.3	322.1	323.1	324.6	320.27	7.7
1970	325.7	326.3	327.5	329.6	330.3	331.2	332.2	333.9	335.5	336.9	330.91	11.2
1980	338.7	340.0	341.1	342.8	344.4	345.9	347.1	349.0	351.4	352.9	345.33	14.2
1990	354.2	355.6	356.4	357.1	358.9	360.9	362.6	363.8	366.6	368.3	360.44	14.1
2000	369.5	371.0	373.1	375.6	377.4	379.6	379.6	381.2	382.8	384.4	377.42	14.9

9. There are k=13 samples (decades) of n=10 observations (years) each.

$\overline{\overline{x}} = (\Sigma\overline{x})/k = 4150.38/13 = 319.26$

$\overline{R} = (\Sigma R)/k = 87.0/13 = 6.692$

For the R chart: $UCL = D_4\overline{R} = 1.777(6.692) = 11.89$

$LCL = D_3\overline{R} = 0.223(6.692) = 1.49$

For the $\overline{x}$ chart: $UCL = \overline{\overline{x}} + A_2\overline{R} = 319.26 + 0.308(6.692) = 319.26 + 2.06 = 321.32$

$LCL = \overline{\overline{x}} - A_2\overline{R} = 319.26 - 0.308(6.692) = 319.26 - 2.06 = 317.20$

11. The R chart is given below. The process variation is not within statistical control because there is an upward trend, there are 8 consecutive points below the centerline, and there are points above the upper control limit.

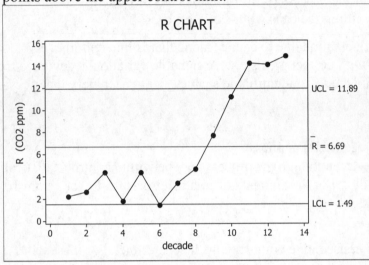

13. There are k=8 samples (years) of n=6 observations (bi-monthly readings) each.

$\overline{\overline{x}} = (\Sigma\overline{x})/k = 22253.67/8 = 2781.71$

$\overline{R} = (\Sigma R)/k = 13835/8 = 1729.375$

For the R chart: $UCL = D_4\overline{R} = 2.004(1729.375) = 3465.67$

$LCL = D_3\overline{R} = 0.000(1729.375) = 0$

For the $\overline{x}$ chart: $UCL = \overline{\overline{x}} + A_2\overline{R} = 2781.71 + 0.483(1729.375)$

$= 2781.71 + 835.29 = 3617.00$

$LCL = \overline{\overline{x}} - A_2\overline{R} = 2781.71 - 0.483(1729.375)$

$= 2781.71 - 835.29 = 1946.42$

year	$\overline{x}$	R
1	3077.67	1345
2	2816.83	2175
3	2640.33	1302
4	2779.00	1020
5	2809.33	3214
6	2706.50	1311
7	2771.67	2342
8	2652.33	1306

15. The $\overline{x}$ chart is given below. The process mean is within statistical control. None of the three criteria for being out of statistical control apply.

17. There are k=8 samples (weeks) of n=5 observations (weekdays) each.

$\overline{\overline{x}} = (\Sigma\overline{x})/k = 989.30/8 = 123.66$

$\overline{R} = (\Sigma R)/k = 3.0/8 = 0.375$

For the R chart: UCL $= D_4\overline{R} = 2.114(0.375) = 0.793$

LCL $= D_3\overline{R} = 0.000(0.375) = 0$

For the $\overline{x}$ chart: UCL $= \overline{\overline{x}} + A_2\overline{R} = 123.66 + 0.577(0.375)$

$= 123.66 + 0.22 = 123.88$

LCL $= \overline{\overline{x}} - A_2\overline{R} = 123.66 - 0.577(0.375)$

$= 123.66 - 0.22 = 123.45$

week	$\overline{x}$	R
1	123.66	0.6
2	123.44	0.3
3	123.68	0.4
4	123.94	0.4
5	123.68	0.6
6	123.36	0.1
7	123.74	0.4
8	123.80	0.2

19. The run chart is given below. There appears to be a pattern of long range cycles which suggest that the process may not be within statistical control. Several consecutive day of lower voltage are followed by several consecutive days of higher voltage, followed by several consecutive days of lower voltage, followed by several consecutive days of higher voltage.

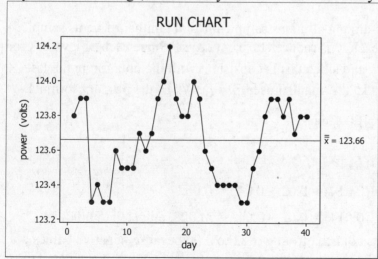

21. There are k=13 samples of size n=10 each.

$\bar{s} = \Sigma s/k = 1.56029/13 = 0.1200$

$UCL = B_4\,\bar{s} = 1.716(0.1200) = 0.2060$

$LCL = B_3\,\bar{s} = 0.284(0.1200) = 0.0341$

This is very similar to the R chart given in the text, and both charts indicate that the process variation is within statistical control. For relatively small n's in the sub-samples (n=10 in this example), there will be a very high correlation between the values of R and s – and so the two charts will be almost identical, but with different labels on the vertical axis.

	s	$\bar{x}$
1	0.164415	13.819
2	0.121033	13.696
3	0.133870	13.741
4	0.157113	13.788
5	0.079889	13.906
6	0.081131	14.016
7	0.077431	14.052
8	0.110156	13.983
9	0.138948	13.938
10	0.124740	14.014
11	0.114232	14.264
12	0.159109	14.396
13	0.098229	14.636

14-3 Control Charts for Attributes

1. No. Control charts determine whether a process is within statistical control relative to the values it is actually producing, not relative to the values it is supposed to be producing.

NOTE: The text defines $\bar{p}$ as the pooled estimate of the proportion of successes among all items sampled (i.e., the weighted mean of the k values for p) and gives

$\bar{p} = (\Sigma x)/(\Sigma n)$

= (total number of "successes" among all items samples)/(total number of items sampled).

When there are k samples of size n each, this reduces to $\bar{p} = (\Sigma p)/k$. Since each of k values for p must be calculated to make the chart, and since this is consistent with the notation in the last section that $\bar{\bar{x}} = (\Sigma \bar{x})/k$ and $\bar{R} = (\Sigma R)/k$, this manual uses $\bar{p} = (\Sigma p)/k$ as the primary formula.

3. There are k=5 samples of size n=100 each.

$\bar{p} = (\Sigma p)/k = 0.10/5 = 0.02$

$\sqrt{\bar{p}\,\bar{q}/n} = \sqrt{(0.02)(0.98)/100} = 0.014$

$UCL = \bar{p} + 3\sqrt{\bar{p}\,\bar{q}/n} = 0.02 + 3(0.014) = 0.02 + 0.042 = 0.062$

$LCL = \bar{p} - 3\sqrt{\bar{p}\,\bar{q}/n} = 0.02 - 3(0.014) = 0.02 - 0.042 = -0.022$, adjusted to 0.00

Yes. The lower control limit was adjusted from -0.022 to 0.00 because negative values are not possible for proportions. Since the process is not within statistical control when data fall outside the control limits (i.e., not on the control limits), a process with the above control limits cannot fail to be within statistical control by having a sample proportion that is too low.

5. The process is within statistical control; none of the three criteria applies. In this case, however, there are warning signs that suggest continuing monitoring of the process is in order – two of the data points are dangerously close to the upper control limit, and the final n=7 consecutive data points below the centerline (along with the general pattern of the data) are dangerously close to indicating the presence of long term cycles.

7. The process is not within statistical control – and in this case, all three criteria apply. The data end with an upward trend. There is a data point above the upper control limit. There are at least 8 consecutive data points above the centerline.

9. There are k=20 samples of n=10,000 each.

$\overline{p} = (\Sigma p)/k = (0.0020 + 0.0014 + 0.0022 + ... + 0.0020)/20 = 0.0400/20 = 0.00200$

$\sqrt{\overline{p}\,\overline{q}/n} = \sqrt{(0.0020)(0.9980)/10000} = 0.0004468$

$UCL = \overline{p} + 3\sqrt{\overline{p}\,\overline{q}/n} = 0.00200 + 3(0.0004468) = 0.00200 + 0.00134 = 0.00334$

$LCL = \overline{p} - 3\sqrt{\overline{p}\,\overline{q}/n} = 0.00200 - 3(0.0004468) = 0.00200 - 0.00134 = 0.00066$

The p chart is given below. The process is within statistical control. None of the three criteria for being out of statistical control applies.

11. There are k=15 samples of n=1000 each.

$\overline{p} = (\Sigma p)/k = (0.601 + 0.625 + 0.619 + ... + 0.667)/15 = 9.523/15 = 0.6349$

$\sqrt{\overline{p}\,\overline{q}/n} = \sqrt{(0.6349)(0.3651)/1000} = 0.015225$

$UCL = \overline{p} + 3\sqrt{\overline{p}\,\overline{q}/n} = 0.6349 + 3(0.015225) = 0.6349 + 0.0457 = 0.6805$

$LCL = \overline{p} - 3\sqrt{\overline{p}\,\overline{q}/n} = 0.6349 - 3(0.015225) = 0.6349 - 0.0457 = 0.5892$

The p chart is given below. The process is within statistical control. None of the three criteria for being out of statistical control applies. Whether or not the enrollments are high enough depends on the desired values – but control charts determine whether a process is within statistical control relative to the values it is actually producing, not relative to the values desired.

13. There are k=23 samples of n=1000 each.

$\overline{p} = (\Sigma p)/k = (0.585 + 0.454 + 0.577 + ... + 0.514)/23 = 10.199/23 = 0.4434$

$\sqrt{\overline{p}\,\overline{q}/n} = \sqrt{(0.4434)(0.5566)/1000} = 0.0157099$

$UCL = \overline{p} + 3\sqrt{\overline{p}\,\overline{q}/n} = 0.4434 + 3(0.0157099) = 0.4434 + 0.0471 = 0.4906$

$LCL = \overline{p} - 3\sqrt{\overline{p}\,\overline{q}/n} = 0.4434 - 3(0.0157099) = 0.4434 - 0.0471 = 0.3963$

The p chart is given below. The process is not within statistical control because there is an

alternating trend, there are points above the upper control limit, and there are points below the lower control limit. The alternating trend is caused by the fact that the odd-numbered samples represent presidential election years and the even-numbered samples represent non-presidential election years. In addition, there may be a long-term cycle: it appears that voting in both types of elections declined, reached a low point, and is beginning to increase again.

15. There are k=20 samples of size n=10000 each. The example calculated $\overline{p} = 0.001$.
The np chart is literally the p chart multiplied by n. It is based on x=np instead of on p.
$$n \cdot \overline{p} = 10000(0.001) = 10$$
$$[OR \quad n \cdot \overline{p} = n \cdot (\Sigma p)/k = (\Sigma np)/k = (\Sigma x)/k = (15+12+14+...+7)/20 = 200/20 = 10]$$
$$n \cdot \sqrt{\overline{p}\,\overline{q}/n} = \sqrt{n\,\overline{p}\,\overline{q}} = \sqrt{10000(0.001)(0.999)} = 3.1607$$
$$UCL = n \cdot \overline{p} + 3\sqrt{n\,\overline{p}\,\overline{q}} = 10.00 + 3(3.1607) = 10.0 + 9.48 = 19.48$$
$$LCL = n \cdot \overline{p} - 3\sqrt{n\,\overline{p}\,\overline{q}} = 10.00 - 3(3.1607) = 10.0 - 0.48 = 0.52$$

The np control chart that follows is identical to the p control chart in the text except for the labeling on the vertical axis. The values on the vertical axis of the np chart are n=10000 times the values on the p chart – e.g., the np UCL is 10000(0.001948) = 19.48, the np center line is 10000(0.001) = 10, and the np LCL is 10000(0.000052) = 0.52.

Statistical Literacy and Critical Thinking

1. Statistical process control is the monitoring of data over time in order to detect trends or problems that indicate the production process is out of statistical control and/or problems that lead to the production of a product that does not meet prescribed specifications.

2. It is important to monitor the manufacturing process over time because conditions affecting the product may occur over time – i.e., the hiring of new employees, the wearing out of the machinery, etc. A possible adverse consequence of not monitoring the manufacturing process would be the financial and/or public relations loss in dealing with products not meeting the proper specifications leaving the plant and entering the market. In addition, the product could become so far out of specification that only a temporary shutdown (and subsequent loss of revenue) could remedy the problem.

3. In general, a process can go out of control because of a change in mean or a change in variation. The fact that the mean is within statistical control does not indicate that the variation is within statistical control, and vice-versa. For this reason, one should monitor a process with both an $\overline{x}$ chart and an R chart.

4. No. Control charts determine whether a process is within statistical control relative to the values it is actually producing, not relative to the values it is supposed to be producing.

Chapter Quick Quiz

1. Process data are values in chronological sequence that measure a characteristic of a controllable activity.

2. Random variation is the natural fluctuation due to chance alone that is inherent in every process which does not function with perfect precision. Assignable variation is deviance from perfect precision that can be attributed to an identifiable source – e.g., machinery that needs adjustment.

3. The test states a process is out of statistical control if any of the following three criteria applies.
   (1) There is a data pattern that is obviously not random.
   (2) There is a data value above the upper control limit or below the lower control limit.
   (3) There are 8 consecutive data points either all above or all below the centerline.

4. An R chart monitors the variation of a process, while an $\bar{x}$ chart monitors the mean of a process.

5. No. The process variation is not within statistical control because the R chart indicates a data point that is above the upper control limit.

6. The R chart indicates that $\bar{R} = 21.2$. In this context, $\bar{R}$ is the mean of the k=20 values for R obtained from the 20 subsamples of size n.

7. No. The process mean is not within statistical control because the X-bar chart indicates that there are at least 8 consecutive data points below the centerline.

8. The X-bar chart indicates that $\bar{\bar{x}} = 6.45$. In this context, $\bar{\bar{x}}$ is the mean of the k=20 values for $\bar{x}$ obtained from the 20 subsamples of size n.

9. A p chart is a control chart to monitor the proportion of items having a particular attribute over time. It is used to determine whether the corresponding process is within statistical control.

10. False. A process is out of control when it appears to be under the influence of some factor or force that causes other than the usual expected random deviation. That factor or force could be causing a smaller proportion of defects, or it could be acting in some way (e.g., increasing the variation) that does not alter the overall proportion of defects.

Review Exercises

1. The run chart is given below. There is not a pattern suggesting that the process is not within statistical control.

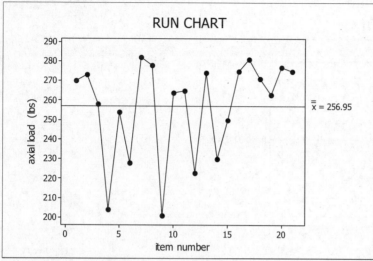

2. There are k=3 samples of n=7 observations each. For the R chart,

$\bar{R} = (\Sigma R)/k = (78+77+31)/3 = 62.0$

$UCL = D_4\bar{R} = 1.924(62.0) = 119.3$

$LCL = D_3\bar{R} = 0.076(62.0) = 4.7$

3. The process variation is within statistical control because none of the three criteria for being out of statistical control applies.

4. There are k=3 samples of n=7 observations each. For the $\bar{x}$ chart,

$\bar{\bar{x}} = (\Sigma\bar{x})/k = (252.71+247.86+270.29)/3 = 256.95$

$\bar{R} = (\Sigma R)/k = (78+77+31)/3 = 62.0$

$UCL = \bar{\bar{x}} + A_2\bar{R} = 256.95 + 0.419(62.0) = 256.95 + 25.98 = 282.93$

$LCL = \bar{\bar{x}} - A_2\bar{R} = 256.95 - 0.419(62.0) = 256.95 - 25.98 = 230.97$

5. The process mean is within statistical control because none of the three criteria for being out of statistical control applies.

6. There are k=15 samples of n=1,000,000 each.

$\bar{p} = (\Sigma p)/k = (0.000085 + 0.000087 + 0.000094 + ... + 0.000056)/15$

$= 0.001153/15 = 0.0000769$

$\sqrt{\bar{p}\,\bar{q}/n} = \sqrt{(0.0000769)(0.9999231)/1000000} = 0.000008767$

$UCL = \bar{p} + 3\sqrt{\bar{p}\,\bar{q}/n} = 0.0000769 + 3(0.000008767) = 0.0000769 + 0.0000263 = 0.0001032$

$LCL = \bar{p} - 3\sqrt{\bar{p}\,\bar{q}/n} = 0.0000769 - 3(0.000008767) = 0.0000769 - 0.0000263 = 0.0000506$

The p chart is given below. The process is not within statistical control because there is a downward trend, there are 8 consecutive points above the centerline. The control chart suggests that the downward trend in the proportion of homicides may have ended and is leveling off.

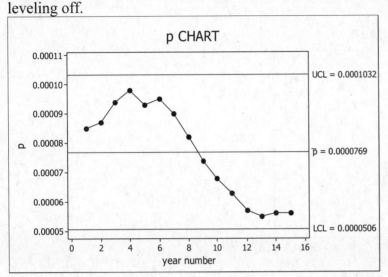

7. There are k=15 samples of n=100 each.

$\overline{p} = (\Sigma p)/k = (0.05 + 0.04 + 0.03 + ... + 0.01)/15 = 0.81/15 = 0.054$

$\sqrt{\overline{p}\,\overline{q}/n} = \sqrt{(0.054)(0.946)/100} = 0.0226$

$UCL = \overline{p} + 3\sqrt{\overline{p}\,\overline{q}/n} = 0.054 + 3(0.0226) = 0.054 + 0.068 = 0.122$

$LCL = \overline{p} - 3\sqrt{\overline{p}\,\overline{q}/n} = 0.054 - 3(0.0226) = 0.054 - 0.068 = -0.014$, truncated at 0.000

The p chart is given below. The process is not within statistical control because there is a pattern of increasing variation.

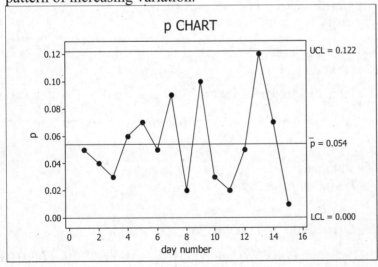

Cumulative Review Exercises

1. The following summary information applies to all parts of the exercise.
   Let $CO_2 = x$ and temperature $= y$.

   $n = 10$

   $\Sigma x = 3774.2$ $\bar{x} = (\Sigma x)/n = 3774.2/10 = 377.42$

   $\Sigma y = 146.36$ $\bar{y} = (\Sigma y)/n = 146.36/10 = 14.636$

   $\Sigma xy = 55242.261$ $n(\Sigma xy) - (\Sigma x)(\Sigma y) = 30.698$

   $\Sigma x^2 = 1424685.94$ $n(\Sigma x^2) - (\Sigma x)^2 = 2273.76$

   $\Sigma y^2 = 2142.2118$ $n(\Sigma y^2) - (\Sigma y)^2 = 0.8684$

   $r = [n(\Sigma xy) - (\Sigma x)(\Sigma y)]/[\sqrt{n(\Sigma x^2) - (\Sigma x)^2}\sqrt{n(\Sigma y^2) - (\Sigma y)^2}]$

   $= 30.698/[\sqrt{2273.76}\sqrt{0.8684}]$

   $= 0.6908$

   a. H_o: $\rho = 0$
      H_1: $\rho \neq 0$
      $\alpha = 0.05$ and df $= 8$
      C.V. $t = \pm t_{\alpha/2} = \pm t_{0.025} = \pm 2.306$ [or $r = \pm 0.632$]
      calculations:

      $t_r = (r - \mu_r)/s_r$

      $= (0.691 - 0) / \sqrt{[1 - (0.691)^2]/8}$

      $= 0.691/0.2556$

      $= 2.703$

      P-value $= 2 \cdot$ tcdf$(2.703,99,8) - 0.0270$
      conclusion:

      Reject H_o; there is sufficient evidence to conclude that $\rho \neq 0$ (in fact, that $\rho > 0$). Yes;
      there is sufficient evidence to support the claim of a linear correlation between carbon
      dioxide concentration and temperature.

   b. No. Linear correlation measures association, not causation. And if there does happen to be
      an cause-and-effect relationship between the two variables, there is nothing in the
      mathematics that would allow a person to say which variable is the cause and which variable
      is the effect.

   c. $b_1 = [n(\Sigma xy) - (\Sigma x)(\Sigma y)]/[n(\Sigma x^2) - (\Sigma x)^2] = 30.698/2273.76 \doteq 0.0135$
      $b_o = \bar{y} - b_1\bar{x} = 14.636 - 0.0135(377.42) = 9.540$
      $\hat{y} = b_o + b_1 x = 9.540 + 0.0135x$

   d. $\hat{y}_{290.7} = 9.540 + 0.0135(290.7) = 13.46\ °C$

      The predicted temperature does not appear to be very close to the actual temperature of
      13.88 °C. But a carbon dioxide concentration of 290.7 is well outside the range of x values
      used to determine the regression line. Since 290.7 is $(290.7 - 377.42)/5.02633 = -17.253$
      standard deviations from the mean of the x scores used to determine the line, the regression
      line should probably not be used for that prediction.

2. There are k=10 samples (weeks) of n=200 belts each.

$\overline{p} = (\Sigma p)/k = (0.010 + 0.015 + 0.005 +...+ 0.085)/10 = 0.335/10 = 0.0335$

$\sqrt{\overline{p}\,\overline{q}/n} = \sqrt{(0.0335)(0.9665)/200} = 0.01272$

$UCL = \overline{p} + 3\sqrt{\overline{p}\,\overline{q}/n} = 0.0335 + 3(0.01272) = 0.0335 + 0.0382 = 0.0717$

$LCL = \overline{p} - 3\sqrt{\overline{p}\,\overline{q}/n} = 0.0335 - 3(0.01272) = 0.0335 - 0.0382 = -0.0047$, truncated at 0.0000

The p chart is given below. The process is not within statistical control because there is a pattern of increasing variability and there are points above the upper control limit.

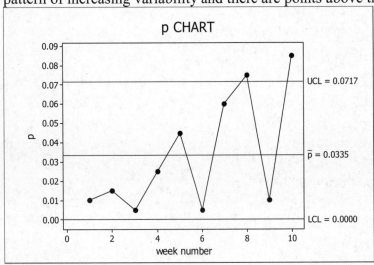

3. $\alpha = 0.05$ and $z_{\alpha/2} = z_{0.025} = 1.96$; $\hat{p} = x/n = 67/2000 = 0.0335$

$\hat{p} \pm z_{\alpha/2}\sqrt{\hat{p}\hat{q}/n}$

$0.0335 \pm 1.96\sqrt{(0.0335)(0.9665)/2000}$

0.0335 ± 0.0079

$0.0256 < p < 0.0414$

We have 95% confidence that the interval from 0.0256 to 0.0414 contains the true proportion of defective seat belts produced by this process.

4. original claim: p > 0.03

$\hat{p} = x/n = 67/2000 = 0.0335$

H₁: p > 0.03

H_o: p = 0.03
H_1: p > 0.03
$\alpha = 0.05$
C.V. $z = z_\alpha = z_{0.05} = 1.645$
calculations:

$z\hat{p} = (\hat{p} - \mu\hat{p})/\sigma\hat{p}$

$= (0.0335 - 0.03)/\sqrt{(0.03)(0.97)/2000}$

$= 0.0035/0.003814$

$= 0.92$

P-value = P(z>0.92) = 1 - 0.8212 = 0.1788

conclusion:

Do not reject H_o; there is not sufficient evidence to support the claim that p > 0.03. There is not sufficient evidence to support the claim that the proportion of defects is greater than 3%.

5. a. P(all 8 above) = P(A$_1$ and A$_2$ and A$_3$ and...and A$_8$)
$$= P(A_1) \cdot P(A_2) \cdot P(A_3) \cdot ... \cdot P(A_8)$$
$$= (1/2)(1/2)(1/2)...(1/2)$$
$$= (1/2)^8 = 1/256 \text{ or } 0.0039$$
   b. P(all 8 below) = P(B$_1$ and B$_2$ and B$_3$ and...and B$_8$)
$$= P(B_1) \cdot P(B_2) \cdot P(B_3) \cdot ... \cdot P(B_8)$$
$$= (1/2)(1/2)(1/2)...(1/2)$$
$$= (1/2)^8 = 1/256 \text{ or } 0.0039$$
   c. P(all above OR all below) = P(all above) + P(all below) – P(all above AND all below)
$$= 1/256 + 1/256 - 0$$
$$= 2/256 = 1/128 \text{ or } 0.0078$$

6. normal distribution: $\mu = 6.0$, $\sigma = 1.0$
   a. P(x>7.0) = P(z>1.00)
$$= 1 - 0.8413$$
$$= 0.1587 \text{ or } 15.87\%$$
   Yes; if the helmets were made to fit only males with head breadths less than 7.0 inches, too many men would be excluded.
   b. For the smallest 5%,   For the largest 5%,
      A = 0.0500 and z = -1.645   A = 0.9500 and z = 1.645
   $x = \mu + z\sigma$ $x = \mu + z\sigma$
      $= 6.0 + (-1.645)(1)$ $= 6.0 + (1.645)(1)$
      $= 6.0 - 1.645$ $= 6.0 + 1.645$
      $= 4.355$ inches $= 7.645$ inches
   The helmets will men with head breadths from 4.355 inches to 7.645 inches.

7. No. Since the sample is a voluntary response sample, which is not necessarily representative of the general population, it is not appropriate to use these results to make any conclusion about the general population

8. No; the values do not appear to come from a normal distribution. The histogram indicates a distribution that is positively skewed and far from bell-shaped, and the normal probability plot is not close to a straight line.

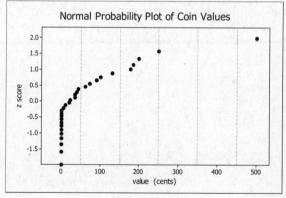

9. Listed in order, the values are as follows.

0 0 0 0 0 0 0 0 0 0 0 0 0 5 10 20 23 35 35 40 43 62 73 90 100 130 178 185 200 250 500

The n=30 have $\Sigma x = 1979$ and $\Sigma x^2 = 469535$.

a. $\bar{x} = (\Sigma x)/n = 1979/30 = 66.0$ cents

b. $\tilde{x} = (20+23)/2 = 21.5$ cents

c. $s^2 = [n(\Sigma x^2) - (\Sigma x)^2]/[n(n-1)]$

$= [30(469535) - (1979)^2]/[30(29)] = 10169609/870 = 11689.2$

$s = \sqrt{11689.2} = 108.1$ cents

10. original claim: $p > 0.50$

$\hat{p} = x/n = 18/30 = 0.60$

$H_o: p = 0.50$

$H_1: p > 0.50$

$\alpha = 0.05$

C.V. $z = z_\alpha = z_{0.05} = 1.645$

calculations:

$z\hat{p} = (\hat{p} - \mu\hat{p})/\sigma\hat{p}$

$= (0.60 - 0.50)/\sqrt{(0.50)(0.50)/30}$

$= 0.10/0.09128$

$= 1.10$

P-value $= P(z>1.10) = 1 - 0.8643 = 0.1357$

conclusion:

Do not reject H_o; there is not sufficient evidence to support the claim that $p > 0.50$. There is not sufficient evidence to support the claim that most students have change in their possession.

FINAL NOTE: Congratulations! You have completed statistics – the course that everybody likes to hate. I trust that this manual has helped to make the course a little more understandable – and that you leave the course with an appreciation of broad principles, and mot merely memories of manipulating formulas. I wish you well in your continued studies, and that you achieve your full potential wherever your journey of life may lead.